Control Allocation for Spacecraft Under Actuator Faults

Qinglei Hu · Bo Li · Bing Xiao · Youmin Zhang

Control Allocation for Spacecraft Under Actuator Faults

Springer

Qinglei Hu
School of Automation Science
and Electrical Engineering
Beihang University
Beijing, China

Bo Li
Institute of Logistics Science
and Engineering
Shanghai Maritime University
Shanghai, China

Bing Xiao
School of Automation
Northwestern Polytechnical University
Xi'an, Shaanxi, China

Youmin Zhang
Department of Mechanical
Industrial and Aerospace Engineering
Concordia University
Montreal, Canada

ISBN 978-981-16-0438-6 ISBN 978-981-16-0439-3 (eBook)
https://doi.org/10.1007/978-981-16-0439-3

This Springer imprint is published by the registered company Springer Nature Singapore Pte Ltd.
The registered company address is: 152 Beach Road, #21-01/04 Gateway East, Singapore 189721, Singapore

To our families, teachers and friends for their love and support.

Preface

Over the last several decades, human's exploration of deep space never stops, due to the great significance and promotion for the human development and technological progress. As the corn carrier of the aerospace missions, spacecraft system techniques have attracted much more attention and many interesting results have been reported in literature. The stability, reliability, autonomy, and intelligence of the spacecraft Attitude Control System (ACS) play a crucial and critical role during the various aerospace missions. As is known, the dynamics of the spacecraft is strong nonlinear physically, and it is vulnerable to the severe and various disturbances from the deep space including gravitational perturbations, aerodynamic torque, radiation torque, and other environmental or non-environmental torques. As a result, faults and uncertainties do occur inevitably in spacecraft actuators and other components (such as sensors, or electrical or software modules) during their long-time working in deep space, which deteriorate the stability, safety, and reliability of the spacecraft ACS. And finally the on-orbiting missions may fail completely for just these reasons. More specifically, actuator fault is one type of the most serious and common faults in spacecraft ACS, accounting for about 44% of all faults. These potential faults and uncertain effects derived from internal or external causes will lead to the degradation of system performance, and even may result in the complete collapse, immeasurable losses, or catastrophic consequences. Therefore, it has significant practical values and theoretical significance to promote the stable and reliable operation abilities of spacecraft especially in the presence of external disturbances, actuator uncertainties, or faults. To achieve this goal, one of the main approaches is to design Fault-Tolerant Control (FTC) system under actuators' faults by utilizing some advanced control techniques. On the other hand, the modern spacecraft are generally equipped with redundant actuators (the number of the actuators is greater than the degree of freedom of the ACS) to improve the reliability and safety of the spacecraft ACS. It introduces the problem of how to distribute appropriately the designed/commanded control signals to the individual actuator, which will generate the related control force/torque to accomplish specified aerospace control missions. Thus, it is also important to design a Control Allocation (CA) approach under some actuator constraints and attitude control performance indexes.

The techniques of FTC and CA have been investigated and taken into application widely in the aeronautics and astronautics fields due to their superiorities to improve the reliability, safety, and control performances for the over-actuated control systems. Although there exist many results to deal with these practical issues in the presence of external disturbances, uncertainties, actuator faults, and constraints, the following problems should be addressed reasonably:

(1) Online FDD strategies are still necessary to be developed and improved to reconstruct the actuator fault information for the spacecraft ACS under various uncertainties and constraints.
(2) The FTC approaches can be improved by promoting the properties of fast convergence and robustness to the unknown uncertainties and disturbances.
(3) The effectiveness and optimal performances of CA schemes could be improved, and the stability of the whole closed-loop control system should also be analyzed theoretically.

In view of the above problems, this book focuses on systematical control system design schemes for the over-actuated spacecraft ACS under external disturbances, uncertainties, actuator faults, and constraints. The main contents of this book are presented in the following aspects:

(1) For the over-actuated spacecraft ACS equipped with redundant Reaction Wheels (RWs), a Saturated Proportional-Derivative (SPD)-based virtual/baseline control law incorporated with a null-space-based optimal CA method is proposed under external disturbances and input saturation.
(2) To improve convergence performance of the virtual control law, a Nonsingular Terminal Sliding Mode Control (NSMC)-technique-based finite-time control law is first proposed to achieve the fast convergence, and then a robust least squares-based CA scheme is developed to suitably distribute the desired virtual control signals into the active actuators.
(3) To consider further the spacecraft attitude stabilization control problem in the presence of external disturbances, input saturation, and inertia uncertainties, a Fast SMC (FSMC)-based attitude stabilization control law is first proposed, and it can be extended to accommodate actuator faults while making use of the remaining active actuating capabilities by utilizing the Pseudo-inverse CA (PICA) method.
(4) To achieve optimal control performances, a finite-time Extended State Observer (ESO) is first investigated to estimate compensation for specified total uncertainty including actuators' misalignments and parametric uncertainties. And then, Control Lyapunov Function (CLF)-approach-based optimal control laws are developed to achieve the spacecraft attitude stabilization control.
(5) Taking the actuators' faults into account, an Iterative Learning Observer (ILO) and Integral-type Sliding Mode Control (ISMC)-based FTC scheme is proposed to achieve the attitude stabilization of rigid spacecraft in the presence of actuator uncertainties, external disturbances, and input saturation. And then a robust CA algorithm is developed to map the virtual control demand onto individual actuator in an optimal manner with fault-tolerant ability.

(6) To analyze the stability of the whole closed-loop spacecraft ACS with CA module, a closed-loop CA scheme is proposed for dealing with the problem of distributing the virtual control command synthesized by the SPD controller into the individual actuator properly by using constrained optimal quadratic programming algorithms. And the asymptotic stability of the closed-loop control allocation system is guaranteed theoretically.
(7) Taking further the actuators' fault in the design of the closed-loop CA system, a Proportional-Derivative (PD)-like control law designed by the dynamic attitude-gain technique is first presented as a virtual or baseline control law to achieve the spacecraft attitude stabilization control. Then, a robust online CA method is developed and an analytical solution is obtained. The asymptotic stability of the closed-loop system with online CA is achieved by a rigorous closed-loop stability analysis.

This book presents systematical and comprehensive descriptions of the CA and FTC issues for over-actuated spacecraft ACS under actuator faults, system uncertainties, and disturbances. It intends to provide a legible platform for reference, and applicability of the FTC and CA for different purposes in aerospace engineering and some other related fields.

Beijing, China — Qinglei Hu
Shanghai, China — Bo Li
Xi'an, China — Bing Xiao
Montreal, Canada — Youmin Zhang
October 2020

Acknowledgements

The contents included in this book are the outgrowth and summary of the authors' academic researches to address the issues of control allocation and fault-tolerant control for over-actuated spacecraft attitude control system under actuator faults, system uncertainties, and disturbances in the past several years. It was partially supported by National Natural Science Foundation of China (61960206011, 61633003, 62073212, 61703272, 61873207, 61833013), Beijing Natural Science Foundation (JQ19017), Major Project of Science and Technology Innovation 2030—"New Generation Artificial Intelligence" (2018AAA0102700), etc. The authors greatly appreciate the above financial supports. The authors would like to thank the cooperators Prof. Guangfu Ma, Prof. Maruthi R. Akella, Prof. Aihua Zhang, Chenliang Wang, and the students Xiao Tan, Guanlin Niu, Wenquan Gong, Yueyang Liu, Wei Chen, Haichao Zhang, Hang Xiong, Biru Chi, et al., for their contributions and efforts dedicated to this book. We appreciate the permission from the John Wiley and Sons, Springer Nature, Elsevier Paris and American Institute of Aeronautics and Astronautics to reuse partly the results published in the relevant journals, such as International Journal of Adaptive Control and Signal Processing (Chap. 3); Nonlinear Dynamics (Chap. 4); Journal of Guidance, Control, and Dynamics (Chaps. 5, 8, and 9); Aerospace Science and Technology (Chaps. 6 and 7). In addition, the authors would also like to thank the editors in Springer Nature for their help to accomplish the publication of this book.

Contents

1 Introduction . . . 1
- 1.1 Background and Motivations . . . 1
- 1.2 Introduction to Control Allocation . . . 2
  - 1.2.1 Control Allocation Methods . . . 2
  - 1.2.2 Applications of Control Allocation in Aeronautics and Astronautics . . . 6
- 1.3 Introduction to Fault-Tolerant Control . . . 9
  - 1.3.1 Fault-Tolerant Control Methods . . . 9
  - 1.3.2 Applications of Fault-Tolerant Control in Aeronautics and Astronautics . . . 10
- 1.4 Organization of the Book . . . 13
- References . . . 16

2 Mathematical Model of the Attitude Control System . . . 23
- 2.1 Spacecraft Attitude Kinematics . . . 23
  - 2.1.1 Euler Angles Based Model . . . 24
  - 2.1.2 Quaternion Based Model . . . 27
  - 2.1.3 Modified Rodrigues Parameters Based Model . . . 28
- 2.2 Spacecraft Attitude Dynamics Model . . . 29
- References . . . 30

3 Null-Space Based Optimal Control Allocation for Spacecraft Attitude Stabilization . . . 33
- 3.1 Introduction . . . 33
- 3.2 Problem Formulation . . . 35
  - 3.2.1 Spacecraft Attitude Dynamics . . . 35
  - 3.2.2 Control Objective . . . 36
- 3.3 Attitude Control Law Design . . . 36
  - 3.3.1 SPD-Based Virtual/Baseline Control Law Design . . . 37
  - 3.3.2 Null-Space-Based Optimal Control Reallocation Scheme . . . 40
- 3.4 Simulation Results . . . 43

3.4.1 Time Response Results Under the Proposed Control Law ... 45
3.4.2 Energy Consumption Analysis ... 49
3.5 Conclusions ... 51
References ... 51

4 Robust Finite-Time Control Allocation for Attitude Stabilization Under Actuator Misalignment ... 55
4.1 Introduction ... 55
4.2 Problem Formulation ... 56
4.2.1 Actuator Configuration with Misalignment ... 56
4.2.2 Control Objective ... 59
4.3 Control Law Design ... 59
4.3.1 Nonsingular Terminal Sliding Mode Based Virtual Finite Time Feedback Controller Design ... 60
4.3.2 Robust Least Squares-Based Control Allocator Design Under Actuator Misalignment ... 62
4.4 Simulation Results ... 64
4.4.1 Simulation Results with Misalignment ... 65
4.4.2 Energy Consumption Analysis ... 67
4.5 Conclusions ... 69
References ... 70

5 Finite-Time Fault-Tolerant Spacecraft Attitude Control with Torque Saturation ... 73
5.1 Introduction ... 73
5.2 Problem Formulation ... 75
5.2.1 Definitions and Lemmas ... 75
5.3 Saturated Finite-Time Fault-Tolerant Control Law Design ... 76
5.3.1 Novel Time-Varying Terminal Sliding Mode Design ... 76
5.3.2 Basic Saturated Finite-Time Controller Design ... 77
5.3.3 Saturated Fault-Tolerant Finite-Time Controller Design ... 79
5.3.4 The Advantage of Modification Matrix $\boldsymbol{A}$ ... 82
5.4 Simulation Results ... 86
5.5 Conclusions ... 92
Appendices ... 96
References ... 107

6 Extended State Observer Based Optimal Attitude Robust Control of Spacecraft ... 109
6.1 Introduction ... 109
6.2 Problem Formulation ... 111
6.3 Robust Optimal Attitude Controller Design ... 111
6.3.1 Extended State Observer Design ... 111

6.3.2 Inverse Optimal Controller Design with Uncertainties ... 114
6.3.3 CLF and Zero Dynamics Based Optimal Controller Design ... 117
6.4 Simulation Results ... 121
6.4.1 Comparison and Analysis Under Different Controllers ... 123
6.4.2 Comparisons and Analyses of Energy Consumptions Under Different Controllers ... 125
6.5 Conclusions ... 130
References ... 130

7 Spacecraft Attitude Fault-Tolerant Control Based on Iterative Learning Observer and Control Allocation ... 133
7.1 Introduction ... 133
7.2 Problem Formulation ... 135
7.3 Iterative Learning Observer Design ... 135
7.4 Control Strategy Design ... 137
7.4.1 Virtual Feedback Controller Design ... 137
7.4.2 Robust Control Allocation Design ... 138
7.5 Simulation Results ... 140
7.5.1 Case 1 ... 141
7.5.2 Case 2 ... 141
7.6 Conclusions ... 145
Appendices ... 149
References ... 154

8 Nonlinear Proportional-Derivative Control Incorporating Closed-Loop Control Allocation for Spacecraft ... 157
8.1 Introduction ... 157
8.2 Problem Formulation ... 159
8.3 Attitude Control Law Design ... 159
8.3.1 SPD-based Virtual/Baseline Control Law Design ... 160
8.3.2 A Closed-Loop Control Allocation Design with Optimization Method ... 162
8.4 Simulation Results ... 170
8.4.1 Attitude Control without Actuator Uncertainties and Disturbances ... 171
8.4.2 Attitude Control with Actuator Uncertainties and Disturbances ... 177
8.5 Conclusions ... 181
References ... 182

9 Closed-Loop Based Control Allocation for Spacecraft Attitude Stabilization with Actuator Faults 185
9.1 Introduction 185
9.2 Problem Formulation 187
9.2.1 System Schematic Diagram 187
9.2.2 Problem Formulation 188
9.3 Control Scheme Design and Stability Analysis 189
9.3.1 Baseline Control Law Design 189
9.3.2 Online Robust Control Allocation Design 189
9.3.3 Online Pseudo-Inverse Control Allocation 192
9.3.4 Closed-Loop Stability with Online Control Allocation Analysis 192
9.4 Simulation Results 194
9.4.1 Performance with FDD Estimation Uncertainty Bound $\eta = 5\%$ 196
9.4.2 Performance with FDD Estimation Uncertainty Bound $\eta = 33\%$ 203
9.5 Conclusions 203
Appendices 210
References 215

10 Conclusions 219
10.1 Conclusions 219
10.2 Open Problems and Challenges 220

Acronyms

ACS	Attitude Control System
FDD	Fault Detection and Diagnosis
FTC	Fault-Tolerant Control
PFTC	Passive Fault-Tolerant Control
AFTC	Active Fault-Tolerant Control
CA	Control Allocation
PICA	Pseudo-Inverse Control Allocation
OLCA	Open-Loop Control Allocation
CLCA	Closed-Loop Control Allocation
SMC	Sliding Mode Control
TSMC	Terminal Sliding Mode Control
FTSMC	Fast Terminal Sliding Mode Control
NTSMC	Nonsingular Terminal Sliding Mode Control
ISMC	Integral Sliding Mode Control
MRPs	Modified Rodrigues Parameters
PID	Proportional-Integration-Derivative
PD	Proportional-Derivative
SPD	Saturated Proportional-Derivative
ESO	Extended State Observer
CLF	Control Lyapunov Function
RW	Reaction Wheel
min	Minimum
max	Maximum
DCM	Direction Cosine Matrix
$\mathbf{R}$	The field of real numbers
$\mathbf{R}^{n \times m}$	The $n \times m$-dimensional real vector or matrix space
$\boldsymbol{I}_n$	An identity unit matrix with corresponding dimensions
$\lvert \cdot \rvert$	The absolute value of a scalar
$\lVert \cdot \rVert_p$	The p-norm of the vector or matrix, and $\lVert \cdot \rVert$ denotes the 2-norm $\lVert \cdot \rVert_2$ for simplify
$(\cdot)^{\times}$	A skew-symmetric matrix of corresponding vector, and it can be also represented as $\boldsymbol{S}(\cdot)$

$(\cdot)^{\mathrm{T}}$	The transposition of a vector or matrix
$(\cdot)^{-1}$	The inverse of a matrix
$(\cdot)^{\dagger}$	The pseudo-inverse of a matrix, $(\cdot)^{\dagger} = (\cdot)^{\mathrm{T}}\left((\cdot)(\cdot)^{\mathrm{T}}\right)^{-1}$
$\mathrm{Tanh}(\cdot)$	The standard hyperbolic tangent function
$\mathrm{diag}(\cdot)$	The diagonal matrix constituting by the vector $(\cdot)$, and the diagonal elements are the elements of the vector correspondingly
$\Delta(\cdot)$	The uncertainty or deviation of the variable $(\cdot)$ (scalar, vector, or matrix)

Chapter 1
Introduction

1.1 Background and Motivations

The exploration of deep space is of great significance and promotion for the human development and technological progress. It has been extensively and will be continuously supported by the USA, Europe, Russia and China, etc. Spacecraft is the core carrier of the space missions, and it is also one of the main and significant components of the space engineering systems. Thus, the growing development of different exploration missions have motivated significant requirements for the guidance, navigation, and control of spacecrafts. As one of the spacecraft's main subsystems, attitude control system (ACS) plays an important role for the on orbit spacecraft. It also determines the stability, reliability, autonomy and intelligence of the spacecraft during the space missions, even whether it can be accomplished successfully. For orbiting spacecraft, dynamics are strongly nonlinear in nature and are also affected by various and harsh disturbances from the environment that influence the mission objectives significantly. What is worse, some payloads or devices (such as sensors, actuators, etc.) would inevitably occur various faults or malfunctions. Whereas, actuator fault is the most serious and common faults in spacecraft ACS, accounting for about 44% of all faults [1]. These potential faults and uncertain effects deriving from internal or external causes will lead to the degradation of system performance, even may result in the completely collapse, immeasurable losses or catastrophic consequences. Therefore, it has significant practical values and theoretical significance to promote the stable and reliable operation abilities of spacecraft in the unknown and various space environments, especially in the presence of actuator uncertainties or faults, and external disturbances.

Recently, there exist some main approaches to improve the stability and reliability of the spacecraft ACS, such as high reliable control system design with redundant devices, fault detection and diagnosis (FDD) based fault tolerant control (FTC) system design and so on. Generally, modern spacecraft often use redundant actuators for enhanced reliability and safety to achieve reliable orbiting operation. In this situation, the number of actuators is usually more than the degree of freedom of

Q. Hu et al., *Control Allocation for Spacecraft Under Actuator Faults*,
https://doi.org/10.1007/978-981-16-0439-3_1

the ACS. That is to say, the spacecraft is then an over-actuated system, and there are freedoms to design the control systems by making use of the existing hardware redundancy due to the over-actuated actuators. To make effective use of these hardware actuator redundancies, the control allocation (CA) technique is one natural and necessary solution for achieving a desired control objective. It will distribute appropriately the lower dimension of the total control signals synthesized by the baseline controller to each individual actuator among the higher-dimension redundant actuators [2–4]. Additionally, the actuator redundancy of the spacecraft ACS can also introduce a design freedom to build an FTC system when some actuators undergo faults [5]. CA is also an alternative way to build a CA-based FTC system by utilizing the remaining active actuators via different strategies or algorithms. Hence, it has important significance to investigate CA strategy in consideration of multi-optimization objectives, various system constraints and actuator faults. On the other hand, FTC is a widely utilized method to deal with spacecraft attitude control in the present of actuator faults [6]. The control systems designed by FTC can achieve the control performances of the stability and reliability using the online compensations of the unknown faults reconstructed by the relevant control strategies. It is very important that the fault tolerance and safety are guaranteed by the designed control system with FTC capability. Therefore, FTC is also one of the most significant issues that should be taken into consideration. The design of the FTC scheme not only requires to ensure the closed-loop stability of the system under actuator faults, but also could maintain a certain degree of robustness against the various uncertainties and disturbances.

This work will present systematical and comprehensive descriptions of the CA and FTC issues for over-acutated spacecraft attitude control system under actuator faults, system uncertainties and disturbances. It intends to provide a legible platform for reference, and applicability of the fault tolerant attitude control and CA for different purposes in aerospace engineering and some related fields. In the following subsections, the introductions and the-state-of-art of CA and FTC methods and their applications in aeronautics and astronautics will be presented in details.

1.2 Introduction to Control Allocation

1.2.1 Control Allocation Methods

In general, there may be more effectors or actuators to generate the relevant control signals than the control degree of freedom for the control system, that is, the number of effectors or actuators than is greater than strictly needed to achieve the objectives of the specified control missions. Thus, the controllability of the chosen states and outputs would also be achieved with less control input signals. Such a control system is the so-called over-actuated control system. According to the survey paper in [7], Johansen et al. have defined three levels of the control algorithm hierarchy of motion control for the over-actuated mechanical systems with redundant effectors and actuators. That is, (1) A high-level motion control algorithm commands a vector of virtual control efforts (i.e. forces and moments) in order to meet the overall motion control objectives. (2) A CA algorithm coordinates the different effectors such that they together produce the desired virtual control efforts, if possible. (3) Low-level control algorithms may be used to control each individual effector via its actuators. Obviously, CA offers the advantage of a modular design where the high-level motion control algorithm can be designed without detailed knowledge about the effectors and actuators. And that, the actuators' constraints, faults and uncertainties, optimization objective for the control performances, could be take into consideration explicitly during the design of the CA schemes. More respectively, if the actuators/effectors occur some faults (time-varying control effectiveness, or other actuator faults), the control system reconfiguration could be performed by utilizing CA module without having to redesign the control schemes. It is an effective and significant way to improve the reliability and safety of the control system with some superiorities. Therefore, the CA methods have been widely investigated and applied in many over-actuated control systems.

For the following over-actuated control system,

$$\begin{aligned} \dot{\boldsymbol{x}} &= \boldsymbol{f}(\boldsymbol{x}, t) + \boldsymbol{u} \\ \boldsymbol{u} &= h(\boldsymbol{\tau}, x, t) \end{aligned} \tag{1.1}$$

where $\boldsymbol{x} \in \mathbf{R}^n$ is the state vector of the system; $\boldsymbol{f}$ is a function related with the states; $\boldsymbol{u} \in \mathbf{R}^n$ is defined as the virtual control or baseline control input vector; $\boldsymbol{\tau} \in \mathbf{R}^m$ is the actual control vector produced by the actuators subject to some constraints, such as limited position and rate magnitude, actuators' uncertainties or faults. $h(\tau, x, t)$ denotes a mapping function from the actual control signals to the virtual control in the over-actuated control system. Commonly, the liner case of the mapping function $h(\tau, x, t)$ in (1.1) is investigated extensively, and it yields to

$$\boldsymbol{u} = \boldsymbol{D}\boldsymbol{\tau} \tag{1.2}$$

where $\boldsymbol{D} \in \mathbf{R}^{n\times m}$ is the so-called CA matrix or distribution matrix, which is used to describe control signal distribution of the physical actuators for representing the influence of each actuator by the virtual control. If dimensions satisfy the constraint $n < m$, (1.1) is an over-actuated control system. In [7], Johansen et al. presented that the control system modularity allows the high-level motion control law to be designed without detailed knowledge about the effector/actuator system. To coordinating the effect of the different actuators/effectors in the over-actuated control system, the attractive issues such as effector/actuator fault tolerance, redundancy, and control constraints could be typically handled within the CA module. In general, the CA model $h(\tau, x, t)$ in (1.1) is usually selected to be static, so the low-level actuator control should handle the dynamic control of the actuators. The actuators/effectors utilization driven by CA module to produce relevant control signals can be treated independently and optimized for practical engineering applications under some constraints. However, it should also be mentioned that the CA scheme design and the high-level motion control law design cannot always be independent. In [8], the author has concluded that the control scheme design of the over-actuated control systems (such as aerospace control system, marine vessel control system, etc.) generally uses the principle of virtual controller based CA framework with the following two steps: (1) design a control law specifying which total control effort to be produced, known as virtual control law ($\boldsymbol{u}$ in (1.1)); and (2) design a CA scheme (or control allocator) that maps the total control demand determined by the virtual control law onto individual actuator settings, enforcing the actuators produce relevant control effect to accomplish the control missions.

According to the CA problem mentioned above, some constraints should be considered [8] to achieve more reliable and stable control performances. The actuator position constraints due to the physical limitations can be described in the form of

$$\tau_{\min} \leq \tau_{\mathrm{i}} \leq \tau_{\max}, i = 1, 2, \ldots, m \tag{1.3}$$

where $\tau_{\min}$ and $\tau_{\max}$ are the lower and upper actuator amplitude limitations. If the rate constraints of the actuators are also exist, one should construct the following inequalities

$$\eta_{\min} \leq \dot{\tau}_{\mathrm{i}} \leq \eta_{\max}, i = 1, 2, \ldots, m \tag{1.4}$$

where $\eta_{\min}$ and $\eta_{\max}$ denote the minimal and maximal actuator rate limitations, respectively. Since the control allocator is part of a digital control system, it is reasonable to approximate the time derivative as

$$\dot{\tau}_{\mathrm{i}}(t) \approx \frac{\tau_{\mathrm{i}}(t) - \boldsymbol{\tau}(t-T)}{T}$$

where T is the sampling time, and $\tau_{\mathrm{i}}(t-T)$ is the control torques in the previous sampling instant. Then one can get the overall actuator constraints as

$$\underline{\tau}_{\mathrm{i}}(t) \leq \tau_{\mathrm{i}}(t) \leq \bar{\tau}_{\mathrm{i}}(t)$$

with $\underline{\tau}_{\mathrm{i}}(t) = \max\{\tau_{\mathrm{i}}(t)_{\min}, \tau_{\mathrm{i}}(t-T) + T\min\{\eta_{\min}\}\}$ and $\bar{\tau}_{\mathrm{i}}(t) = \min\{\tau_{\mathrm{i}}(t)_{\max}, \tau_{\mathrm{i}}(t-T) + T\max\{\eta_{\max}\}\}$.

CA techniques have attracted extensive attentions in both theoretical study and practical applications [8]. In view of its applications, there exist abundant research results focus on aerospace engineering (spacecraft, aircraft, etc.), marine applications, and other over-actuated vehicles [9–16]. By analysing and studying the existing overview articles and research papers [6–8, 17, 18], the CA algorithms can be classified into the following categories: (1) The traditional CA methods; (2) Optimization based CA methods; (3) Dynamic CA methods; (4) Fault tolerant CA methods. The detailed introductions and analyses are presented as follows.

(1) **Traditional CA methods**. This category of CA includes daisy-chain allocation [19, 20], direct allocation [21] and so on. However, these methods cannot meet some specified control system performance requirements due to single functional capability or actuator saturation problem or loss of actuator effectiveness. Thus, the traditional CA methods have gradually become obsolete.

(2) **Optimization based CA methods**. The optimization based CA methods are attracted widely attention along with the development of the computational performances of the embedded or on-board computers. And it can take the physical constraints and control effectiveness into account by using the optimization algorithms. Reference [22] investigated and analysed contrastively the optimal control and control allocation for the actuator redundancy managements of the over-actuated nonlinear systems. The common method to solve the CA problem subject to $\boldsymbol{u} = \boldsymbol{D}\boldsymbol{\tau}$ is the pseudo-inverse CA (PICA) method. The solution of the CA problem by using PICA is $\boldsymbol{\tau} = \boldsymbol{D}^{\mathrm{T}}(\boldsymbol{D}\boldsymbol{D}^{\mathrm{T}})^{-1}\boldsymbol{u}$ [3, 23]. In the optimization based CA, it is converted into an optimization problem by minimizing the specified cost functions subject to constraints. To improve the CA efficiency, a redistributed pseudo inverse is proposed by using genetic algorithm to find the best pseudo-inverse matrix [24]. When the cost function is established by l_p-norm (1-norm, 2-norm or ∞-norm), linear programming and quadratic programming based algorithms can be utilized to solve the CA problems [25–28]. Nonlinear programming [29] has been also utilized to solve the CA formulation with nonlinear effector model, non-quadratic cost function and constraint sets that are not polyhedral [7]. Consider the multiple-objective of the cost function subject to constraints, a multiple-object nonlinear CA method was presented in [30]. For the over-actuated in-wheel-motor-driven electric vehicle, a multiple-object optimal CA scheme is proposed by combining a pre-allocation law and a control reallocation law, to improve the energy saving and driving performance [16]. In order to further deal with over-actuated control system where parameters affine in the effector model may be assumed unknown, an adaptive approach based CA method has been investigated [31] to achieve the asymptotic stability of the control system. And the proposed adaptive CA scheme in [18] enhanced the universality without requiring the estimation of uncertainties for the over-actuated control system under actuator uncertainties and constraints. The adaptive CA method is a new idea of combining optimization and adaptive technique, which will become a hot-spot in the field of CA. Whereas, the algorithm structure and the computation become more complex and larger along with the increasing of optimization objectives and various constraints. Additionally, the intelligent algorithms based CA schemes have also attracted scholars' attention due to the strong nonlinear function approximation performance of intelligent algorithms, such as neural networks, genetic algorithms, etc. [32, 33].

(3) **Dynamic CA methods**. In the practical engineering, the requirements of stability and accuracy of the CA-based control system are increasing. Although static CA methods are very popular, the dynamic CA technique has already received extensive attention [3]. The most difference between the dynamic CA and static CA is that the former can take the dynamic response characteristics of the actuator into consideration during the control system design. The dynamic characteristics

generally contain the actuators' position and rate constrains, and actuator dynamics. Ref. [34] has compared the static CA and dynamic CA techniques, and verified that introduce of CA could enhance control performance and stability of the over-actuated control system. In [35], a dynamic CA approach was proposed for the over-actuator control system with actuators' magnitude and rate limits. Considering actuators' dynamics, the CA problem was converted into a LMI problem in [36], which was solved by a straightforward methodology.

(4) **Fault tolerant CA methods**. In view of the possible uncertainties and faults of actuators in practical applications, the CA techniques have been already utilized to design the FTC system. One of the advantages of the CA is that the redundant actuators provide the ability of fault tolerance, by utilizing the available actuators without reconfiguring the control structure or scheme. In the presence of actuator faults, an effective reallocation of the virtual control signals to the remaining healthy actuators is needed to maintain system performance, which is referred to as a reconfigurable CA problem [37]. In the context of reconfigurable FTC, Zhang et al. [37, 38] present the concept of CA and reallocation for aircraft with redundant control effectors. Furthermore, for the sake of the overall system performance and stability, a high-level virtual controller is needed to provide the desired virtual control signals for the low-level CA module [39]. Consider an over-actuated control system under matched disturbance, model uncertainties and unknown actuator nonlinearities, a robust fault tolerant CA has been proposed by utilizing a weighted PICA method to distribute the virtual control signal to the actuators [40]. In [41], an online sliding mode CA scheme was proposed to distribute the demanded control effect to the available actuators without the reconfiguration of the designed controller when some actuator faults occur. Consider the unmanned aerial vehicle control system with the actuator stuck fault, Zhou et al. [42] proposed two CA schemes by utilizing the cascade pseudo-inverse method and the fixed point method respectively, and the effectiveness of the designed schemes was demonstrated by numerical simulations.

1.2.2 Applications of Control Allocation in Aeronautics and Astronautics

CA techniques have been investigated and taken into application widely in the aeronautics and astronautics fields due to their superiority to improve the performances for the over-actuated control systems [7]. For the aircraft being equipped with multiple control effectors, there exist abundant theoretical study and engineering application results of the CA methods for flight control systems. For the modern spacecraft, the redundant actuators (such as RWs, control moment gyros, thrusters and so on) are generally utilized to ensure the requirements of high precision, reliability and safety of ACSs. For the reentry vehicle during the flight phase from exoatmospheric to endoatmospheric, a mixed integer linear programming based quantized CA was utilized to distribute the actuators [43]. In [10], a hierarchical multiple-objective CA based control strategy was presented for different flight phases, and an active set based dynamic programming method was utilized to solve the real-time CA optimization problem. Consider the spacecraft ACS equipped with hybrid actuators (single-gimbal control moment gyros and RWs), an online dynamic CA was developed to distribute the control effects in [11], and the singularity problem was avoided effectively. In order to obtain better control performances for the on-ground aircraft runway centerline tracking, a weighted pesudo-inverse method based dynamic weighted CA scheme was presented in [44], to distribute the baseline control signal to the available actuators under some constraints. In [45], a robust constrained control law incorporating with off-line CA was proposed for multivariate over-actuated systems, and it was verified by numerical simulations using two fighter aircraft models under uncertainties of the normal control effectiveness matrix. Actually, the uncertain control effectiveness matrix does not take into consideration mostly in the design process of CA schemes. To deal with this problem, a new robust least-square CA method was proposed in [46] for the flight control system in the presence of disturbances. By considering the constrained attitude angular velocity and actuators' misalignments of the spacecraft ACS, a combination of finite-time control and robust least squares-based CA with uncertain configuration matrix was further proposed in [47].

In aerospace engineering, the ACS design under actuator faults is still an open problem and a challenge urgently to be addressed. As it is known, it is of great significance to introduce the CA technique in the flight control system and spacecraft ACS to improve the performance and robust fault tolerant capability of the control systems. In [48], a continuous integral sliding mode and a fixed CA method were utilized to design the FTC system, which was verified by the numerical simulations of the benchmark B747-100/200 civil aircraft model. In [49], a novel linear parameter varying (LPV) system representation based sliding mode FTC approach has been established by incorporating with a CA law. It is able to allocate the available actuators to execute the control command under actuator faults. Alwi et al. [50, 51] have carried out many researches on the CA and CA-based FTC approaches, and numerical simulations were performed on the flight control system or flight simulator. A novel adaptive sliding-mode-based online CA scheme to deal with simultaneous actuator faults in [52], which was applied and verified by a modified unmanned multi-rotors helicopter. For the over-actuated satellite ACS, Ref. [53] constructed the CA problem with an index function of minimum energy, and the optimal CA vector was corrected by using the zero-space technique. Consider the attitude tracking control issue of an over-actuated spacecraft under actuator faults and external disturbances, Shen et al. proposed a novel online CA based FTC scheme incorporating with an adaptive law to deal with the unknown inertia and actuator faults [54], and the uniformly ultimately bounded stability of the ACS was guaranteed; in addition, a robust CA method was investigated to redistribute control effects of the available actuators in [55], which was optimized by minimizing system energy consumption. In [56], a finite-time attitude compensation control scheme was firstly developed for spacecraft subject to actuator faults, misalignments, external disturbances and uncertain inertia parameters, and then a Karush–Kuhn–Tucker condition based optimal CA method was proposed by minimizing the energy consumption. For the attitude stabilization control of the combined spacecraft in the presence of inertia uncertainties and external disturbances, a dynamic infinity-norm constrained CA scheme was designed in [57], which was solved by using linear programming method; by further considering the uncertain control effectiveness matrix, actuator faults and saturation, a novel robust CA method incorporating with a disturbance observer based dynamic surface control law has been developed in [58]. In [59], Gui et al. proposed an online CA and an adaptive fault-tolerant pose tracking control law to deal with actuator faults, mass and inertia uncertainties and external disturbances for a rigid spacecraft. A novel block backstepping sliding mode and ESO based integrated guidance and control method was developed in [60] for the near space interceptor with aerodynamic fins and reaction jets, and a new dynamic CA algorithm was presented to distribute the baseline control signal among the actuators with position and rate constraints.

In order to analyse the various CA methods, Page et al. [61] compared the performances of 16 different CA algorithms in the open-loop and closed-loop. As it is known, the over-actuated control system design can be divided into two processes by introducing CA techniques, namely, the baseline/virtual control law design and the CA method design for the redundant actuators. It is generally recognized that the stability of the closed-loop over-actuated control system is guaranteed by the baseline/virtual control law. Whether the introduction of the CA component will affect the stability of the system, it needs to be analyzed and proved. Zhang et al. [62] designed a feedback control law and a cascaded generalized inverse CA scheme to achieve the high-precision pointing of satellites, and then provided a sufficient and necessary condition for the stability of the closed-loop control system. In order to ensure the stability of the closed-loop aircraft flight control system after introducing CA technique, Gai et al. [63] designed a class of closed-loop control schemes based on nonlinear dynamic inversion and dynamic CA, and the stability of the entire closed-loop system was analyzed in the presence of actuators' partial and stuck faults. For the spacecraft attitude stabilization control system with position and rate constraints of the redundant RWs, Hu et al. have proposed a closed-loop constrained optimal CA in [4], which was solved by the quadratic programming approach; by further taking actuator faults, actuator saturation, and external disturbances into consideration, an FDD based FTC scheme was developed incorporating with a novel online closed-loop constrained optimal fault tolerant CA scheme in [64], and the stability of the closed-loop fault tolerant CA process was analyzed utilizing the discrete-time feedback control system theory. In [65], a modified PD-type virtual controller incorporating with the online robust CA and online PICA has been proposed for attitude stabilization in the presence of actuator faults, and the stability of the closed-loop spacecraft ACS with online CA was guaranteed utilizing a Lyapunov-like analysis. In [66], an anti-disturbance attitude control law was first proposed for combined spacecraft, and then a modified closed-loop CA was modeled by minimizing the energy consumption and CA errors in the presence of actuator uncertainties. For the over-actuated aircraft flight control system under uncertainties and actuator faults, an adaptive closed-loop CA based super-twisting sliding mode controller was developed in [67], and the convergence of the adaptive closed-loop CA and the stability of the FTC system were analyzed by utilizing Lyapunov theory.

1.3 Introduction to Fault-Tolerant Control

1.3.1 Fault-Tolerant Control Methods

The IFAC SAFEPROCESS Technical Committee has given the definition of a fault in 1997, which is an unpermitted deviation of at least one characteristic property or parameter of the system from the acceptable/usual/standard condition [68, 69]. When the control systems occur some faults derived from actuators, sensors and other components, an FTC system can guarantee the continual safety and reliability or acceptable performance of the system utilizing its specified control structure to restrain or compensate the negative influences of faults [70]. Moreover, the fine performances of the control system can be guaranteed by the FTC scheme in normal situation as well. Thus, the need of the FTC systems in practical engineering applications has motivated significant researches in the last decades [6, 69, 71–77]. According to the redundant results of FTC, the design methods of a fault-tolerant controller are usually divided into the following two categories: Passive FTC (PFTC) and Active FTC (AFTC).

Generally, the PFTC is known as a fixed control structure to tolerate a limited types of faults, including some foreseeable faults. It does not need to design an FDD module to observe the fault in real time. This type of method owns strong robustness to system faults, and usually requires the system to be equipped with redundant hardware [78]. It can also attempt to ensure the stability and safety of the closed-loop control system under certain faults by utilizing some robust control techniques. By introducing the advanced control methods, such as adaptive control, sliding mode control, intelligent control, data-based control methods and so on, the FTC systems can not only guarantee the stability when the system components occurring faults, but it can also deal with the influences caused by system uncertainties and external disturbances. However, the realization of robustness and stability by PTFC is only at the expense of decreased nominal performances. And it is difficult to maintain the system performances when the PFTC system suffers from some severe unforeseen and sudden faults. Thus, it can be seen that the PFTC approaches have a certain conservative property.

As an alternative, AFTC has attracted much more attention. The biggest difference or superiority of AFTC is that the structure of the controllers are not fixed. In the design process of AFTC strategies, the real-time information about faults can be reconstructed and provided by the FDD module, which is usually utilized to reconfigure the control gain or structure. Figure 1.1 shows an example of the AFTC system block diagram for spacecraft attitude FTC system with CA. The FDD plays a very significant role in AFTC systems, aiming to reconstruct the fault information and achieve real-time decision making and controller reconfiguration [6]. It is an effective way to reduce the drawbacks of PFTC mentioned above, such that the control systems can achieve an acceptable performance. This kind of control methods has a strong property of autonomy, and can ensure that the system still be stable when the foreseen or unforeseen faults occur, by using the available hardware and

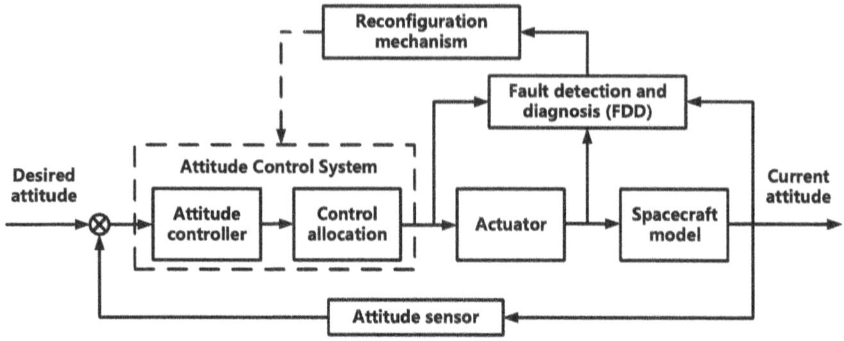

Fig. 1.1 Block diagram for spacecraft attitude fault tolerant control system with CA

software reasonably. Therefore, the advantages of the AFTC method can be described as follows [6, 79]: (1) Make full use of analytical redundancy; (2) Reconfigure the control scheme based on FDD; (3) Obtain fine performance under some faults; (4) The controller is with less conservativeness. Due to the aforementioned advantages and superiorities, AFTC has attracted extensive researches and applications.

1.3.2 Applications of Fault-Tolerant Control in Aeronautics and Astronautics

FTC is particularly significant for the safety-critical systems, such as aircraft, spacecraft, and so on. Zhang et al. presented an overall picture of historical, current, and future developments in the FTC field in [72]. A robust regional eigenvalue assignment technique based control law was presented for an over-actuated dynamic system in [80], which was also verified by utilizing an example of the aircraft bank angle control problem. For the spacecraft formation flying control system in the presence of model uncertainties and external disturbances, a PFTC based control scheme was proposed to guarantee the stability subject to thruster and sensor faults [81]. Due to the fine estimation performance for uncertainties, the adaptive control method has been utilized to damage estimations, and a robust integrated FTC was proposed for a flight control system under actuator faults, control effector damage and disturbances in [82]. Consider a spacecraft ACS subject to uncertain inertia moment and actuator faults, authors in [83] proposed an adaptive robust FTC scheme to guarantee the uniformly ultimately bounded stability of the attitude tracking system. Busan et al. [84] proposed an FTC method based on SMC and adaptive technique, to ensure the uniformly ultimately bounded stability. Hu et al. have done many researches and developed some interesting FTC approaches for the spacecraft ACS under uncertainties and actuator faults, by utilizing some advanced control techniques such as adaptive control and SMC to achieve stable, fast and reliable attitude control [85–

91]. An adaptive FTC based on a command filter was designed in [92] for rigid-body attitude tracking system subject to actuator faults and angular velocity constraints. For the spacecraft rendezvous and docking mission in the presence of input saturation and actuator faults, a smooth adaptive FTC based on the adaptive nonlinear state feedback method was presented in [93]. In [94], a novel adaptive fuzzy FTC has been designed to deal with the rigid spacecraft attitude tracking control problem under unavailable velocities, external disturbances, actuator faults and saturation; the fuzzy control technique was utilized in both the design of fuzzy adaptive observer and adaptive FTC scheme. In [95], an FTC based on fuzzy logic and backstepping method was designed for spacecraft ACS, subject to unknown inertia moment, disturbances, actuator faults and constraints. Ma et al. [73] proposed a discontinuous adaptive FTC scheme by using a common approximate function to restrain the chattering phenomenon, and adaptive laws to estimate the unknown parameters. In addition, SMC has also attracted extensive attention and application in FTC scheme design, due to its superior properties such as fast response, high steady-state precision and strong robustness to uncertainties and disturbances [96, 97]. For the satellite ACS with unknown and slow-varying satellite mass distribution and solar parameter, a robust FTC based on SMC and adaptive technique was proposed in [98]. To deal with the problem of satellite attitude maneuvering under disturbacnes and actuator faults, Lee et al. proposed a robust and fault-tolerant SMC scheme to achieve fast attitude maneuver in [99]. Bustan et al. proposed a time-varying SMC based attitude FTC for the attitude tracking of a spacecraft subject to saturation and actuator faults, and the global asymptotic stability of the ACS was guaranteed in [100]. For the spacecraft ACS with actuator faults and external disturbances, Shen et al. investigated the attitude FTC schemes utilizing SMC and adaptive technique in [101]; by further considering the actuator constraints, a fast terminal SMC (FTSMC) based finite-time FTC was presented by using adaptive methods to handle the uncertainties and input saturation [102]. Furthermore, some intelligent algorithms have also been used to deal with FTC problems in recent years. By utilizing wavelet analysis and neural network technique, Liang et al. [103] proposed an FTC for the satellite ACS under actuator faults, and the effectiveness and fine performance of the proposed scheme was demonstrated by numerical simulations. In [104], an FTC scheme for the spacecraft rendezvous mission under actuator faults and uncertainties was proposed based on genetic algorithm, and the asymptotic stability and robustness of the control system was guaranteed by using Lyapunov method. Consider the spacecraft attitude stabilization control system under actuator faults and constraints, a novel neural network based adaptive fixed-time FTC scheme has been developed to handle the catastrophic actuator faults and uncertainties in [105]; and then the authors improved the control law with fully adaptive-gain to enhance the performances of robustness and anti-saturation.

On the other hand, the FDD-based AFTC with favourable and powerful fault tolerance capabilities has attracted more and more attention from scholars and engineers. Generally, FDD is involved in the AFTC system as an essential module, such as the control system in Fig. 1.1. It refers to the process of judging whether some certain type of faults have occurred in the control system through detecting or diagnosing methods, and further determining the approximate location and degree of the faults [6, 106]. The approaches of FDD design can be divided into two main categories: The analytical model based FDD and the knowledge based FDD. The former FDD methods can obtain faults information by model and states of the control systems. In [107], a robust model-based fault detection and isolation scheme was presented for the atmospheric reentry vehicle under actuator faults. Chen et al. [108–110] investigated the iterative learning observer to achieve the fault detection for the nonlinear systems, and then proposed an iterative learning observer based on the attitude information in the previous sampling instant, to estimate the time-varying actuator faults for the satellite ACS in the presence of thruster faults. In [111], a novel FDD scheme was proposed to reconstruct the occurance time and amplitude of the actuator faults, by utilizing a sliding mode based fault detection observer and networks-based fault estimation observer. On the other hand, the knowledge based FDD methods include statistical methods, data based methods, qualitative methods and neural network, etc. Due to the complexity of spacecraft system, the accurate mathematical model of the control system and related devices or actuators can't be established. Thus, the model-free knowledge based FDD approaches have been attracted widely attention and applications [112–114]. Jiang et al. have developed an adaptive neural network observer to deal with control effector damage of the near space vehicle flight control system [115]. And then his research team has published a monograph [116] to introduce the FDD and FTC design of the nonlinear systems and relevant applications, by using fuzzy logic system, command filter, adaptive technique, neural network and so on. For spacecraft formation flight control system equipped with RWs as actuator, the authors in [117] developed a decentralized dynamic neural network based FDD scheme.

Base on the aforementioned FDD approaches, the FDD-based AFTC and related applications in aeronautics and astronautics have been investigated extensively in recent decades. By designing a modified iterative learning law to reconstruct the actuator faults and disturbances, an FTC scheme was developed in [118] for a satellite ACS. Zhang et al. [119] proposed an iterative learning observer to reconstruct the faults information of the RWs, and a robust fault-tolerant attitude controller was designed by utilizing the observation information. By further considering actuator uncertainties and constraints, an improved iterative learning observer was first designed to observe the torque deviations in [120], and then integral-type SMC (ISMC) law incorporating with robust CA method was developed to ensure the stability and performance of the closed-loop control system. In view of the spacecraft ACS in the presence of external disturbances, sensor measurement errors and actuator faults, the authors in [121] studied the problems of fault detection, isolation and control reconfiguration. For the sample data based satellite ACS, Shen et al. [122] designed a class of linear matrix inequality (LMI) based observers to estimate the states and faults at the same time, and the fault tolerance capability of the control system was analyzed rigorously. For spacecraft attitude stabilization control system under model uncertainties, actuator failures, and external disturbances simultaneously, an ESO based finite-time robust FTC scheme has been developed in [123]. In [124], a SMC and adaptive techniques based FDD scheme was first proposed to estimate the multiple actuator faults, and then a non-singular fast terminal SMC strategy was designed for a rigid spacecraft ACS with inertia uncertainties, external disturbances and input saturation. Consider the spacecraft ACS with unavailable attitude angular velocity information, actuator faults and constraints, a super-twisting observer and an adaptive law were designed to deal with the unknown angular velocity and the loss of effectiveness respectively [125], then an FTC scheme was developed by using neural networks method to handle the unknown nonlinear function. Shen et al. proposed an AFTC scheme for spacecraft ACS with actuator faults and limitations in [126]. The authors proposed the FDD to reconstruct the lumped faults containing multiplicative and additive actuator faults, and then a backstepping method based FTC scheme was developed by taking the fault estimation errors and actuator constraints into account.

1.4 Organization of the Book

The above sections present the background and motivations, the introduction of CA and FTC of the over-actuated spacecraft ACS under actuator faults, and then introduce briefly the applications of CA and FTC methods in aeronautics and astronautics in recent decades. Although there exist many results to deal with the issues of CA and FTC in the presence of external disturbances, uncertainties, actuator faults and constraints, it is necessary and significant to investigate the reasonable and effective approaches to handle the following problems.

(1) Online FDD strategies are still necessary to be developed and improved to reconstruct the actuator faults information for the spacecraft ACS under various uncertainties and constraints. In practical aerospace engineering, the unknown faults and uncertainties may result in the instability of the control system, even lead to the failure of the specified control missions. Thus, the design of a simple and effective FDD scheme with important engineering application value, is the significant fundamental issue of the FTC scheme design of the spacecraft ACS.

(2) The FTC approaches can be improved by promoting the properties of fast convergence and robustness to the unknown uncertainties and disturbances. The traditional FTC schemes generally can achieve the asymptotic or exponential stability of the control systems with infinite time. Such infinite time control methods cannot meet the requirements of some control performances, such as convergence, accuracy and robustness, in the presence of system faults and uncertainties. In the sense of optimal control, the finite-time control is the time-optimal control method. The finite-time control can not only promote the response speed, but also has strong robustness to the uncertainties and disturbances. Thus, utilizing the advanced control techniques, such as the finite-time control, intelligent algorithms, adaptive control, to design a class of FDD based AFTC schemes with the characteristics of strong fault tolerance and less conservativeness, is also an open and significant issue in the field of aeronautics and astronautics.

(3) The effectiveness and optimal performances of CA schemes could be improved, and the stability of whole closed-loop control system should also be analysed rigorously theoretically. In fact, the traditional CA methods cannot meet the specified control system performance requirements. In addition, the analysis of the stability of the closed-loop control system, including the modules of FDD, FTC and CA, is still a difficult problem to be well solved reasonably. Therefore, it is very necessary and important to investigate the CA scheme of the over-actuated spacecraft ACS under actuator faults and constraints (such as position and rate limitations), uncertainties and disturbances, in order to achieve some optimal indexes (such as minimizing energy consumption, minimizing CA errors, and so on) meeting practical engineering application requirements.

To address the above problems, this book will present systematical control system design schemes for the over-actuated spacecraft ACS, including optimal CA for spacecraft stabilization control, robust finite-time CA under actuator misalignment, finite-time fault-tolerant control with torque saturation, extended state observer based robust attitude control with input saturation, fault-tolerant control based on iterative learning observer and CA, nonlinear PD control incorporating closed-loop CA, closed-loop-based CA with actuator faults. The remaining chapters of this book are organized as follows.

This chapter has presented the background and motivations of this book, the introduction and the-state-of-art of CA and FTC methods and their applications in aeronautics and astronautics.

Chapter 2 introduces dynamic and kinematic models of the spacecraft ACS, then presents related attitude control and CA problem formulation.

Chapter 3 addresses the problem of spacecraft attitude stabilization control and optimal CA in the presence of disturbance and input saturation. Firstly, a SPD based virtual baseline nonlinear controller is developed to guarantee the globally asymptotic stability under input constraints and external disturbances. This is achieved with inexpensive online computations by dynamically adjusting a single parameter to ensure the desired performance. Then, a novel null-space-based optimal CA method is employed to map the specified virtual control command to the redundant actuators. The optimal control solution is obtained by penalizing the CA errors at a lower power/energy cost using quadratic programming algorithm.

Chapter 4 investigates the finite-time control and CA problems of the over-actuated spacecraft attitude stabilization control system with uncertain CA matrix (configuration/distribution matrix). The virtual control law is designed based on the nonsingular terminal sliding mode surface, and the finite-time stability of the attitude stabilization control system is guaranteed in the presence of external disturbances. Then, a robust least squares-based CA is developed to deal with the problem of distributing the three axis torques over the available actuators with constraints and uncertain CA matrix deriving from actuators misalignments.

In Chap. 5, the finite-time FTC problem of spacecraft attitude stabilization control system equipped with redundant thrusters is designed under external disturbances, input saturation, and inertia uncertainties. In order to improve the convergence rate of the spacecraft ACS, a novel time-varying fast terminal sliding-mode surface is first developed in conjunction with a judiciously designed control signal that rigorously obeys a known maximum-torque constraint. The controller can be viewed as a continuous analog to a variable structure approach, with the smoothness of the control permitted to vary with time according to a set of user-defined parameters. And then the control algorithm is extended to accommodate actuator faults while making use of the remaining active actuating capabilities by utilizing the PICA method. A Lyapunov-based analysis of the proposed formulation is provided, and finite-time stability of the closed-loop system is guaranteed.

Chapter 6 addresses the problem of robust optimal control of spacecraft attitude stabilization in the presence of external disturbances, actuator misalignments and constraints simultaneously. As a stepping stone, an ESO is developed to reconstruct the specified uncertainties, and the uniformly ultimately bounded stability of the estimation errors is ensured in the sense of finite-time stable. Utilizing the reconstructed information, an inverse optimal CLF approach is developed to guarantee asymptotic stability of the closed-loop system, such that an optimal/minimum performance index can be achieved. The associated stability proof is constructive and accomplished by the development of a novel Lyapunov function candidate. Furthermore, with the concept of input–output linearization of dynamics transverse to zero dynamic manifold, a rapid exponential stabilization CLF based optimal control scheme is investigated by utilizing the quadratic programming technique to restrain the actuator saturation. And the optimal analytical solution can provide the final control signals for the redundant actuators of the spacecraft ACS, to achieve the attitude stabilization control mission with the specified optimal performance index.

With further consideration to the actuator faults, uncertainties and constraints, Chap. 7 proposes an ILO based FTC scheme for the spacecraft attitude stabilization control system under external disturbances. More specifically, an ILO is firstly developed to estimate the torque deviation and steer the estimation errors into some small residual sets. And also the detailed derivations of the observer are provided, along with a thorough analysis for the associated ultimate bounded stability and estimation error convergence property. Then, an integral-type sliding mode control law is designed to produce the three-axis virtual control signals with the desired performance for being distributed among the individual actuators. Under this, a robust CA algorithm is developed to map the virtual control demand onto individual actuator in an optimal manner, which takes into account the estimation uncertainties and ensures some fault-tolerant ability. The key feature of the proposed strategies is that the whole closed-loop FTC system can be guaranteed theoretically to be stable by the development of Lyapunov methodology.

Chapter 8 attempts to solve the problem of closed-loop CA system design for spacecraft ACS under actuator constraints with respect to the actuator amplitude and rate constraints, and the stabilization analysis of the closed-loop control system under combination of CA with the virtual controller is provided by using discrete-time control system theory. More specifically, a SPD-based virtual nonlinear controller is designed to guarantee the globally asymptotic stability under control input/signal constraints. Then, a closed-loop constrained optimal CA scheme is employed to distribute the moments synthesized by the baseline controller over the redundant actuators with constraints due to actuator amplitude and rate constraints. The optimal control solution is to be found by penalizing the CA errors and rates using optimal quadratic programming algorithms.

Further taking the actuator faults into consideration during the design of the closed-loop CA system, Chap. 9 investigates the closed-loop CA problem for spacecraft attitude stabilization control system with actuator faults. Firstly, a PD-type controller is modified as a virtual control law by utilizing by the dynamic attitude-gain technique, to regulate the spacecraft maneuver and provide robustness to bounded fault-estimation uncertainty. Then, a robust CA method is developed under actuator faults in this chapter, and an analytical solution is obtained. Furthermore, the asymptotic stability of the whole closed-loop ACS including online CA mechanism is authenticated via a Lyapunov-like analysis.

In Chap. 10, some discusses and conclusions are presented.

References

1. Robertson, B., Stoneking, E. (2003) Satellite GN and C anomaly trends. Advances in the Astronautical Sciences 113: 531–542
2. Guillaume, J.J.D. (2009) Fault-tolerant flight control and guidance systems. Springer London
3. Härkegård, O. (2004) Dynamic control allocation using constrained quadratic programming. Journal of Guidance, Control, and Dynamics 27(6): 1028–1034
4. Hu, Q.L., Li, B., Zhang, Y.M. (2014) Nonlinear proportional-derivative control incorporating closed-loop control allocation for spacecraft. Journal of Guidance, Control, and Dynamics 37(3): 799–812
5. Argha, A., Su, S.W., Celler, B.G. (2019) Control allocation-based fault tolerant control. Automatica 103: 408–417
6. Yin, S., Xiao, B., Ding, S.X., Zhou, D.H. (2016) A review on recent development of spacecraft attitude fault tolerant control system. IEEE Transactions on Industrial Electronics 63(5): 3311–3320
7. Johansen, T.A., Fossen, T.I. (2013) Control allocation—A survey. Automatica 49(5): 1087–1103
8. Härkegård, O. (2003) Backstepping and control allocation with applications to flight control. PhD thesis, Linkoping Univ, Linkoping, Sweden
9. Burken, J.J., Lu, P., Wu, Z.L., Bahm, C. (2012) Two reconfigurable flight-control design methods: Robust servomechanism and control allocation. Journal of Guidance, Control, and Dynamics 24(3): 482–493
10. Sun, L.G., Zhou, Q., Jia, B.X., Tan, W.Q., Li, H.G. (2020) Effective control allocation using hierarchical multi-objective optimization for multi-phase flight. Chinese Journal of Aeronautics 33(7): 2002–2013
11. Hu, Q.L., Tan, X (2020) Dynamic near-optimal control allocation for spacecraft attitude control using a hybrid configuration of actuators. IEEE Transactions on Aerospace and Electronic Systems 56(2): 1430–1443
12. Lindegaard, K.P., Fossen, T.I. (2003) Fuel-efficient rudder and propeller control allocation for marine craft: Experiments with a model ship. IEEE Transactions on Control Systems Technology 11(6): 850–862
13. Feemster, M.G., Esposito, J.M. (2011) Comprehensive framework for tracking control and thrust allocation for a highly overactuated autonomous surface vessel. Journal of Field Robotics 28(1): 80–100

14. Tjonnas, J., Johansen, T.A. (2010) Stabilization of automotive vehicles using active steering and adaptive brake control allocation. IEEE Transactions on Control Systems Technology 18(3): 545–558
15. Chen, Y., Wang, J.M. (2014) Design and experimental evaluations on energy efficient control allocation methods for overactuated electric vehicles: Longitudinal motion case. IEEE/ASME Transactions on Mechatronics 19(2): 538–548
16. Jing, H.H., Jia, F.J., Liu, Z.Y. (2018) Multi-objective optimal control allocation for an over-actuated electric vehicle. IEEE Access 6: 4824–4833
17. Ma, J.J. (2009) Research and application of control allocation theory for overactuated system. PhD thesis, National University of Defense university, Hunan, China
18. Tohidi, S.S., Yildiz, Y., Kolmanovsky, I. (2020) Adaptive control allocation for constrained systems. Automatica 121: 109161
19. Adams, R.J., Buffington, J.M., Banda, S.S. (1994) Design of nonlinear control laws for high-angle-of-attack flight. Journal of Guidance, Control, and Dynamics 17(4): 737–746
20. Buffington, J.M., Enns, D.F. (1996) Lyapunov stability analysis of daisy chain control allocation. Journal of Guidance, Control, and Dynamics 19(6): 1226–1230
21. Durham, W.C. (1994) Constrained control allocation - three-moment problem. Journal of Guidance, Control, and Dynamics 17(2): 330–336
22. Härkegård, O., Glad, S.T. (2005) Resolving actuator redundancy—optimal control vs. control allocation. Automatica 41(1): 137–144
23. Durham, W. C. (1993) Constrained control allocation. Journal of Guidance, Control, and Dynamics 16(4): 717–725
24. Shi, J.P., Zhang, W.G., Li, G.G., Liu, X.X. (2010) Research on allocation efficiency of the redistributed pseudo inverse algorithm. Science China Information Science 53(2): 271–277
25. Bodson, M. (2002) Evaluation of optimization methods for control allocation. Journal of Guidance, Control, and Dynamics 25(4): 703–711
26. Bodson, M., Frost, S. (2010) Control allocation with load balancing. Journal of Guidance, Control, and Dynamics 34: 380–387
27. Petersen, J., Bodson, M. (2005) Interior-point algorithms for control allocation. Journal of Guidance, Control, and Dynamics 28(3): 471–480
28. Petersen, J., Bodson, M. (2006) Constrained quadratic programming techniques for control allocation. IEEE Transactions on Control Systems Technology 14(1): 91–98
29. Johansen, T.A., Fossen, T.I., Berge, S.P. (2004) Constrained nonlinear control allocation with singularity avoidance using sequential quadratic programming. IEEE Transactions on Control Systems Technology 12(1): 211–216
30. Yang, E.Q., Gao, J.Y., Li, W.Q. (2008) Research on multi-object nonlinear control allocation method. Acta Aeronautica et Astronautica Sinica 29(4): 995–1001
31. Tjønnås, J., Johansen, T.A. (2008) Adaptive control allocation. Automatica 44(10): 2754–2765
32. Luo, Y., Serrani, A., Yurkovich, S., Oppenheimer, M.W. and Doman, D.B. (2007) Model-predictive dynamic control allocation scheme for reentry vehicles. Journal of Guidance, Control, and Dynamics 30(1): 100–113
33. He, G., Zhou, J., Hu, W. (2010) A novel control allocation algorithm based on genetic algorithm GA and quadratic programming. Journal of Northwestern Polytechnical University 28(1): 23–26
34. Tavasoli, A., Naraghi, M. (2011) Comparison of static and dynamic control allocation techniques for integrated vehicle control. In: International Federation of Automatic Control World Congress, Milano, Italy
35. Zaccarian, L. (2009) Dynamic allocation for input redundant control systems. Automatica 45(6): 1431–1438
36. Kishore, W.C.A., Sen, S., Ray, G., Ghoshal, T.K. (2008) Dynamic control allocation for tracking time-varying control demand. Journal of Guidance, Control, and Dynamics 31(4): 1150–1157

37. Zhang, Y.M., Suresh, V.S., Jiang, B., Theilliol, D. (2007) Reconfigurable control allocation against aircraft control effector failures. In Proc. IEEE Int. Conf. Control Appl., Oct. 2007, Singapore, IEEE, pp. 1085–1992
38. Zhang, Y.M., Rabbath, C.A., Su, C.Y. (2008) Reconfigurable control allocation applied to an aircraft benchmark model. In Proc. Amer. Control Conf., Seattle, WA, USA, 11–13 June 2008, IEEE, pp. 1052–1057
39. Wang, B., Zhang, Y.M. (2018) An adaptive fault-tolerant sliding mode control allocation scheme for multirotor helicopter subject to simultaneous actuator faults. IEEE Transactions on Industrial Electronics 65(5): 4227–4236
40. Zhi, J.H., Chen, Y., Dong, X.M., Liu, Z.C., Shi, C. (2017) Robust adaptive FTC allocation for over-actuated systems with uncertainties and unknown actuator non-linearity. IET Control Theory and Applications 12(2): 273–281
41. Alwi, H., Edwards, C. (2008) Fault tolerant control using sliding modes with online control allocation. Automatica 44(7): 1859–1866
42. Zhou, Q.L., Zhang, Y.M., Rabbath, C.A., Apkarian, J. (2010) Two reconfigurable control allocation schemes for unmanned aerial vehicle under stuck actuator failures. In: AIAA Guidance, Navigation, and Control Conference, Toronto, Canada
43. Doman, D.B., Gamble, B.J., Ngo, A.D. (2009) Quantized control allocation of reaction control jets and aerodynamic control surfaces. Journal of Guidance, Control, and Dynamics 32(1): 13–24
44. Sadien, E., Roos, C., Birouche, A., Carton, M., Grimault, C., Romana,L.E., Basset, M. (2020) A simple and efficient control allocation scheme for on-ground aircraft runway centerline tracking. Control Engineering Practice 95: 104228
45. De Almeida, F.A. (2016) Robust off-line control allocation. Aerospace Science and Technology 52: 1–9
46. Cui, L., Yang, Y. (2011) Disturbance rejection and robust least-squares control allocation in flight control system. Journal of Guidance, Control, and Dynamics 34(6): 1632–1643
47. Hu, Q.L., Li, B., Zhang, A.H. (2013) Robust finite-time control allocation in spacecraft attitude stabilization under actuator misalignment. Nonlinear Dynamics 73(1–2): 53–71
48. Rios, H., Kamal, S., Fridman, L.M., Zolghadri, A. (2015) Fault tolerant control allocation via continuous integral sliding-modes: A HOSM-observer approach. Automatica 51: 318–325
49. Chen, L.J., Edwards, C., Alwi, H., Sato, M. (2020) Flight evaluation of a sliding mode online control allocation scheme for fault tolerant control. Automatica 114: 18829
50. Edwards, C., Alwi, H., Tan, C. (2012) Sliding mode methods for fault detection and fault tolerant control with application to aerospace systems. International Journal of Applied Mathematics and Computer Science 22(22): 109–124
51. Hamayun, M.T., Edwards, C., Alwi, H., Bajodah, A. (2015) A fault tolerant direct control allocation scheme with integral sliding modes. International Journal of Applied Mathematics and Computer Science 25(1): 93–102
52. Wang, B., Zhang, Y.M. (2018) An adaptive fault-tolerant sliding mode control allocation scheme for multirotor helicopter subject to simultaneous actuator faults. IEEE Transactions on Industrial Electronics 65(5): 4227–4236
53. Tsiotras, P., Shen, H., Hall, C. (2001) Satellite attitude control and power tracking with energy/momentum wheels. Journal of Guidance, Control, and Dynamics 24(1): 23–34
54. Shen, Q., Wang, D., Zhu, S., Poh, E.K. (2015) Inertia-free fault-tolerant spacecraft attitude tracking using control allocation. Automatica 62: 114–121
55. Shen, Q., Wang, D., Zhu, S., Poh, E.K. (2017) Robust control allocation for spacecraft attitude tracking under actuator faults. IEEE Transactions on Control Systems Technology 25(3): 1068–1075
56. Zhang, A.H., Hu, Q.L., Friswell, M.I. (2013) Finite-time fault tolerant attitude control for over-activated spacecraft subject to actuator misalignment and faults. IET Control Theory and Applications 7(16): 2007–2020
57. Huang, X.W., Duan, G.R. (2019a) Dynamic infinity-norm constrained control allocation for attitude tracking control of overactuated combined spacecraft. IET Control Theory and Applications 13: 1692–1703

58. Huang, X.W., Duan, G.R. (2019b) Robust control allocation in attitude fault-tolerant control for combined spacecraft under measurement uncertainty. IEEE Access 7: 156191–156206
59. Gui, H.C., De Ruiter, A.H.J. (2019) Adaptive fault-tolerant spacecraft pose tracking with control allocation. IEEE Transactions on Control Systems Technology 27(2): 479–494
60. Guo, C., Liang, X.G. (2015) Integrated guidance and control based on block backstepping sliding mode and dynamic control allocation. Proceedings of the Institution of Mechanical Engineers, Part G: Journal of Aerospace Engineering 229(9): 1559–1574
61. Page, A., Steinberg, M. (2012) A closed-loop comparison of control allocation methods. In: AIAA Guidance, Navigation, and Control Conference and Exhibit, Dever, USA
62. Zhang, Y., Chen, Z. (2012) A closed-loop control allocation method for satellite precision pointing. In: The 10th International Conference on Industrial Informatics, Beijing, China
63. Gai, W., Wang, H. (2013) Closed-loop dynamic control allocation for aircraft with multiple actuators. Chinese Journal of Aeronautics 26(3): 676–686
64. Li, B., Hu, Q.L., Ma, G.F., Yang, Y.S. (2019) Fault-tolerant attitude stabilization incorporating closed-loop control allocation under actuator failure. IEEE Transactions on Aerospace and Electronic Systems 55(4): 1989–2000
65. Hu, Q.L., Tan, X., Akella, M.R. (2018) Closed-loop-based control allocation for spacecraft attitude stabilization with actuator fault. Journal of Guidance, Control, and Dynamics 41(4): 943–952
66. Qiao, J., Liu, Z., Li, W. (2018) Anti-disturbance attitude control of combined spacecraft with enhanced control allocation scheme. Chinese Journal of Aeronautics 31(8): 1741–1751
67. Lu, B. , Ma, J., Zheng, Z. (2019) Adaptive closed-loop control allocation-based fault tolerant flight control for an overactuated aircraft. IEEE Access 7: 179505–179516
68. Isermann, R., Ballé, P. (1997) Trends in the application of model-based fault detection and diagnosis of technical processes. Control Engineering Practice 5(5): 709–719
69. Jain, T., Yamé, J.J., Sauter, D. (2017) Active Fault-Tolerant Control Systems:A Behavioral System Theoretic Perspective, 1st edn. Springer
70. Blanke, M., Izadi Zamanabadi, R., Bøgh, S.A., Lunau, C.P. (1997) Fault-tolerant control systems - A holistic view. Control Engineering Practice 5(5): 693–702
71. Blanke, M., Kinnaert, M., Lunze, J., Staroswiecki, M. (2006) Diagnosis and Fault-Tolerant Control. Springer Berlin Heidelberg
72. Zhang, Y.M., Jiang, J. (2008) Bibliographical review on reconfigurable fault-tolerant control systems. Annual Reviews in Control 32(2): 229–252
73. Ma, Y.J., Jiang, B., Tao, G., Cheng, Y.H. (2014) Actuator failure compensation and attitude control for rigid satellite by adaptive control using quaternion feedback. Journal of the Franklin Institute 351(1): 296–314
74. Guo, B., Chen, Y. (2019) Adaptive fast sliding mode fault tolerant control integrated with disturbance observer for spacecraft attitude stabilization system. ISA Transactions 94: 1–9
75. Shi, X.N., Zhou, Z.G., Zhou, D. (2020) Adaptive fault-tolerant attitude tracking control of rigid spacecraft on lie group with fixed-time convergence. Asian Journal of Control 22(1): 423–435
76. Yao, X.M., Wu, L.G., Guo, L. (2020) Disturbance-observer-based fault tolerant control of high-speed trains: A markovian jump system model approach. IEEE Transactions on Systems, Man, and Cybernetics: Systems 50(4): 1476–1485
77. Mao, Z.H., Yan, X.G., Jiang, B., Chen, M. (2020) Adaptive fault-tolerant sliding-mode control for high-speed trains with actuator faults and uncertainties. IEEE Transactions on Intelligent Transportation Systems 21(6): 2449–2460
78. Jiang, J., Yu, X. (2012) Fault-tolerant control systems: A comparative study between active and passive approaches. Annual Reviews in Control 36(1): 60–72
79. Zhang, Y.M., Jiang, J. (2001) Fault tolerant control systems design with consideration of performance degradation. In: American Control Conference, Arlington, USA
80. Jiang, J, Zhao, Q (2000) Design of reliable control systems possessing actuator redundancies. Journal of Guidance, Control, and Dynamics 23(4): 709–718

81. Godard, Kumar, K.D. (2010) Fault tolerant reconfigurable satellite formations using adaptive variable structure techniques. Journal of Guidance, Control, and Dynamics 33(3): 95–99
82. Boskovic, J.D., Bergstrom, S., Mehra, R.K. (2005) Robust integrated flight control design under failures, damage, and state-dependent disturbances. Journal of Guidance, Control, and Dynamics 28(5): 902–917
83. Cai, W.C., Liao, X.H., Song, D.Y. (2008) Indirect robust adaptive fault-tolerant control for attitude tracking of spacecraft. Journal of Guidance, Control, and Dynamics 31(5): 1456–1463
84. Bustan, D., Sani, S.K.H., Pariz,N. (2014) Adaptive fault-tolerant spacecraft attitude control design with transient response control. IEEE/ASME Transactions on Mechatronics 19(4): 1404–1411
85. Hu, Q.L. (2010) Robust adaptive sliding-mode fault-tolerant control with L2-gain performance for flexible spacecraft using redundant reaction wheels. IET Control Theory and Applications 4(6): 1055–1070
86. Xiao, B., Hu, Q.L., Zhang, Y.M. (2011) Adaptive sliding mode fault tolerant attitude tracking control for flexible spacecraft under actuator saturation. IEEE Transactions on Control Systems Technology 20(6): 1605–1612
87. Hu, Q.L., Xiao, L., Wang, C.L. (2019) Adaptive fault-tolerant attitude tracking control for spacecraft with time-varying inertia uncertainties. Chinese Journal of Aeronautics 32(3): 674–687
88. Hu, Q.L., Shao, X.D., Chen, W.H. (2018) Robust fault-tolerant tracking control for spacecraft proximity operations using time-varying sliding mode. IEEE Transactions on Aerospace and Electronic Systems 54(1): 2–17
89. Dong, H.Y., Hu, Q.L., Friswell, M. I., Ma, G.F. (2017) Dual-quaternion-based fault-tolerant control for spacecraft tracking with finite-time convergence. IEEE Transactions on Control Systems Technology 25(4): 1231–1242
90. Hu, Q.L., Shao, X.D., Guo, L. (2018) Adaptive fault-tolerant attitude tracking control of spacecraft with prescribed performance. IEEE/ASME Transactions on Mechatronics 23(1): 331–341
91. Shao, X.D., Hu, Q.L., Shi, Y., Jiang, B.Y. (2020) Fault-tolerant prescribed performance attitude tracking control for spacecraft under input saturation. IEEE Transactions on Control Systems Technology 28(2): 574–582
92. Shen, Q., Yue, C.F., Goh, C.H., Wu, B.L., Wang, D.W. (2018) Rigid-body attitude tracking control under actuator faults and angular velocity constraints. IEEE/ASME Transactions on Mechatronics 23(3): 1338–1349
93. Sun, L. (2019) Constrained adaptive fault-tolerant attitude tracking control of rigid spacecraft. Advances in Space Research 63(7): 2229–2238
94. Huo, B.Y., Xia, Y.Q., Yin, L.J., Fu, M.Y. (2017) Fuzzy adaptive fault-tolerant output feedback attitude-tracking control of rigid spacecraft. IEEE Transactions on Systems, Man, and Cybernetics: Systems 47(8): 1898–1908
95. Zou, A.M., Kumar, K.D. (2011) Adaptive fuzzy fault-tolerant attitude control of spacecraft. Control Engineering Practice 19(1): 10–21
96. Xu, S.D., Chen, C.C., Wu, Z.L. (2015) Study of nonsingular fast terminal sliding-mode fault-tolerant control. IEEE Transactions on Industrial Electronics 62(6): 3906–3913
97. Van, M., Mavrovouniotis, M., Ge, S.S. (2019) An adaptive backstepping nonsingular fast terminal sliding mode control for robust fault tolerant control of robot manipulators. IEEE Transactions on Systems, Man, and Cybernetics: Systems 49(7): 1448–1458
98. Varma, S., Kumar, K.D. (2010) Fault tolerant satellite attitude control using solar radiation pressure based on nonlinear adaptive sliding mode. Acta Astronautica 66(3–4): 486–500
99. Lee, H., Kim, Y. (2010) Fault-tolerant control scheme for satellite attitude control system. IET Control Theory and Applications 4(8): 1436–1450
100. Bustan, D., Pariz, N., Sani, S.K.H. (2014) Robust fault-tolerant tracking control design for spacecraft under control input saturation. ISA Transactions 53(4): 1073–1080
101. Shen, Q., Wang, D.W., Zhu, S.Q., Poh, E.K. (2015a) Integral-type sliding mode fault-tolerant control for attitude stabilization of spacecraft. IEEE Transactions on Control Systems Technology 23(3): 1131–1138

102. Shen, Q., Wang, D.W., Zhu, S.Q., Poh, E.K. (2015b) Finite-time fault-tolerant attitude stabilization for spacecraft with actuator saturation. IEEE Transactions on Aerospace and Electronic Systems 51(3): 2390–2405
103. Liang, J., Dong, C., Wang, Q. (2008) A fault-tolerant attitude control system for a satellite based on fuzzy global sliding mode control algorithm. In: The 7th International Symposium on Instrumentation and Control Technology: Measurement Theory and Systems and Aeronautical Equipment, Beijing, China
104. Li, Z., Yang, X.B., Gao, H.J. (2013) Autonomous impulsive rendezvous for spacecraft under orbital uncertainty and thruster faults. Journal of the Franklin Institute 350(9): 2455–2473
105. Zhou, N., Cheng, X.D., Xia, Y.Q., Liu, Y.J. (2020) Fully adaptive-gain-based intelligent failure-tolerant control for spacecraft attitude stabilization under actuator saturation. IEEE Transactions on Cybernetics pp 1–13 (Early Access)
106. Hwang, I., Kim, S., Kim, Y., Seah, C.E. (2010) A survey of fault detection, isolation, and reconfiguration methods. IEEE Transactions on Control Systems Technology 18(3): 636–653
107. Falcoz, A., Henry, D., Zolghadri, A. (2010) Robust fault diagnosis for atmospheric reentry vehicles. IEEE Transactions on Systems, Man, and Cybernetics, Part A: Systems and Humans 40(5): 886–899
108. Chen, W., Saif, M. (2001) An iterative learning observer-based approach to fault detection and accommodation in nonlinear systems. In: IEEE Conference on Decision and Control, pp 4469–4474
109. Chen, W., Chen, W.T., Saif, M., Li, M.F. (2013) Simultaneous fault isolation and estimation of lithium-ion batteries via synthesized design of luenberger and learning observers. IEEE Transactions on Control Systems Technology 22(1): 290–298
110. Chen, W., Saif, M. (2007) Observer-based fault diagnosis of satellite systems subject to time-varying thruster faults. Journal of Dynamic Systems Measurement & Control 129(3): 352–356
111. Gao, Z.F., Cheng, P., Qian, M.S., Jiang, G.P., Lin, J.X. (2018) Active fault-tolerant control approach design for rigid spacecraft with multiple actuator faults. Proceedings of the Institution of Mechanical Engineers 232(10): 1365–1378
112. Talebi, H.A., Khorasani, K. (2013) A neural network-based multiplicative actuator fault detection and isolation of nonlinear systems. IEEE Transactions on Control Systems Technology 21(3): 842–851
113. Yin, S., Li, X.W., Gao, H.J., Kaynak, O. (2015) Data-based techniques focused on modern industry: An overview. IEEE Transactions on Industrial Electronics 62(1): 657–667
114. Lu, K.F., Li, T.Y., Zhang, L.J. (2019) Active attitude fault-tolerant tracking control of flexible spacecraft via the chebyshev neural network. Transactions of the Institute of Measurement and Control 41(4): 925–933
115. Jiang, B., Xu, D.Z., Shi, P., Lim, C.C. (2014) Adaptive neural observer-based backstepping fault tolerant control for near space vehicle under control effector damage. IET Control Theory and Applications 8(9): 658–666
116. Shen, Q.K., Jiang, B., Shi, P. (2017) Fault Diagnosis and Fault-Tolerant Control Based on Adaptive Control Approach. Springer Berlin Heidelberg
117. Mousavi, S., Khorasani, K. (2014) Fault detection of reaction wheels in attitude control subsystem of formation flying satellites. International Journal of Intelligent Unmanned Systems 2(1): 2–26
118. Lee, H.J., Kim, Y.D., Kim, H.S. (2010) Satellite attitude control with a modified iterative learning law for the decrease in the effectiveness of the actuator. International Journal of Aeronautical and Space Sciences 11(2): 87–97
119. Zhang, A.H., Hu, Q.L., Zhang, Y.M. (2015) Observer-based attitude control for satellite under actuator fault. Journal of Guidance, Control, and Dynamics 38(4): 806–811
120. Hu, Q.L., Niu, G.L., Wang, C.L. (2018) Spacecraft attitude fault-tolerant control based on iterative learning observer and control allocation. Aerospace Science and Technology 75: 245–253

121. Baldi, P., Castaldi, P., Simani, S., Bertoni, G. (2010) Fault diagnosis and control reconfiguration for satellite reaction wheels. In: 2010 Conference on Control and Fault-Tolerant Systems, Nice, France
122. Shen, Y., Wang, Z.H., Zhang, X.L. (2013) Fault diagnosis and fault-tolerant control for sampled-data attitude control systems: An indirect approach. Proceedings of the Institution of Mechanical Engineers, Part G: Journal of Aerospace Engineering 228(7): 1047–1057
123. Li, B., Hu, Q.L., Yu, Y.B., Ma, G.F. (2017) Observer-based fault-tolerant attitude control for rigid spacecraft. IEEE Transactions on Aerospace and Electronic Systems 53(5): 2572–2582
124. Zhang, X.B., Zhou, Z.P. (2019) Integrated fault estimation and fault tolerant attitude control for rigid spacecraft with multiple actuator faults and saturation. IET Control Theory and Applications 13: 2365–2375
125. Wang, X.H., Tan, C.P. (2019) Fault-tolerant spacecraft attitude control under actuator saturation and without angular velocity. International Journal of Robust and Nonlinear Control 29(18): 6483–6506
126. Shen, Q., Yue, C.F., Goh, C.H., Wang, D.W. (2019) Active fault-tolerant control system design for spacecraft attitude maneuvers with actuator saturation and faults. IEEE Transactions on Industrial Electronics 66(5): 3763–3772

Chapter 2
Mathematical Model of the Attitude Control System

This chapter gives a brief introduction to the attitude kinematics and dynamics of a rigid spacecraft, as well as some related basic knowledge.

2.1 Spacecraft Attitude Kinematics

The attitude of the spacecraft refers to the relative azimuth or direction in a predetermined spatial reference coordinate system. In order to accurately describe the attitude or position, the appropriate reference coordinate systems are necessary to be established [1]. Generally speaking, the geocentric inertial coordinate system is usually selected as the reference coordinate system of a spacecraft. In addition, some auxiliary coordinates such as the body-fixed coordinate system and the orbital coordinate system are also required for the spacecraft orbit or attitude control systems. Please see the definitions of coordination reference frames for spacecraft attitude control system in Fig. 2.1.

(1) Geocentric inertial reference coordinate system $O_I X_I Y_I Z_I$. The origin O_I is placed at the center of the earth; the $O_I X_I$ axis points to the vernal equinox along the intersection of the earth's equatorial plane and the ecliptic plane; the $O_I Z_I$ axis is aligned with the earth's rotation axis pointing to the north pole of the earth. The $O_I Y_I$ axis lies in the equatorial plane to form a right-handed rotating coordinate system with the $O_I X_I$ and $O_I Z_I$ axes.

(2) Body coordinate system $O_B X_B Y_B Z_B$. It is a frame joined to the body with an origin attached to a point of the body (the rigid spacecraft's center of mass). The three axes $O_B X_B$, $O_B Y_B$ and $O_B Z_B$ are formed as a right-handed Cartesian system, and they are generally aligned with the three directions in the geometry characteristics of the spacecraft body.

Q. Hu et al., *Control Allocation for Spacecraft Under Actuator Faults*,
https://doi.org/10.1007/978-981-16-0439-3_2

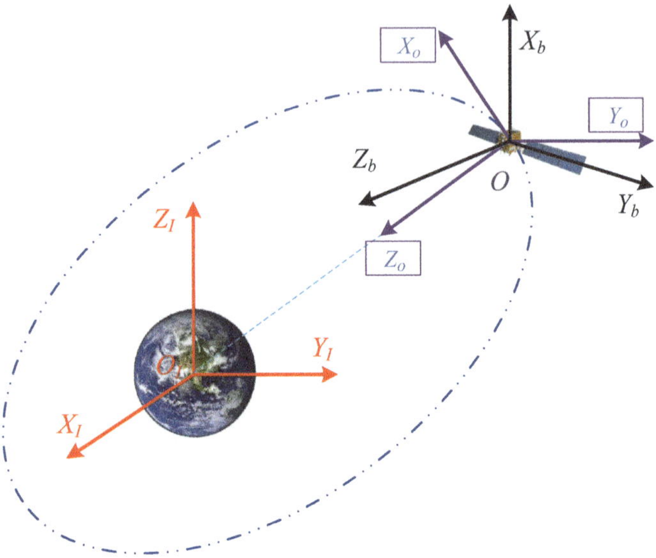

Fig. 2.1 Definitions of three coordination reference frames for spacecraft attitude control system

(3) Orbital coordinate system $OX_OY_OZ_O$. The origin is the rigid spacecraft's center of mass. The triad axes depend on the spacecraft's position and velocity. The OZ_O axis points to the center of the earth and it is formed the orbital plane coincides with the OX_O; the OX_O axis points to the flight direction of the spacecraft and is perpendicular to the OZ_O axis in the orbital plane. The OY_o axis satisfies the standard right-hand coordinate system law with the OZ_O and OX_O axis, and is parallel to the normal direction of the orbital plane.

The descriptions of the spacecraft attitude kinematics mainly include Euler angles, Quaternion and Modified Rodrigues Parameters (MRPs). The Euler angles (roll angle ϕ, pitch angle θ, and yaw angle ψ) based attitude description have the advantages of simple and intuitive, easy to implement with clear physical meanings. But, the singularity phenomenon should be taken into consideration during the application of Euler angles during the design of ACS. Thus, the quaternion and MRPs are utilized to present the attitude kinematics and design the ACS in this book, and the related definitions will be given as follows.

2.1.1 *Euler Angles Based Model*

According to [2], the direction cosine matrix (DCM) which describes the orientation of B relative to A, can be described in the form of

$$C^{B/A} = \begin{bmatrix} b_1 \cdot a_1 & b_1 \cdot a_2 & b_1 \cdot a_3 \\ b_2 \cdot a_1 & b_2 \cdot a_2 & b_2 \cdot a_3 \\ b_3 \cdot a_1 & b_3 \cdot a_2 & b_3 \cdot a_3 \end{bmatrix} = \boldsymbol{B}^{\mathrm{T}} \cdot \boldsymbol{A} \tag{2.1}$$

where $\boldsymbol{B} = [b_1, b_2, b_3]$ and $\boldsymbol{A} = [a_1, a_2, a_3]$ are the right-hand set of three orthogonal unit vectors, and they also present two reference frames. The DCM $\boldsymbol{C}^{B/A}$ is also called the rotation matrix or coordinate transformation matrix to $\boldsymbol{B}$ from $\boldsymbol{A}$. Similarly, one can easily obtain the DCM $\boldsymbol{C}^{A/B}$ presenting a coordinate transformation matrix to $\boldsymbol{A}$ from $\boldsymbol{B}$ in the following formulation

$$C^{A/B} = \begin{bmatrix} a_1 \cdot b_1 & a_1 \cdot b_2 & a_1 \cdot b_3 \\ a_2 \cdot b_1 & a_2 \cdot b_2 & a_2 \cdot b_3 \\ a_3 \cdot b_1 & a_3 \cdot b_2 & a_3 \cdot b_3 \end{bmatrix} = \boldsymbol{A}^{\mathrm{T}} \cdot \boldsymbol{B} \tag{2.2}$$

It should be noted that the DCM is an orthonormal matrix that satisfies $\boldsymbol{C}\boldsymbol{C}^{\mathrm{T}} = \boldsymbol{I} = \boldsymbol{C}^{\mathrm{T}}\boldsymbol{C}$. And then, three elementary rotations respectively about the first, second, and third axes of the reference frame $\boldsymbol{A}$ are described by the following rotation matrices

$$\boldsymbol{C}_1(\theta_1) = \begin{bmatrix} 1 & 0 & 0 \\ 0 & \cos\theta_1 & \sin\theta_1 \\ 0 & -\sin\theta_1 & \cos\theta_1 \end{bmatrix} \tag{2.3}$$

$$\boldsymbol{C}_2(\theta_2) = \begin{bmatrix} \cos\theta_2 & 0 & -\sin\theta_2 \\ 0 & 1 & 0 \\ \sin\theta_2 & 0 & \cos\theta_2 \end{bmatrix} \tag{2.4}$$

$$\boldsymbol{C}_3(\theta_3) = \begin{bmatrix} \cos\theta_3 & \sin\theta_3 & 0 \\ -\sin\theta_3 & \cos\theta_3 & 1 \\ 0 & 0 & 1 \end{bmatrix} \tag{2.5}$$

in which $\boldsymbol{C}_1(\theta_1)$, $\boldsymbol{C}_2(\theta_2)$ and $\boldsymbol{C}_3(\theta_3)$ represent the DCM $\boldsymbol{C}$ of an elementary rotation about the ith axis of $\boldsymbol{A}$ with the relative angles θ_1, θ_2 and θ_3, respectively.

In the practical aerospace missions, it is possible to bring a rigid body into an arbitrary orientation by performing three successive rotations that involve the axes fixed in an inertial reference frame. This scheme will then provide another 12 sets of Euler angles for the so-called space-axis rotations. Here we also introduce a particular sequence that presents the three successive rotations from the reference frame $\boldsymbol{B}$ to $\boldsymbol{A}$ according to [2]. Define two other intermediate reference frames representing in the form of $\boldsymbol{A}' = [a_1', a_2', a_3']$ and $\boldsymbol{A}'' = [a_1'', a_2'', a_3'']$. Consider the following rotation sequence

$$\begin{aligned} \boldsymbol{C}_3(\theta_3) &: \boldsymbol{A}' \leftarrow \boldsymbol{A} \\ \boldsymbol{C}_2(\theta_2) &: \boldsymbol{A}'' \leftarrow \boldsymbol{A}' \\ \boldsymbol{C}_1(\theta_1) &: \boldsymbol{B} \leftarrow \boldsymbol{A}'' \end{aligned} \tag{2.6}$$

And the corresponding rotation is presented as follows

$$\begin{bmatrix} a_1' \\ a_2' \\ a_3' \end{bmatrix} = \boldsymbol{C}_3(\theta_3) \begin{bmatrix} a_1 \\ a_2 \\ a_3 \end{bmatrix}$$

$$\begin{bmatrix} a_1'' \\ a_2'' \\ a_3'' \end{bmatrix} = \boldsymbol{C}_2(\theta_2) \begin{bmatrix} a_1' \\ a_2' \\ a_3' \end{bmatrix}$$

$$\begin{bmatrix} b_1 \\ b_2 \\ b_3 \end{bmatrix} = \boldsymbol{C}_1(\theta_1) \begin{bmatrix} a_1'' \\ a_2'' \\ a_3'' \end{bmatrix}$$

And then the three rotation angles θ_1, θ_2 and θ_3 are called Euler angles. By combining the above three rotations according to the sequence in (2.6), one can obtain that

$$\begin{aligned} \begin{bmatrix} b_1 \\ b_2 \\ b_3 \end{bmatrix} &= \boldsymbol{C}_1(\theta_1) \begin{bmatrix} a_1'' \\ a_2'' \\ a_3'' \end{bmatrix} = \boldsymbol{C}_1(\theta_1)\boldsymbol{C}_2(\theta_2) \begin{bmatrix} a_1' \\ a_2' \\ a_3' \end{bmatrix} \\ &= \boldsymbol{C}_1(\theta_1)\boldsymbol{C}_2(\theta_2)\boldsymbol{C}_3(\theta_3) \begin{bmatrix} a_1 \\ a_2 \\ a_3 \end{bmatrix} \end{aligned} \tag{2.7}$$

Define the Euler rates as $\dot{\theta}_1$, $\dot{\theta}_2$ and $\dot{\theta}_3$, and angular velocity vector as $\boldsymbol{\omega} = [\omega_1, \ \omega_2, \ \omega_3]^{\mathrm{T}}$. One can get [2]

$$\begin{aligned} \begin{bmatrix} \omega_1 \\ \omega_2 \\ \omega_3 \end{bmatrix} &= \begin{bmatrix} \dot{\theta}_1 \\ 0 \\ 0 \end{bmatrix} + \boldsymbol{C}_1(\theta_1) \begin{bmatrix} 0 \\ \dot{\theta}_2 \\ 0 \end{bmatrix} + \boldsymbol{C}_1(\theta_1)\boldsymbol{C}_2(\theta_2) \begin{bmatrix} 0 \\ 0 \\ \dot{\theta}_3 \end{bmatrix} \\ &= \begin{bmatrix} 1 & 0 & -\sin\theta_2 \\ 0 & \cos\theta_1 & \sin\theta_1 \cos\theta_2 \\ 0 & -\sin\theta_1 & \cos\theta_1 \cos\theta_2 \end{bmatrix} \begin{bmatrix} \dot{\theta}_1 \\ \dot{\theta}_2 \\ \dot{\theta}_3 \end{bmatrix} \end{aligned} \tag{2.8}$$

Then, the kinematic model represented by Euler angles can be written in the form of

$$\begin{bmatrix} \dot{\theta}_1 \\ \dot{\theta}_2 \\ \dot{\theta}_3 \end{bmatrix} = \frac{1}{\cos\theta_2} \begin{bmatrix} \cos\theta_2 & \sin\theta_1 \sin\theta_2 & \cos\theta_1 \sin\theta_2 \\ 0 & \cos\theta_1 \cos\theta_2 & -\sin\theta_1 \cos\theta_2 \\ 0 & \sin\theta_1 & \cos\theta_1 \end{bmatrix} \begin{bmatrix} \omega_1 \\ \omega_2 \\ \omega_3 \end{bmatrix} \tag{2.9}$$

It should be noted that the above kinematic equation will be singular, when $\theta_2 = \frac{\pi}{2}$. Thus, we will introduce the quaternion based attitude representation without singularity in the following subsection.

2.1.2 Quaternion Based Model

Quaternion based attitude representation is widely utilized for the spacecraft ACS. According to the Euler theorem, four attitude parameters composing by the Euler axis/angles can be presented in the form of

$$\boldsymbol{q} = \left[\cos\frac{\Phi}{2}, e_x \sin\frac{\Phi}{2}, e_y \sin\frac{\Phi}{2}, e_z \sin\frac{\Phi}{2}\right]^{\mathrm{T}} \tag{2.10}$$

where $\boldsymbol{e} = \left[e_x, e_y, e_z\right]^{\mathrm{T}} \in \mathbf{R}^{3\times 1}$ is the Euler rotation axis and satisfying $e_x^2 + e_y^2 + e_z^2 = 1$; $\Phi \in \mathbf{R}$ is the rotation angle around the Euler rotation axis. By denoting $\boldsymbol{q} = [q_0, q_1, q_2, q_3]^{\mathrm{T}} \in \mathbf{R}^{4\times 1}$ the above four parameters apparently satisfy the following relationship

$$q_0^2 + q_1^2 + q_2^2 + q_3^2 = 1 \tag{2.11}$$

Then, the unit quaternion $\boldsymbol{q}$ formed by the above four parameters can be defined as

$$\begin{aligned} \boldsymbol{q} &= q_0 + iq_1 + jq_2 + kq_3 \\ &= [q_0, q_1, q_2, q_3]^{\mathrm{T}} \\ &= \left[q_0, \boldsymbol{q}_v\right]^{\mathrm{T}} \end{aligned} \tag{2.12}$$

where q_0 and $\boldsymbol{q}_v = [q_1, q_2, q_3]^{\mathrm{T}} \in \mathbf{R}^{3\times 1}$ represent the scalar and vector parts of the quaternion $\boldsymbol{q}$ respectively; i, j, k are imaginary units. Define the conjugate or inverse $\boldsymbol{q}^*$ of the quaternion $\boldsymbol{q}$ as follows

$$\begin{aligned} \boldsymbol{q}^* &= q_0 - iq_1 - jq_2 - kq_3 \\ &\triangleq [q_0, -q_1, -q_2, -q_3]^{\mathrm{T}} \\ &\triangleq \left[q_0, -\boldsymbol{q}_v\right]^{\mathrm{T}} \end{aligned} \tag{2.13}$$

Then, one can obtain that the 2-norm of unit quaternion $\|\boldsymbol{q}\| \equiv \sqrt{\boldsymbol{q}\boldsymbol{q}^*} = \sqrt{\boldsymbol{q}^*\boldsymbol{q}} = \sqrt{q_0^2 + q_1^2 + q_2^2 + q_3^2} = 1$.

To avoid the singularity phenomenon deriving from Euler angles based attitude representation, the spacecraft attitude kinematics described by quaternion can be expressed as

$$\dot{\boldsymbol{q}} = \frac{1}{2}\Xi(\boldsymbol{q})\boldsymbol{\omega} = \frac{1}{2}\begin{bmatrix} -\boldsymbol{q}_v^{\mathrm{T}} \\ q_0\boldsymbol{I}_3 + \boldsymbol{q}_v^{\times} \end{bmatrix}\boldsymbol{\omega} \tag{2.14}$$

where $\boldsymbol{\omega} = [\omega_1, \omega_2, \omega_3]^{\mathrm{T}} \in \mathbf{R}^{3\times 1}$ denotes the attitude angular velocity vector. $\boldsymbol{I}_n$ represents an identity unit matrix with corresponding dimensions, and $\boldsymbol{I}_3 = \mathrm{diag}([1, 1, 1]) \in \mathbf{R}^{3\times 3}$ is defined in (2.14). The operator $(\cdot)^{\times}$ represents the skew-symmetric matrix satisfying

$$\boldsymbol{x}^{\times} = \begin{bmatrix} 0 & -x_3 & x_2 \\ x_3 & 0 & -x_1 \\ -x_2 & x_1 & 0 \end{bmatrix}, \quad \boldsymbol{x} = [x_1, x_2, x_3]^{\mathrm{T}} \in \mathbf{R}^{3\times 1} \tag{2.15}$$

and it can be also represented as $\boldsymbol{S}(\boldsymbol{x})$.

2.1.3 Modified Rodrigues Parameters Based Model

Modified Rodrigues parameters based attitude presentation is a three-parameters and non-redundant attitude description method derived from unit quaternion. The MRPs are defined as

$$\begin{aligned} \rho_i &= \frac{q_i}{1+q_0}, \quad i = 1,2,3 \quad \text{or} \\ \boldsymbol{\rho} &= \boldsymbol{e}\tan\left(\frac{\Phi}{4}\right), \quad \Phi \in \left[0^\circ, 360^\circ\right) \end{aligned} \tag{2.16}$$

where $\boldsymbol{\rho} = [\rho_1, \rho_2, \rho_3]^{\mathrm{T}} \in \mathbf{R}^{3\times 1}$ is the attitude vector described by MRPs.

On the basis of the above attitude presentation, the kinematics equation of the spacecraft attitude described by MRPs is in the form of

$$\dot{\boldsymbol{\rho}} = \boldsymbol{G}(\boldsymbol{\rho})\boldsymbol{\omega} \tag{2.17}$$

where the matrix $\boldsymbol{G}(\boldsymbol{\rho})$ is given by

$$\begin{aligned} \boldsymbol{G}(\boldsymbol{\rho}) &= \frac{1}{2}\left(\boldsymbol{I} - \boldsymbol{\rho}^{\times} + \boldsymbol{\rho}\boldsymbol{\rho}^{\mathrm{T}} - \left(1 + \boldsymbol{\rho}^{\mathrm{T}}\boldsymbol{\rho}\right)\boldsymbol{I}/2\right) \\ &= \frac{1}{4}\boldsymbol{A} \end{aligned} \tag{2.18}$$

with

$$\boldsymbol{A} = \begin{bmatrix} 1+\sigma_1^2-\sigma_2^2-\sigma_3^2 & 2\left(\sigma_1\sigma_2-\sigma_3\right) & 2\left(\sigma_3\sigma_1+\sigma_2\right) \\ 2\left(\sigma_1\sigma_2+\sigma_3\right) & 1+\sigma_2^2-\sigma_3^2-\sigma_1^2 & 2\left(\sigma_2\sigma_3-\sigma_1\right) \\ 2\left(\sigma_3\sigma_1-\sigma_2\right) & 2\left(\sigma_2\sigma_3+\sigma_1\right) & 1+\sigma_3^2-\sigma_1^2-\sigma_2^2 \end{bmatrix}$$

In addition, the matrix $\boldsymbol{G}\left(\boldsymbol{\rho}\right)$ has the following properties

$$\begin{aligned} \boldsymbol{G}^{-1}(\boldsymbol{\rho}) &= \frac{16}{\left(1+\boldsymbol{\sigma}^{\mathrm{T}}\boldsymbol{\rho}\right)^2}\boldsymbol{G}^{\mathrm{T}}(\boldsymbol{\rho}) \\ &= \left(1-\boldsymbol{\rho}^{\mathrm{T}}\boldsymbol{\rho}\right)\boldsymbol{I}_3 - 2\left[\boldsymbol{\rho}^{\times}\right] + 2\boldsymbol{\rho}\boldsymbol{\rho}^{\mathrm{T}} \\ &= \frac{4}{\left(1+\boldsymbol{\rho}^{\mathrm{T}}\boldsymbol{\rho}\right)^2}\boldsymbol{A}^{\mathrm{T}} \end{aligned} \tag{2.19a}$$

$$\boldsymbol{\rho}^{\mathrm{T}}\boldsymbol{G}(\boldsymbol{\rho})\boldsymbol{\omega} = \left(\frac{1+\boldsymbol{\rho}^{\mathrm{T}}\boldsymbol{\rho}}{4}\right)\boldsymbol{\rho}^{\mathrm{T}}\boldsymbol{\omega} \tag{2.19b}$$

$$\boldsymbol{G}^{\mathrm{T}}(\boldsymbol{\rho})\boldsymbol{G}(\boldsymbol{\rho}) = \left(\frac{1+\boldsymbol{\rho}^{\mathrm{T}}\boldsymbol{\rho}}{4}\right)^2 \boldsymbol{I}_3 \tag{2.19c}$$

2.2 Spacecraft Attitude Dynamics Model

In general, the spacecraft is the so-called over-actuated control system due to the equipment with redundant actuators (such as reaction wheels, control moment gyros, thrusters and so on), to improve the safety and reliability of the aerospace missions. That is to say, the number of available or working actuators is more than the control freedom. For the spacecraft ACS, the control freedom is equal to $n = 3$, which is the number of the control signals to be designed and distributed to redundant actuators with specified configuration. Suppose that the number of the actuators is defined as m. When $m > 3$, the spacecraft ACS is an over-actuated control system, and it is very necessary and effective to develop the control schemes by introducing CA. Consider the CA module, the dynamic equations of the spacecraft ACS can be presented in the form of

$$\begin{aligned} \boldsymbol{J}\dot{\boldsymbol{\omega}} &= -\boldsymbol{\omega}^{\times}\boldsymbol{J}\boldsymbol{\omega} + \boldsymbol{u} + \boldsymbol{d} \\ \boldsymbol{u} &= \boldsymbol{D}\boldsymbol{\tau} \end{aligned} \tag{2.20}$$

where $\boldsymbol{J} \in \mathbf{R}^{3\times3}$ denotes the moments of the inertia matrix, which is symmetric positive. $\boldsymbol{u} = [u_1, u_2, u_3]^{\mathrm{T}} \in \mathbf{R}^{3\times1}$ is the control torque acting on the spacecraft. $\boldsymbol{d} = [d_1, d_2, d_3]^{\mathrm{T}} \in \mathbf{R}^{3\times1}$ is the external disturbances, which are stemming from the complex and various deep-space environments. $\boldsymbol{\tau} = [\tau_1, \tau_2, \ldots, \tau_m]^{\mathrm{T}} \in \mathbf{R}^{m\times1}$ denotes the actual control torques produced by the redundant actuators. $\boldsymbol{D} \in \boldsymbol{R}^{3\times m}$ presents the CA matrix or distribution matrix, which is used to describe the distribution of the designed control signal $\boldsymbol{u} \in \mathbf{R}^{3\times1}$ to the m actuators, such that each actuator could generate the demanded control torque or effect τ_i under some specified constraints according to the solution of CA. In addition, the physical position limitations, configuration deviations and various faults of the actuators will also be taken into consideration in the following chapters.

In view of Eq. (2.20), $\boldsymbol{\tau}$ denotes the actual control torques produced by the redundant actuators. In this book, reaction wheels (RWs) are mainly equipped in the spacecraft ACS as the actuators. According to principle of conservation of angular momentum, the attitude control of the spacecraft is realized by using the angular momentum exchange between the RWs and the spacecraft itself to provide the attitude control torques. A RW is a kind of momentum exchange device, which generally has two working modes: torque mode and speed mode. It is vulnerable to the effects derived from the harsh space environment, friction torque and other severe conditions, during the RWs' long time on-orbiting operation in the deep space. As described in [3–5], RWs are vulnerable to four main sources of failures: Failure 1 (F1), decreased reaction torque; Failure 2 (F2), continuous generation of reaction torque; Failure 3 (F3), increased bias torque; and Failure 4 (F4), failure to respond to control signals. Accordingly, the actual output torque generated by the above failures can be mathematically modeled for each RW by

$$\tau_i = e_{ii} v_i + \bar{v}_i \tag{2.21}$$

where v_i ($i = 1,2,\ldots,m$) is the desired control torque allocating to the individual actuator; $\bar{v}_i$ represents additive fault induced by F2 or F3, and $e_{ii} \in [0, 1]$ characterizes the actuator control effectiveness. Then, the actuators' faults can be categorised as the following five cases: (1) $e_{ii} = 1, \bar{v}_i = 0$ implies that the ith RW is working perfectly without any failure; (2) $0 < e_{ii} < 1, \bar{v}_i = 0$ indicates the ith RW undergoes F1; (3) $e_{ii} = 0, \bar{v}_i \neq 0$ indicates the ith RW undergoes F2; (4) $e_{ii} = 1, \bar{v}_i \neq 0$ shows the occurrence of F3; (5) $e_{ii} = 0, \bar{v}_i = 0$ implies completely fault of actuator.

Accordingly, the spacecraft dynamics under failures can be established as

$$\begin{aligned} \boldsymbol{J}\dot{\boldsymbol{\omega}} &= -\boldsymbol{\omega}^{\times}\boldsymbol{J}\boldsymbol{\omega} + \boldsymbol{D}\boldsymbol{E}\boldsymbol{v} + \boldsymbol{D}\bar{\boldsymbol{v}} + \boldsymbol{d} \\ &\triangleq -\boldsymbol{\omega}^{\times}\boldsymbol{J}\boldsymbol{\omega} + \boldsymbol{D}\boldsymbol{E}\boldsymbol{v} + \bar{\boldsymbol{d}} \end{aligned} \tag{2.22}$$

with $\bar{\boldsymbol{d}} = \boldsymbol{D}\bar{\boldsymbol{v}} + \boldsymbol{d}$. $\boldsymbol{E} = \mathrm{diag}\,(e_{11}, e_{22}, \ldots, e_{mm}) \in \mathbf{R}^{m \times m}$ is characterizing the health condition of the actuators, called the actuator effectiveness matrix. It is noted that the number of actuators satisfying $e_{ii} \neq 0$ must be greater than three for the requirement of the controllability. Additionally, $\boldsymbol{v} = [v_1, v_2, \ldots, v_m]^{\mathrm{T}}$ and $\bar{\boldsymbol{v}} = [v_1, v_2, \ldots, v_m]^{\mathrm{T}}$ denote the desired torque vector and additive fault output vector, respectively.

Assumption 2.1 Due to physical limitations on the actuators (such as thruster, reaction wheel, control moment gyro and so on) installed in spacecraft, it is assumed that actuators' output torques have the same constraint value for simplicity, that is $|\tau_i| \leq \tau_{max}$. Additionally, the additive fault $\bar{v}_i$ is bounded with a small value because of the constraints of the actuators.

Assumption 2.2 The variable $\boldsymbol{d}$ in (2.20) including the gravitational perturbation, aerodynamic torque, radiation torque, and other environmental or non-environmental torques, is assumed to be bounded and differentiable. Additionally, with respect to the physical limitations of the actuators and the actuator faults defined above, the accidental actuator output $\bar{v}_i$ is also bounded. Thus, it is reasonable to assume that there always exists a constant $\bar{d}_1$ such that $\|\bar{\boldsymbol{d}}\| \leq \bar{d}_1$.

References

1. Mazzini, L. (2016) Flexible Spacecraft Dynamics Control and Guidance. Springer, Berlin
2. Wie, B. (2008) Space Vehicle Dynamics and Control, Second Edition. AIAA

3. Murugesan, S., Goel, P.S. (1987) Fault-tolerant spacecraft attitude control system. Sadhana-Academy Proceeding in Engineering Sciences 11(12): 233–261
4. Xiao, B., Hu, Q.L., Zhang, Y.M., Huo, X. (2014) Fault-tolerant tracking control of spacecraft with attitude-only measurement under actuator failures. Journal of Guidance, Control, and Dynamics 37(3): 838–849
5. Li, B., Hu, Q.L., Ma, G.F., Yang, Y.S. (2019) Fault-tolerant attitude stabilization incorporating closed-loop control allocation under actuator failure. IEEE Transactions on Aerospace and Electronic Systems 55(4): 1989–2000

Chapter 3
Null-Space Based Optimal Control Allocation for Spacecraft Attitude Stabilization

3.1 Introduction

For the on-orbiting spacecraft, the accurate and reliable ACS is very important and necessary for the accomplishment of the aerospace missions, such as attitude stabilization, high-precision pointing, attitude tracking or maneuver. Attitude stabilization control of a rigid spacecraft is one of the most significant problems and has attracted a great deal of interests nowadays [1–5]. It should be noted that the dynamics of the spacecraft control system is strong nonlinear in nature. And it is also affected by various disturbances including gravitational torque, aerodynamic torque, radiation torque, and other environmental and non-environmental torques, which influence the aerospace missions inevitably if the appropriate methods are not used to deal with the unknown disturbances. Thus, it is a difficult and challenging issue to design an available ACS scheme which meets the requirement of high-precision stabilization, and obtains the desired control performance during the spacecraft control mission.

During the past decades, many results have been put forward to deal with the attitude control problems based on several inspiring techniques, such as optimal control [6], nonlinear feedback control [7], adaptive control [8], and robust control or their integrations [9–11]. Generally speaking, most or part of the previous works consider the situation that the spacecraft ACS performs a mission under external disturbances and uncertainties in a certain degree, but very few have explicitly dealt with these issues concurrently. In view of its simplicity and good performance for practical applications, Proportional-Integral-Derivative (PID) or PID-like controllers have been extensively studied as well [12–14]. However, the actuators are required to produce large and fast enough joint torques in these control methods, while this will hardly be achieved due to limited magnitude and slew rate constraints in practice. This can lead to substantial performance deterioration and even to instability of the system. Hence, global stabilization algorithms that take control input saturation explicitly into account have drawn considerable attention. Such control algorithms bring about the control objectives under limited control authority which is more

Q. Hu et al., *Control Allocation for Spacecraft Under Actuator Faults*,
https://doi.org/10.1007/978-981-16-0439-3_3

applicable in practice. Ali et al. [15] designed a nonlinear bounded control using a backstepping method and an inverse tangent-based tracking function. Su et al. [16] presented a simple saturated PD control by introducing the standard hyper tangent function, and the globally asymptotic stabilization of a rigid spacecraft can be guaranteed. In [17], Xiao et al. presented an observer based control scheme for the tracking control of uncertain systems with external disturbances in which the globally exponential stability of the closed-loop system is guaranteed. In addition, Hu [18, 19], Tsiotras [20], Bang [21] and the references therein have developed some inspired and creative controller to effectively handle the actuator output saturation issue.

Due to the spacecraft ACS is generally an over-actuated control system, the problem of CA for spacecraft with redundant actuators should also be considered during the control scheme design. CA is a technique that distribute the virtue control input to each of the actuators in the best way while taking the constraints into consideration. CA techniques have been extensively investigated since last decades, such as: daisy chaining [22], linear or nonlinear programming-based optimization algorithms [23, 24], direct allocation [25], and dynamic control allocation [10, 24, 26–28]. However, most or partial previous works study linear CA by mathematical programming algorithms, which can be iteratively conducted to minimize the error between the commands produced by virtual control law and the moments produced by practical actuator combinations. Recently, a robust least-squares-based control allocation scheme applied to flight/spacecraft control system with an uncertain control effectiveness matrix is investigated in [29, 30]. In order to make full use of the existing redundancy due to the over-actuated actuators, the actuator constraints and some specified optimization performances should be considered explicitly during the design of an actuation algorithm to map the desired control command. To realize this objective, power minimizing based torque-distribution strategies were presented to reduce the maneuver energy requirements [4, 31]. In [4, 32], the null-space based control allocation scheme has been used to yield a power-optimal torque solution for reaction wheels at the instant, which does not consider control allocation error, only the traditional minimum-torque optimization problem.

In this chapter, a simple saturated proportional-derivative (SPD) control integrated with a novel null-space based optimal control reallocation is developed for spacecraft attitude stabilization, in the presence of the external disturbances and control input constraints. Firstly, a simple SPD control law will be proposed to achieve the desired rotation maneuvers under input saturation. Then, the null-space based optimal control reallocation technique will be utilized to distribute the total virtual control command into each individual actuator properly. Finally, the superiority of the proposed scheme will be verified via numerical simulations. The key features/achievements of the designed control scheme are: (1) global asymptotical stability of the closed-loop system; and (2) simple design procedure and structure of the proposed baseline control and control allocation scheme with inexpensive online computations.

(Remark: The main control schemes and contents of this chapter are from the published journal paper "Bo Li, Qinglei Hu and Juntong Qi, Null-space-based optimal control reallocation for spacecraft stabilization under input saturation, Interna-

tional Journal of Adaptive Control and Signal Processing, 2015, 29(6): 705–724". The authors appreciate the permission from the John Wiley and Sons to reuse partly the results published in the relevant Journal.)

3.2 Problem Formulation

3.2.1 Spacecraft Attitude Dynamics

Let us consider the spacecraft attitude dynamic equipped with m reaction wheels as actuators, then the total angular momentum of the spacecraft system is defined as

$$\boldsymbol{H} = \boldsymbol{H}_s + \boldsymbol{H}_w \tag{3.1}$$

where $\boldsymbol{H}_s$ denotes the angular momentum of the spacecraft about body-fixed axis, $\boldsymbol{H}_w$ is the total angular momentum of the reaction wheels (RWs) cluster. Whereas, the angular momentum of the spacecraft is defined as

$$\boldsymbol{H}_s = \boldsymbol{J}\boldsymbol{\omega} \tag{3.2}$$

where $\boldsymbol{J} \in \mathbf{R}^{3\times 3}$ is the total inertia matrix of the spacecraft with $\boldsymbol{J} = \boldsymbol{J}^{\mathrm{T}}$, which is a positive definite symmetrical matrix. In an analogous manner, the angular momentum of the reaction wheels cluster is given by

$$\boldsymbol{H}_w = \boldsymbol{D}\boldsymbol{J}_w\boldsymbol{\omega}_w,\ \boldsymbol{h}_w = \boldsymbol{J}_w\boldsymbol{\omega}_w \tag{3.3}$$

where $\boldsymbol{D} \in \mathbf{R}^{3\times m}$ is used to describe distribution of the physical actuators for representing the influence of each actuator on the angular acceleration of the spacecraft, which is the so-called control configuration/projection matrix, in other words, it maps the angular momentum of the RWs cluster described in the wheel axes to the spacecraft body-based axes. In addition, $\boldsymbol{J}_w = \mathrm{diag}\,(J_{w1}, J_{w2} \ldots, J_{wm})$ is the diagonal inertial matrix of RWs, and $\boldsymbol{\omega}_w = \mathrm{diag}\,(\omega_{w1}, \omega_{w2}, \ldots \omega_{wm})$ denotes the angular rate vector.

Then the dynamic equation of the rigid spacecraft is typically written as

$$\dot{\boldsymbol{H}} = \boldsymbol{d} - \boldsymbol{\omega}^{\times}\boldsymbol{H} \tag{3.4}$$

which yields to

$$\boldsymbol{J}\dot{\boldsymbol{\omega}} = -\boldsymbol{\omega}^{\times}\boldsymbol{J}\boldsymbol{\omega} - \boldsymbol{\omega}^{\times}\boldsymbol{D}\boldsymbol{h}_w - \boldsymbol{D}\dot{\boldsymbol{h}}_w + \boldsymbol{d} \tag{3.5}$$

where $\boldsymbol{d} = [d_1, d_2, d_3]^{\mathrm{T}} \in \mathbf{R}^3$ denotes the external disturbance torque vector induced from the environment, which is assumed to be unknown but bounded for all time, i.e., $\boldsymbol{d} \in S_d = \left\{\boldsymbol{d} \mid |d_1| + |d_2| + |d_3| \le \bar{d}\right\}$ for a constant $\bar{d}$. More particularly, $\boldsymbol{u} =$

$-\boldsymbol{D}\dot{\boldsymbol{h}}_w = [u_1, u_2, u_3]^{\mathrm{T}} \in \mathbf{R}^3$ is defined as the combined control torque produced by the actuators. Then the spacecraft dynamics can be rewritten as

$$\boldsymbol{J}\dot{\boldsymbol{\omega}} = -\boldsymbol{\omega}^{\times}\boldsymbol{J}\boldsymbol{\omega} - \boldsymbol{\omega}^{\times}\boldsymbol{D}\boldsymbol{h}_w + \boldsymbol{u} + \boldsymbol{d} \tag{3.6}$$

For the synthesis of the control law design, the following assumption is made.

Assumption 3.1 Due to physical limitations on the actuators, for simplicity, it is assumed that actuator output torques have the same constraint value $\tau_{\max}$, i.e.

$$\boldsymbol{\tau}(t) \in \boldsymbol{\Omega}_{\tau} = \left\{ \boldsymbol{\tau} \in \mathbf{R}^m \,\middle|\, |\tau_i| \le \tau_{\max}, i = 1, 2, \ldots m \right\} \tag{3.7}$$

where m is the number of actuators considered. That is to say, the virtual control demand designed from the control law ought to be subject to the following constraint

$$\boldsymbol{u}(t) \in \boldsymbol{\Omega}_{u} = \left\{ \boldsymbol{u} \in \mathbf{R}^3 \,\middle|\, |u_i| \le u_{\max}, i = 1, 2, 3 \right\} \tag{3.8}$$

where $u_{\max}$ is a common maximum magnitude of the control signals, and relative to the actuator configuration matrix $\boldsymbol{D}$ and $\tau_{\max}$.

3.2.2 *Control Objective*

Based on the above statement and assumption, the control objective of this work can be stated as: design a control law for the plant represented by Eqs. (2.14) and (3.6) under the constraints represented by Eqs. (3.7) or (3.8), such that, for all physically realizable initial conditions, the states of the closed-loop system can be stabilized asymptotically, which can be expressed as

$$\lim_{t\to\infty} \boldsymbol{q}(t) = \boldsymbol{0}, \ \lim_{t\to\infty} \boldsymbol{\omega}(t) = \boldsymbol{0}.$$

3.3 Attitude Control Law Design

In view of exploiting available actuator redundancy, attitude control law design in this chapter will use the principle of baseline/virtual controller and CA framework with the following two steps [33]: (1) design a control law specifying which total control effort to be produced, known as virtual control law or baseline control law; and (2) design a CA scheme (or control allocator) that maps the total control demand onto individual actuator settings (such as reaction wheels, etc.).

3.3.1 SPD-Based Virtual/Baseline Control Law Design

From a practical point of view, the design of efficient and low-cost attitude controllers is an important issue which is of great interest for aerospace industry for instance. In this section, on the one hand, due to either cost limitations (implementation constraints), a nonlinear control law with control input constraint is considered for the attitude stabilization problem of the rigid spacecraft described by Eqs. (2.14) and (3.6). On the other hand, this nonlinear control law is simple and user friendly in that it does not involve a time-consuming design procedure and demands little redesigning or reprogramming during orbiting operation. With these in mind, a PID-like controller for attitude stabilization with control input saturation is discussed in the following. According to [5], the control solution to the underlying spacecraft attitude stabilization problem is summarized:

Theorem 3.1 *Given the sapcecraft attitude stabilization control system modeled by Eqs. (2.14) and (3.6), there exists controller gains* $k_1 > 0$, $k_2 > 0$, $\gamma > 0$ *and* $\delta > 0$ *to ensure the system is globally asymptotically stable under the following SPD control law and time-vary adjusting parameter* k*:*

$$u_i = -k_2 \frac{\omega_i}{|\omega_i| + k^2\delta} - k_1 q_i, \ i = 1,\ 2,\ 3 \tag{3.9}$$

$$\dot{k} = -\gamma \sum_{i=1}^{3} \frac{k_2 |\omega_i| k\delta}{|\omega_i| + k^2\delta} \tag{3.10}$$

Proof The proof of Theorem 3.1 consists of the following two parts.
Step 1. Globally asymptotically stable of the spacecraft system
Consider the following Lyapunov function candidate

$$V = \frac{1}{2} k_1 \left[(1 - q_0)^2 + \boldsymbol{q}^{\mathrm{T}} \boldsymbol{q} \right] + \frac{1}{2} \boldsymbol{\omega}^{\mathrm{T}} \boldsymbol{J} \boldsymbol{\omega} + \frac{1}{2\gamma} k^2 \tag{3.11}$$

Differentiation of V with respect to time and substitution of the attitude dynamics from Eq. (3.6) and the control law Eqs. (3.9) and (3.10) yield

$$\begin{aligned} \dot{V} &= k_1 \boldsymbol{q}^{\mathrm{T}} \boldsymbol{\omega} + \boldsymbol{\omega}^{\mathrm{T}} \left(-\boldsymbol{\omega}^{\times} \boldsymbol{J} \boldsymbol{\omega} - \boldsymbol{\omega}^{\times} \boldsymbol{D} \boldsymbol{h}_w + \boldsymbol{u} + \boldsymbol{d} \right) + \frac{1}{\gamma} k \dot{k} \\ &\le k_1 \sum_{i=1}^{3} q_i \omega_i + \sum_{i=1}^{3} \omega_i \left(-k_2 \frac{\omega_i}{|\omega_i| + k^2\delta} - k_1 q_i + d_i \right) + \frac{1}{\gamma} k \dot{k} \end{aligned} \tag{3.12}$$

It is next noted that the following equation holds

$$\frac{\omega_i}{|\omega_i| + k^2\delta} = |\omega_i| - \frac{|\omega_i| k^2 \delta}{|\omega_i| + k^2\delta}$$

Hence,

$$
\begin{aligned}
\dot{V} &\le k_1 \sum_{i=1}^{3} q_i\omega_i + \sum_{i=1}^{3}\left(-k_2\frac{\omega_i^2}{|\omega_i|+k^2\delta} - k_1\omega_i q_i + \omega_i d_i\right) \\
&\quad - \frac{1}{\gamma}k\gamma \sum_{i=1}^{3}\frac{k_2|\omega_i|k\delta}{|\omega_i|+k^2\delta} \\
&\le k_1 \sum_{i=1}^{3} q_i\omega_i + \sum_{i=1}^{3}\left(-k_2\left(|\omega_i| - \frac{|\omega_i|k^2\delta}{|\omega_i|+k^2\delta}\right)\right) \\
&\quad - k_1\sum_{i=1}^{3}\omega_i q_i + \sum_{i=1}^{3}\omega_i d_i - \sum_{i=1}^{3}\frac{k_2|\omega_i|k^2\delta}{|\omega_i|+k^2\delta} \\
&\le \sum_{i=1}^{3}(-k_2|\omega_i| + |\omega_i d_i|) \le -|\boldsymbol{\omega}|\left(k_2 - \bar{d}\right)
\end{aligned}
\tag{3.13}
$$

where $|\boldsymbol{\omega}| = |\omega_1| + |\omega_2| + |\omega_3|$.

From the control law Eq. (3.9), one can obtain that $|u_i| \le k_2 + k_1$, and here k_2 is one of the dominant terms for the amplitude of the control torques. For the stabilization of the spacecraft attitude and rejection of the external disturbances, one proper value of k_2 should be chosen to satisfy the constraint $k_2 \ge \bar{d}$, and then this yields

$$
\dot{V} = -c|\boldsymbol{\omega}| \le 0,\ c = k_2 - \bar{d} \ge 0 \tag{3.14}
$$

It follows that $\boldsymbol{q}, \boldsymbol{\omega}$ and k are bounded. Integrating both sides of the above equation with respect to time from zero to infinite, one obtains

$$
\int_0^{\infty} \dot{V}(\tau)d\tau \le \int_0^{\infty} -c|\boldsymbol{\omega}(\tau)|d\tau \Rightarrow V(0) - V(\infty) \ge c\int_0^{\infty} |\boldsymbol{\omega}(\tau)|d\tau \tag{3.15}
$$

Because the term on the left-hand side of Eq. (3.14) is bounded, it follows that $\boldsymbol{\omega} \in \mathcal{L}^1$, also, since $\boldsymbol{u}$ and $\boldsymbol{\omega}$ are bounded, it follows that $\dot{\boldsymbol{q}}$ and $\dot{\boldsymbol{\omega}}$ are also bounded. Therefore one can achieve that $\lim_{t\to\infty}\boldsymbol{q}(t) = \boldsymbol{0},\ \lim_{t\to\infty}\boldsymbol{\omega}(t) = \boldsymbol{0}$ from Barbalat's lemma [34] and LaSalle's invariance principle [35], that is to say the globally asymptotically stable is achieved.

In addition, for any given finite $\gamma > 0$, the integral $\int_0^{\infty}|\boldsymbol{\omega}(\tau)|d\tau$ is bounded by some positive function $c(\gamma)$, that is $\int_0^{\infty}|\boldsymbol{\omega}(\tau)|d\tau < c(\gamma)$.

Step 2. Behavior of the adjustment $k(t)$

From previous analysis, it is noted that the asymptotic stabilization of the attitudes $\lim_{t\to\infty}\boldsymbol{q}(t) = \boldsymbol{0}$ and $\lim_{t\to\infty}\boldsymbol{\omega}(t) = \boldsymbol{0}$ are established independently of the behavior of $k(t)$ because $\dot{V} \le 0$ from Eq. (3.14), while, a potential problem is that $k(t)$ might go to zero before $\boldsymbol{\omega}(t)$ converges to zero, which can cause chattering and/or singularity of the control input. To this, the following will show that $k(t)$ will stay above some positive value $\bar{k}$ for all time.

In view of Eq. (3.10), due to $\frac{k_2|\omega_i|k\delta}{|\omega_i|+k^2\delta} \le k_2 k\delta$, then the following inequality holds

$$\dot{k} \ge -3\gamma k_2 \delta k \tag{3.16}$$

Then integrating the above inequality with respect to time, one can obtain

$$k(t) \ge k_0 \exp(-3\gamma k_2 \delta t) \tag{3.17}$$

where $k(0) = k_0 > 0$ is defined. This established a positive lower bound on $k(t) > 0$ for all time. In other words, the value of $k(t)$ cannot cross zero or tend to zero in finite time, but only at infinity.

More specifically, it is noted that

$$\dot{k} \ge -\gamma \frac{k_2}{k} \sum_{i=1}^{3} |\omega_i| \ge -\gamma \frac{k_2}{k} |\boldsymbol{\omega}| \tag{3.18}$$

Then integrating Eq. (3.18) with respect to time from zero to infinity yields

$$\frac{k^2(\infty)}{2} - \frac{k_0^2}{2} \ge -k_2\gamma \int_0^{\infty} |\omega(\tau)| d\tau \ge -k_2 \gamma c_1(\gamma) \tag{3.19}$$

If $\gamma > 0$ exists such as

$$\gamma c_1(\gamma) \le \frac{k_0^2 - \bar{k}^2}{2k_2} \tag{3.20}$$

then $\bar{k} \le k(\infty)$ is achieved. Since $k(t)$ is non-increasing function of time, this implies that $k(t) \ge \bar{k}$ is established for all time $t \in [0, \infty)$. □

Remark 3.1 Inequality (3.20) imposes a condition that γ needs to be satisfied in order to ensure that $k(t) \ge \bar{k}$ for all time. However, it is seen from Eq. (3.20) that this condition depends on $c_1(\gamma)$, that is, there is an implicit condition that γ needs to be guaranteed. Finding an explicit condition for γ is the main technical difficulty associated with the proposed approach. However, in simulation studies, when γ is chosen to be small, that is, $\gamma \ll 1$, convergence of k to some nonzero value is always guaranteed.

Remark 3.2 In view of the analysis above, the states $\boldsymbol{q}, \boldsymbol{\omega}$ and the adaptive parameter $k(t)$ are bounded all the time under the bounded controller gains $k_1 > 0$, $k_2 > 0$ and $\delta > 0$

$$|u_i| = \left| -k_2 \frac{\omega_i}{|\omega_i| + k^2\delta} - k_1 q_i \right| \le k_1 + k_2$$

Then, the control input saturation avoidance can be guaranteed by selecting a proper value of k_1 and k_2, which is determined by the physical limitations on the spacecraft and each available actuator.

3.3.2 *Null-Space-Based Optimal Control Reallocation Scheme*

From above analysis, even if the attitude control mission can be achieved by virtual control law in Eqs. (3.9) and (3.10), other practical issues, such as energy saving (minimizing energy/power consumption), practical control capability constraints due to actuator amplitude and rate saturation limits, or some optimal performance indexes, etc., should be taken into account simultaneously for practical spacecraft control system design. CA technique is then useful for control of over-actuated spacecraft systems, which deals with suitably distributing the desired total control command among the individual over-actuated actuators. By using CA, the actuator selection task is separated from the regulation task in the control design.

In this sense, two steps are split for the controller design using the principle of CA: (1) design of a control law in Eq. (3.9) as a virtual control $\boldsymbol{u}$ which specifies the velocity constraints, and (2) design of a control allocator that maps the virtual control demand onto individual actuator settings τ (such as reaction wheel torques, etc.). In general, the CA problem is formulated as an optimization problem systematically handling: redundant sets of actuators, actuator constraints and power consumption minimization, wearing/tearing and other undesirable effects of actuators. Since it is a practical and powerful approach for solving control problems for systems with over-actuated actuators, a novel null-space-based optimal control reallocation scheme is investigated in the sequel to achieve some optimal performance indexes.

From the principle of CA, an equivalent representation of Eq. (3.6) can be written as

$$\begin{aligned} \boldsymbol{J}\dot{\boldsymbol{\omega}} &= -\boldsymbol{\omega}^{\times}\boldsymbol{J}\boldsymbol{\omega} - \boldsymbol{\omega}^{\times}\boldsymbol{D}\boldsymbol{h}_w + \boldsymbol{u}(t) + \boldsymbol{d}(t) \\ \boldsymbol{u}(t) &= \boldsymbol{D}\boldsymbol{\tau}(t) \end{aligned} \tag{3.21}$$

In view of Eq. (3.21), it can be clearly seen that the developed control law in Eq. (3.9) can be utilized as the virtual control to guarantee the attitude stabilization and external disturbances rejection, like the analyses above. In what follows, we shall develop a novel control allocator for actuators to distribute the virtual control requirements to individual actuator settings in one better possible manner.

3.3.2.1 Pseudo-Inverse Control Allocation

For redundant RWs setups, the typical and frequently-used RW motor torque distribution strategy is to seek the minimum norm solution $\boldsymbol{\tau}$, which leads to the smallest absolute motor torques. Then, the pseudo-inverse algorithm can be transformed to the unconstrained minimum l_2 norm CA problem [33], i.e.,

$$\begin{aligned} \min \quad & J_1 = \frac{1}{2}\|\boldsymbol{\tau}\|_2^2 \\ \text{subject to} \quad & \boldsymbol{D}\boldsymbol{\tau} = \boldsymbol{u} \end{aligned} \tag{3.22}$$

Given the above issue defined in Eq. (3.22), as it is well known, if matrix $\boldsymbol{D}$ is full rank, then the actuators can produce the exact required control torque $\boldsymbol{u}$, if not, then $\boldsymbol{u}$ can only be partially produced. Further, for the specified RWs cluster control issue, $\boldsymbol{D}$ is a constant full-rank matrix. So it has an explicit solution shown as following

$$\boldsymbol{\tau} = \boldsymbol{D}^{\dagger}\boldsymbol{u} \tag{3.23}$$

where $\boldsymbol{D}^{\dagger} = \boldsymbol{D}^{\mathrm{T}}(\boldsymbol{D}\boldsymbol{D}^{\mathrm{T}})^{-1}$ is the pseudo-inverse of the matrix $\boldsymbol{D}$. This solution is convenient when the RW motor toque magnitude limits are out of concern.

If more than three RWs are employed, then the $\boldsymbol{D}$ matrix may contain a nonempty null-space, resulting in an infinity of $\boldsymbol{\tau}$ combinations that produce the required control torque $\boldsymbol{u}$. In view of this, a novel null-space-based optimal control reallocation scheme shall be considered for the optimal salutation further in the sequel.

3.3.2.2 Null-Space-Based Optimal Control Reallocation

Although the RWs control law solution shown in Eq. (3.23) via pseudo-inverse method attempts to minimize the actuator output torque, it may not be the optimal solution for spacecraft with power or energy consumption limitations. To solve this problem, a novel null-space-based optimal control reallocation method is investigated to map the demand control signal $\boldsymbol{u}$ to the individual actuator output torque $\boldsymbol{\tau}$ properly.

Let R be the rank of the control configuration matrix $\boldsymbol{D} \in \mathbf{R}^{3\times m}$, while $M = m - R$ is the degree of redundancy in RWs cluster. Since the RWs control law solution shown in Eq. (3.23) is only one infinity of solutions (which is defined as $\boldsymbol{\tau}^*$ here), the general actuator torque can be expressed as follows [4]

$$\boldsymbol{\tau} = \boldsymbol{\tau}^* + [\boldsymbol{N}]\boldsymbol{l} \tag{3.24}$$

satisfying

$$\boldsymbol{D}[\boldsymbol{N}] = [0_{3\times M}] \tag{3.25}$$

where $[\boldsymbol{N}] = [\boldsymbol{n}_1 \ \ \boldsymbol{n}_2 \ \ \dots \ \ \boldsymbol{n}_m]^{\mathrm{T}} \in \mathbf{R}^{m\times M}$ is the null-space matrix of $\boldsymbol{D}$, which does not affect the attitude motions of spacecraft, and note that $\boldsymbol{n}_i \in \mathbf{R}^{m\times 1}$ are nonzero column vectors. The vector $\boldsymbol{l} \in \mathbf{R}^{M\times 1}$ contains the M null-space scaling parameters through $\boldsymbol{l} = [l_1 \ \ l_2 \ \ \dots \ \ l_M]^{\mathrm{T}}$. Then $\boldsymbol{\tau}$ and $\boldsymbol{\tau}^*$ denote the actual and demand actuator torque obtained by pseudo-inverse CA respectively. To this end, the control reallocation goal is to find the null-space scaling parameters l_i, $i = 1, 2 \dots, M$ so that some optimal performance indexes are achieved for the specified actuator configurations.

In view of the above-mentioned problem and constraints, if the control demand signal produced via Eq. (3.23) is unattainable (i.e. the demand control is not possible to attain in one sampling interval) due to position or rate constraints of actuators, the optimal formulation in Eq. (3.22) is not ideal one. Inspired the null-space technique in [4] and control allocation problem presented in [10, 11], a novel null-space-based optimal control reallocation method is investigated in this work. Then a new performance measure is selected by adding a weighting penalizing term related to CA error ($\boldsymbol{\tau} - \boldsymbol{\tau}^*$), the difference between the demand actuator output and that attained. More specially, the proposed CA algorithm can be formulated as an optimal quadratic programming problem as follows:

$$\begin{aligned} \min \quad & J = \frac{1}{2}\boldsymbol{\tau}^{\mathrm{T}}\boldsymbol{W}_1^2\boldsymbol{\tau} + \frac{1}{2}\left(\boldsymbol{\tau} - \boldsymbol{\tau}^*\right)^{\mathrm{T}} \boldsymbol{W}_2^2 \left(\boldsymbol{\tau} - \boldsymbol{\tau}^*\right) \\ \text{subject to} \quad & \boldsymbol{D}\boldsymbol{\tau}^* = \boldsymbol{u} \end{aligned} \tag{3.26}$$

where $\boldsymbol{W}_1$ and $\boldsymbol{W}_2$ denote diagonal positive-defined weighting matrices. Then using Eq. (3.24), the costing function Eq. (3.26) can be rewritten as

$$\begin{aligned} \min \quad & J = \frac{1}{2}\left(\boldsymbol{\tau}^* + [\boldsymbol{N}]\boldsymbol{l}\right)^{\mathrm{T}} \boldsymbol{W}_1^2 \left(\boldsymbol{\tau}^* + [\boldsymbol{N}]\boldsymbol{l}\right) + \frac{1}{2}\left([\boldsymbol{N}]\boldsymbol{l}\right)^{\mathrm{T}} \boldsymbol{W}_2^2 \left([\boldsymbol{N}]\boldsymbol{l}\right) \\ \text{subject to} \quad & \boldsymbol{D}\boldsymbol{\tau}^* = \boldsymbol{u} \end{aligned} \tag{3.27}$$

A necessary condition for the minimum of the index function defined above with respect to the null-space parameter $\boldsymbol{l}$ is

$$\frac{\partial J}{\partial \boldsymbol{l}} = \left(\boldsymbol{\tau}^* + [\boldsymbol{N}]\boldsymbol{l}\right)^{\mathrm{T}} \boldsymbol{W}_1^2[\boldsymbol{N}] + \left([\boldsymbol{N}]\boldsymbol{l}\right)^{\mathrm{T}} \boldsymbol{W}_2^2[\boldsymbol{N}] = 0 \tag{3.28}$$

and yields to

$$[\boldsymbol{N}]^{\mathrm{T}}\boldsymbol{W}_1^2\boldsymbol{\tau}^* + [\boldsymbol{N}]^{\mathrm{T}}\boldsymbol{W}_1^2[\boldsymbol{N}]\boldsymbol{l} + [\boldsymbol{N}]^{\mathrm{T}}\boldsymbol{W}_2^2[\boldsymbol{N}]\boldsymbol{l} = 0 \tag{3.29}$$

and then further

$$[\boldsymbol{N}]^{\mathrm{T}}\left(\boldsymbol{W}_1^2 + \boldsymbol{W}_2^2\right)[\boldsymbol{N}]\boldsymbol{l} = -[\boldsymbol{N}]^{\mathrm{T}}\boldsymbol{W}_1^2\boldsymbol{\tau}^* \tag{3.30}$$

Because $[N]$ has rank M through its definition as the null-space matrix of D, and the weighting matrices $\boldsymbol{W}_1$ and $\boldsymbol{W}_2$ are relevant diagonal positive-defined, the rank of $[\boldsymbol{N}]^{\mathrm{T}}\left(\boldsymbol{W}_1^2 + \boldsymbol{W}_2^2\right)[\boldsymbol{N}]$ is M, that is to say, the matrix parameter $[\boldsymbol{N}]^{\mathrm{T}}\left(\boldsymbol{W}_1^2 + \boldsymbol{W}_2^2\right)[\boldsymbol{N}]$ is invertible. Therefore, the null-space scaling parameter vector $\boldsymbol{l}$ is obtained as

$$\begin{aligned} \boldsymbol{l} &= -\left([\boldsymbol{N}]^{\mathrm{T}}\left(\boldsymbol{W}_1^2 + \boldsymbol{W}_2^2\right)[\boldsymbol{N}]\right)^{-1}[\boldsymbol{N}]^{\mathrm{T}}\boldsymbol{W}_1^2\boldsymbol{\tau}^* \\ \boldsymbol{\tau}^* &= \boldsymbol{D}^{\dagger}\boldsymbol{u} \end{aligned} \tag{3.31}$$

Since $\frac{\partial J}{\partial l} = 0$ is the only necessary condition for the optimal solution of optimal quadratic programming problem in (3.26), $\frac{\partial J^2}{\partial l^2} > 0$ must be a positive definite matrix in order to guarantee an optimal solution. Differentiating the Eq. (3.28) with respect to $\boldsymbol{l}$ yields to

$$\frac{\partial J^2}{\partial l^2} = [\boldsymbol{N}]^{\mathrm{T}}\left(\boldsymbol{W}_1^2 + \boldsymbol{W}_2^2\right)[\boldsymbol{N}] \tag{3.32}$$

which is a positive defined matrix obviously.

Hence, the optimal solution of the control reallocation problem can be given by

$$\begin{aligned} \tau &= \tau^* + [N]l \\ l &= -\left([N]^{\mathrm{T}}\left(W_1^2 + W_2^2\right)[N]\right)^{-1}[N]^{\mathrm{T}} W_1^2 \tau^* \\ \tau^* &= D^{\dagger} u \end{aligned} \tag{3.33}$$

Remark 3.3 Equation (3.26) is a cost function to be minimized subject to some constraints. From this cost function, the total power consumption is represented by two parts. For the first term, it can be seen that the larger weighting matrix W_1 is chosen, the smaller energy consumption will be achieved; for the second term, it is used to minimize the error of the CA, such that the control torques would be convergent to the desired command quickly when a large weighting matrix W_2 is selected. In fact, it is worth noting that the terms defined in Eq. (3.26) do not contradict each other since the weighting matrixes W_1 and W_2 are used for representing their relevant importance, priority, or weight in this overall optimal objective function during the optimization procedure.

Remark 3.4 Although the previous work reported in Dr. H. Schaub's paper in 2009 from [4] made pioneering contribution to the field of null-space based control allocation, it does not consider CA error (the difference between the demand actuator output and that attained) defined in Eq. (3.26), only the traditional minimum-torque optimization problem. In comparison to [4], in our work, a new performance measure is selected by adding a weighting penalizing term related to CA error ($\tau - \tau^*$), and a different energy measurement manner is considered. As it is shown in Remark 3.3, the most improvement/innovation over the related pioneering works is that the novel CA minimizes the error of the CA, such that the control torques would be convergent to the desired command quickly with a smaller energy consumption using the null-space technique by selecting the proper weighting parameters.

3.4 Simulation Results

To verify the effectiveness and performance of the proposed control scheme (noted as SPD+NSOCA), numerical simulations on a rigid spacecraft with four RWs under various conditions are conducted. The configuration with four RWs is shown in Fig. 3.1, in which the rotation axes of three reaction wheels (such as reaction wheels 1, 2, and 3) are orthogonal to the spacecraft's orthogonal shaft and the fourth one is installed with the equiangular direction with the orthogonal-to-each other three axes. As it is shown in Fig. 3.1, α_4 denotes the angle between the angular momentum vector of the fourth reaction wheel and its projection vector in XOY plane, while β_4 denotes the angle between the projection vector and OX axis. Then for this considered configuration, the configuration matrix D in Eq. (3.21) can be represented as

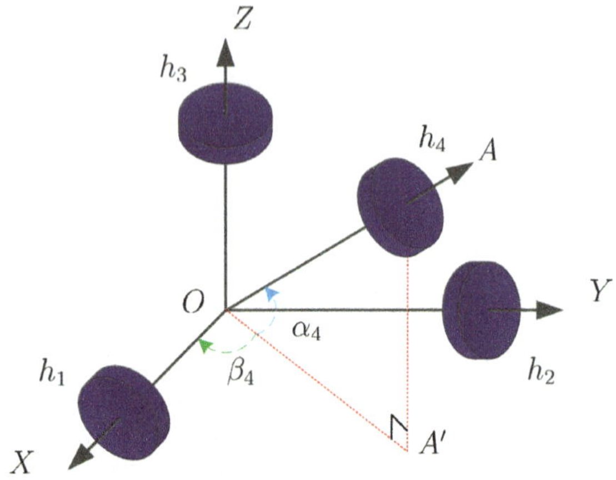

Fig. 3.1 Configuration of four reaction wheels in a spacecraft

Table 3.1 Simulation parameters and initial condition

Model parameters	Inertia moments $\boldsymbol{J} = [20, 0, 0.9; 0, 17, 0; 0.9, 0, 15]$ kg · m^2, initial angular velocity $\boldsymbol{\omega}_0 = [0.1; -0.1; -0.05]$ rad · s^{-1}, initial quaternion $\boldsymbol{Q}_0 = [0.9; -0.3; 0.26; 0.18]$, Inertia moments of RW $\boldsymbol{J}_w = 0.0409$ kg · m^2, assembling angles of the 4th RW $\alpha_4 = 35.26$ deg, $\beta_4 = 45$ deg, maximum torque of RW $\tau_{max} = \pm 0.2$ N.m, maximum rotation speed of RW 3500 rpm
SPD	$k_1 = 2.3,\ k_2 = 0.15,\ \gamma = 0.01,\ \delta = 0.01$
NSOCA	$\boldsymbol{W}_1 = \mathrm{diag}(1, 1, 1),\ \boldsymbol{W}_2 = 0.01 * \boldsymbol{W}_1$

$$\boldsymbol{D} = \begin{bmatrix} 1 & 0 & 0 & \cos\alpha_4 \cos\beta_4 \\ 0 & 1 & 0 & \cos\alpha_4 \sin\beta_4 \\ 0 & 0 & 1 & \sin\alpha_4 \end{bmatrix} \tag{3.34}$$

In general, $\alpha_4 = 35.26$ deg and $\beta_4 = 45$ deg are selected and used practically. For this case, the null-space matrix $[N]$ of $\boldsymbol{D}$ is presented as follows

$$[N] = \left[-\frac{1}{\sqrt{3}}, -\frac{1}{\sqrt{3}}, -\frac{1}{\sqrt{3}}, 1\right]^{\mathrm{T}} \tag{3.35}$$

For all numerical simulations presented in this section, the parameters are provided in Table 3.1 for all simulations carried out in the MATLAB/Simulink software environment, and the external disturbances are assumed to be as follows

$$d(t) = 0.2 \times 10^{-3} \cdot \begin{bmatrix} 3\cos(10t) + 4\sin(3t) - 10 \\ -1.5\sin(2t) + 3\cos(5t) + 15 \\ 3\sin(10t) - 8\sin(4t) + 10 \end{bmatrix} \text{N.m.}$$

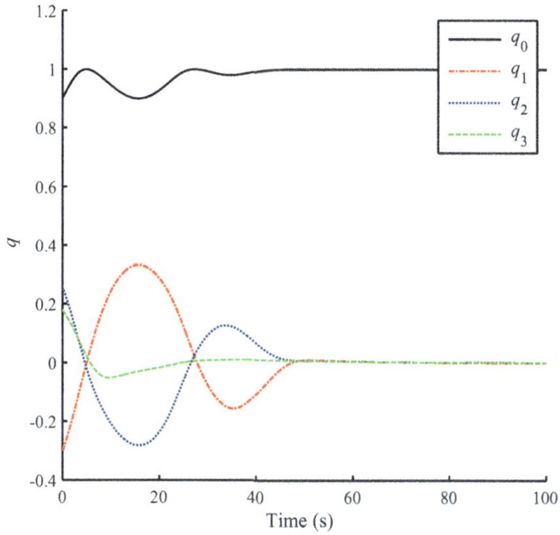

Fig. 3.2 Time responses of the spacecraft attitude quaternion ***q***

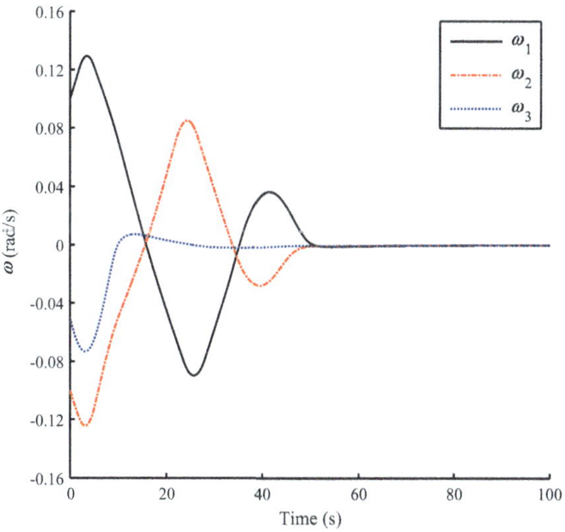

Fig. 3.3 Time responses of the spacecraft attitude angular velocity ***ω***

3.4.1 Time Response Results Under the Proposed Control Law

The results of simulations are shown in Figs. 3.2, 3.3, 3.4, 3.5, 3.6 and 3.7. It can be clearly seen from Figs. 3.2, 3.3, 3.4 and 3.5 that the control objective is achieved despite the presence of external disturbances with a settling time less than 50 s. Moreover, from Figs. 3.6 and 3.7 that the actuator saturation constraint is never

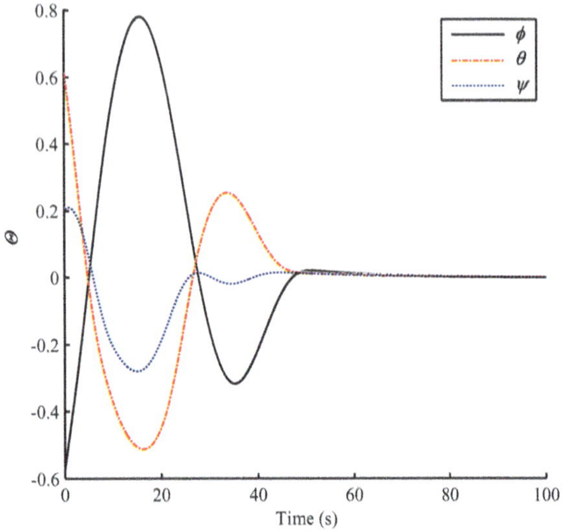

Fig. 3.4 Time responses of the spacecraft attitude Euler angle $\mathbf{\Theta}$

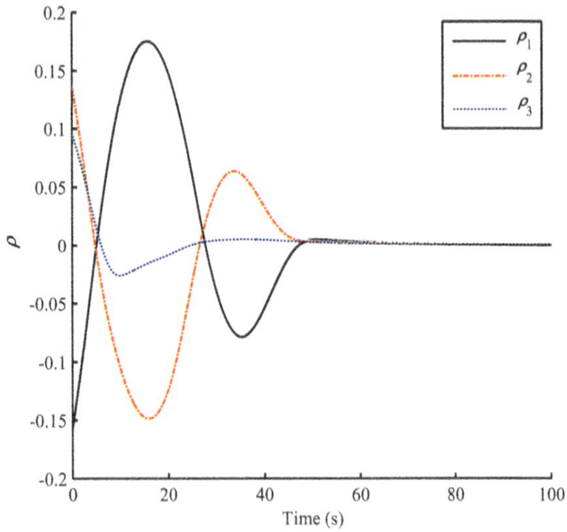

Fig. 3.5 Time responses of the spacecraft attitude MRPs $\boldsymbol{\rho}$

violated. It is also seen in Fig. 3.8 that $k(t)$ decreases during the transient, but then converges to a constant value. This shows further the effectiveness of the proposed control scheme.

In addition, for clearness of the comparison, the spacecraft attitude responses using Euler angles $[\phi, \theta, \psi]^{\mathrm{T}}$ (ϕ, θ and ψ are, respectively, the roll, pitch, and yaw angle of the spacecraft), are shown as Fig. 3.9. It is clearly that results of SPD+NSOCA has the good performance with a short settling time for achieving the rotation maneuvers, which further illustrates the effectiveness and feasibility of

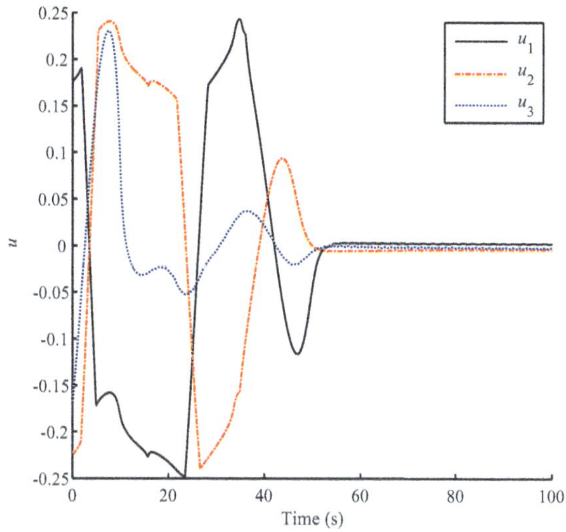

Fig. 3.6 Time responses of the control torques $\boldsymbol{u}$

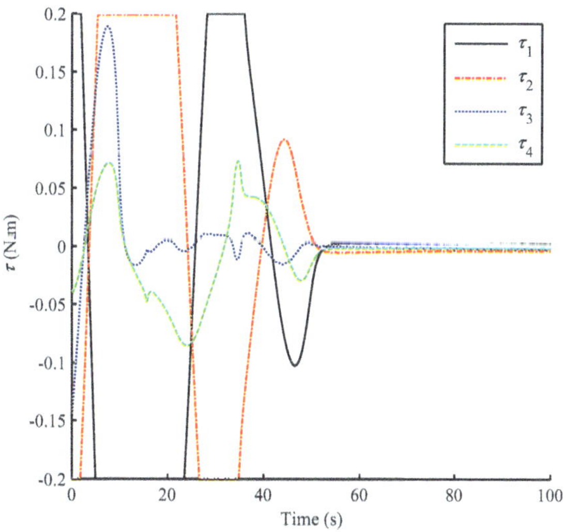

Fig. 3.7 Time responses of the actual actuator control torques $\boldsymbol{\tau}$

the proposed method. Moreover, the time responses of the phase space diagram with Euler angle and angular velocity are shown in Figs. 3.10, 3.11 and 3.12, in which we can obtain that the proposed scheme SPD+NSOCA achieves a smooth and short trajectory.

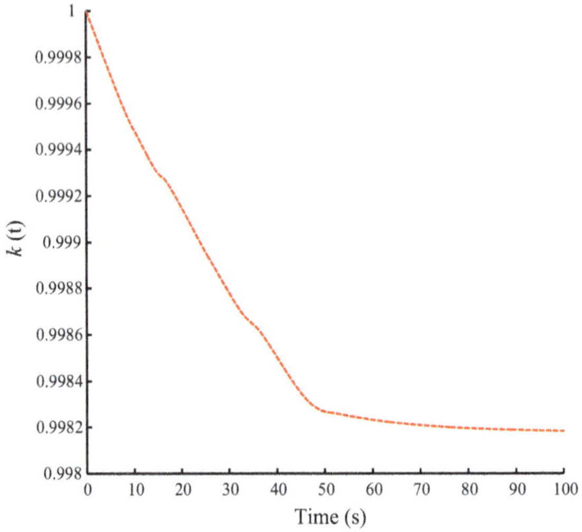

Fig. 3.8 Time responses of $k(t)$

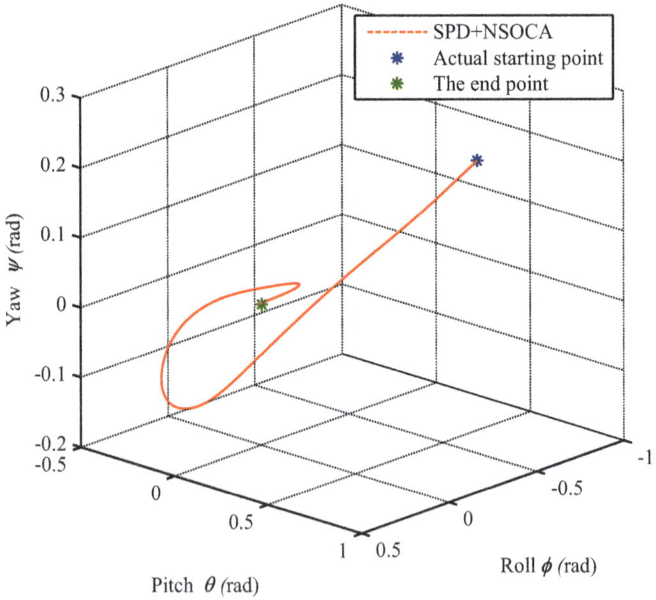

Fig. 3.9 Time responses of the Euler angles

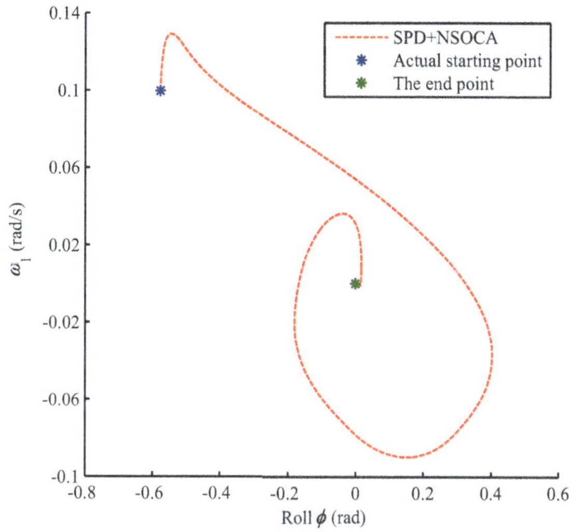

Fig. 3.10 Time responses of the phase space diagram with Euler angle ϕ and angular velocity

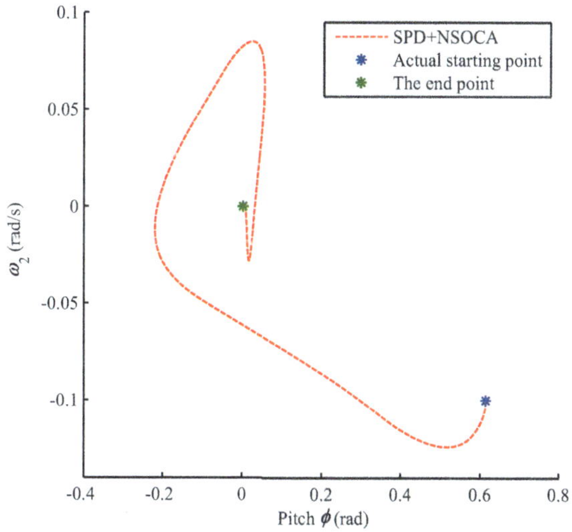

Fig. 3.11 Time responses of the phase space diagram with Euler angle θ and angular velocity

3.4.2 Energy Consumption Analysis

Fuel or electrical energy saving for spacecraft on orbit is of crucial importance to prolong the working life and achieve the target missions, and therefore the energy consumption optimal performance should also be considered. Given the RWs are utilized as the actuator of the rigid spacecraft, the index of instantaneous power consumption of the individual wheel is defined as $P_i = Z_i \tau_i$. Z_i denotes the spin rate (rotation speed) of the ith RW. Figure 3.13 shows the time response of the total instantaneous energy consumption $P = \sum_{i=1}^{m} |P_i|$. The corresponding time response

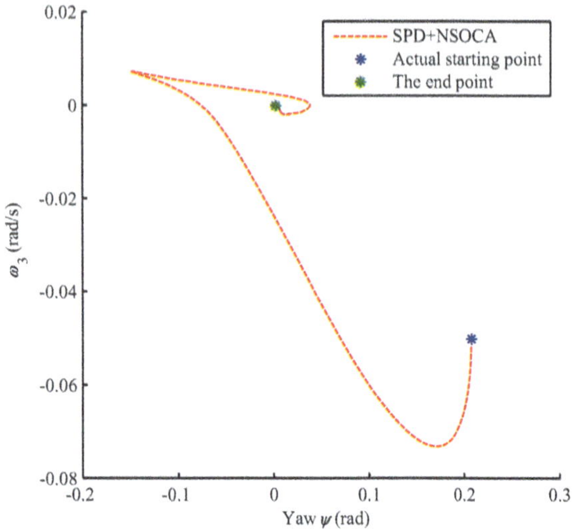

Fig. 3.12 Time responses of the phase space diagram with Euler angle ψ and angular velocity

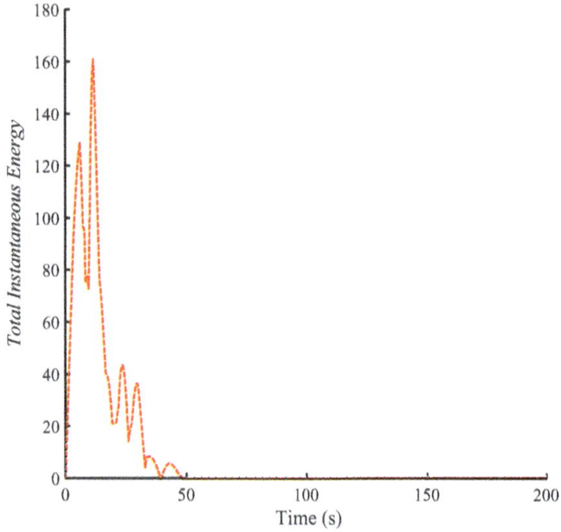

Fig. 3.13 Time response of the total instantaneous energy consumption

of the magnitude of total control torques $|\tau| = \sum_{i=1}^{m} |\tau_i|$, $m = 4$, are illustrated in Fig. 3.14. From these results, a significant energy savings is achieved with the proposed optimal control reallocation solution. It is shown that the proposed control scheme SPD+ NSOCA in this chapter can significantly provide the fine performance for the attitude stabilization control of a rigid spacecraft. Moreover, the proposed scheme provides the theoretical basis for the practical application of the advanced control theory to spacecraft ACS under external disturbances.

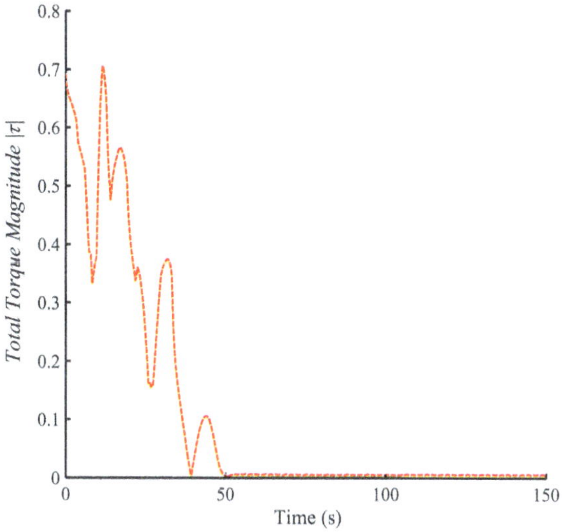

Fig. 3.14 Time response of the magnitude of total control torques

3.5 Conclusions

This chapter presents a novel SPD attitude control law incorporated with null-space based optimal control reallocation scheme for spacecraft under external disturbances and input saturation. By dynamically adjusting a single parameter, the SPD-based virtual controller is designed to guarantee that both attitude and angular velocity errors will tend to zero asymptotically. Then, an inspired null-space-based optimal control reallocation method is employed to distribute this control demands synthesized by the baseline controller into the individual actuator. This optimal control solution is to be found by penalizing the CA errors using quadratic programming algorithms. Numerical simulation results for the orbiting rigid spacecraft model demonstrate advantages and improvements with respect to the time responses and energy consumption saving, which illustrate and validate the effectiveness and feasibility of the proposed scheme. However, it should be pointed out that actuator uncertainties or faults are not taken into account in this chapter, and the fast time response of the control system is also needed for the practical engineering.

References

1. Zhu, S.Q., Wang, D.W., Shen, Q., Poh, E.K. (2017) Satellite attitude stabilization control with actuator faults. Journal of Guidance Control and Dynamics 40(5): 1–11
2. Sun, L., Zheng, Z.W. (2017) Disturbance observer-based robust backstepping attitude stabilization of spacecraft under input saturation and measurement uncertainty. IEEE Transactions on Industrial Electronics 64(10): 7994–8002

3. Hu, Q.L., Jiang, B.Y., Zhang, Y.M. (2017) Observer-based output feedback attitude stabilization for spacecraft with finite-time convergence. IEEE Transactions on Control Systems Technology 27(2): 781–789
4. Schaub, H., Lappas, V.J. (2009) Redundant reaction wheel torque distribution yielding instantaneous l2 power optimal attitude control. Journal of Guidance, Control, and Dynamics 32(4): 1269–1276
5. Boskovic, J.D., Li, S. M., Mehra, R. K. (2004) Robust tracking control design for spacecraft under control input saturation. Journal of Guidance, Control, and Dynamics 27(4): 627–633
6. Park, Y. (2015) Robust and optimal attitude control of spacecraft with disturbances. International Journal of Systems Science 46(7): 1222–1233
7. Liu, C., Sun, Z.W., Shi, K.K., Feng, W. (2017) Robust dynamic output feedback control for attitude stabilization of spacecraft with nonlinear perturbations. Aerospace Science and Technology 64: 102–121
8. Zhu, Z.H., Guo, Y. (2018) Adaptive coordinated attitude control for spacecraft formation with saturating actuators and unknown inertia. Journal of the Franklin Institute 356(2): 1021–1037
9. Joshi, S.M., Kelkar, A.G., Wen, J.T.Y. (1995) Robust attitude stabilization of spacecraft using nonlinear quaternion feedback. IEEE Transactions on Automatic Control 40(10): 1800–1803
10. Hu, Q.L., Li, B., Huo, X., Shi,Z. (2013) Spacecraft attitude tracking control under actuator magnitude deviation and misalignment. Aerospace Science and Technology 28(1): 266–280
11. Hu, Q. L., Li, B., Zhang, Y.M. (2013) Robust attitude control design for spacecraft under assigned velocity and control constraints. ISA Transactions 52(4): 480–493
12. Tsiotras, P. (1996) Stabilization and optimality results for the attitude control problem. Journal of Guidance, Control, and Dynamics 19(4): 772–779
13. Wie, B., Barba, P.M. (1985) Quaternion feedback for spacecraft large angle maneuvers. Journal of Guidance, Control, and Dynamics 8(3): 360–365
14. Song, C., Kim, S.J., Kim, S.H., Nam, H.S. (2006) Robust control of the missile attitude based on quaternion feedback. Control Engineering Practice 14(7): 811–818
15. Ali, I., Radice, G., Kim, J. (2010) Backstepping control design with actuator torque bound for spacecraft attitude maneuver. Journal of Guidance, Control, and Dynamics 33(1): 254–259
16. Su, Y.X., Zheng, C.H. (2011) Globally asymptotic stabilization of spacecraft with simple saturated proportional derivative control. Journal of Guidance, Control, and Dynamics 31(6): 1932–1935
17. Xiao, B., Dong, Q.C., Ye, D., Liu, L., Huo, X. (2017) A general tracking control framework for uncertain systems with exponential convergence performance. IEEE/ASME Transactions on Mechatronics 23(1): 111–120
18. Hu, Q.L. (2008) Adaptive output feedback sliding-mode maneuvering and vibration control of flexible spacecraft with input saturation. IET Control Theory and Applications 2(6): 467–478
19. Hu, Q.L., Xiao, B., Friswell, M.I. (2011) Robust fault-tolerant control for spacecraft attitude stabilisation subject to input saturation. IET Control Theory and Applications 5(2): 271–282
20. Tsiotras, P., Luo, J. (2000) Control of underactuated spacecraft with bounded inputs. Automatica 36(8): 1153–1169
21. Bang, H., Tahk, M.J., Choi, H.D. (2003) Large angle attitude control of spacecraft with actuator saturation. Control Engineering Practice 11(9): 989–997
22. Buffington, J.M., Enns, D.F. (1996) Lyapunov stability analysis of daisy chain control allocation. Journal of Guidance, Control, and Dynamics 19(6): 1226–1230
23. Bodson, M. (2002) Evaluation of optimization methods for control allocation. Journal of Guidance, Control, and Dynamics 25(4): 703–711
24. Härkegård, O. (2004) Dynamic control allocation using constrained quadratic programming. Journal of Guidance, Control, and Dynamics 27(6): 1028–1034
25. Tang, S.Y., Zhang, S.J., Zhang, Y.L. (2011) A modified direct allocation algorithm with application to redundant actuators. Chinese Journal of Aeronautics 24(3): 299–308
26. Kishore, W.A.C., Dasgupta, S., Ray, G., Sen, S. (2013) Control allocation for an over-actuated satellite launch vehicle. Aerospace Science and Technology 28(1): 56–71

27. Schierman, J.D., Ward, D.G., Hull, J.R., Gandhi,N., Oppenheimer,M., Doman,D.B. (2004) Integrated adaptive guidance and control for re-entry vehicles with flight test results. Journal of Guidance, Control, and Dynamics 27(6): 975–988
28. Casavola, A., Garone, E. (2010) Fault-tolerant adaptive control allocation schemes for overactuated systems. International Journal of Robust and Nonlinear Control 20(17): 1958–1980
29. Cui, L., Yang, Y. (2011) Disturbance rejection and robust least-squares control allocation in flight control system. Journal of Guidance, Control, and Dynamics 34(6): 1632–1643
30. Hu, Q.L., Li, B., Zhang, A.H. (2013) Robust finite-time control allocation in spacecraft attitude stabilization under actuator misalignment. Nonlinear Dynamics 73(1–2): 53–71
31. Ivanyukhin, A.V., Petukhov, V.G. (2016) Optimization of the interplanetary trajectories of spacecraft with a solar electric propulsion power plant of minimal power. Solar System Research 50(7): 552–559
32. Tsiotras, P., Shen, H., Hall, C. (2001) Satellite attitude control and power tracking with energy/momentum wheels. Journal of Guidance, Control, and Dynamics 24(1): 23–34
33. Härkegård, O. (2003) Backstepping and control allocation with applications to flight control. Ph d thesis, Linkoping Univ, Linkoping, Sweden
34. Narendra, K.S., Annaswamy, A.M. (1988) Stable Adaptive Systems. Prentice–Hall, New Jersey
35. Slotine, J.J.E., Li, W. (1991) Applied Nonlinear Control. Prentice–Hall, New Jersey

Chapter 4
Robust Finite-Time Control Allocation for Attitude Stabilization Under Actuator Misalignment

4.1 Introduction

Attitude control of a rigid spacecraft has attracted tremendous attention for the capability of achieving operational services, such as remote sensing, communication and variety of space-related research. As mentioned in the last chapter, multiple techniques were investigated to deal with the spacecraft attitude stabilization problem under the model uncertainties and the external disturbances with globally asymptotical convergence. Recently, variable structure control (VSC) was employed with different attitude representations to solve the robust attitude control problem [1–4]. However, these existing schemes can only guarantee the asymptotic stability, which implies that the system trajectories converge to the equilibrium with infinite settling time. Obviously, such an infinite settling time is not an option. Finite-time control is an alternative to achieve faster convergence rate, higher stabilization precision and better disturbance rejection ability. Reference [5] provides a general design principle of terminal sliding mode variable and the corresponding controller to realize the finite-time control. However, the control law contains a singularity problem, which was solved in [6] by utilizing an identical item to compensate this term in the feedback control law. In [7, 8], the authors proposed finite-time nonsingular terminal sliding mode based schemes for spacecraft attitude control systems in the presence of disturbances.

However, these results assumed that the actuators in spacecraft are able to provide any requested joint torque whose axis directions and/or input scaling of the actuators (such as gas jets, reaction fly-wheels) are exactly known. This assumption can rarely be satisfied in practice because of the power limitation and the possible misalignment of the actuators during installation. In order to deal with these problems, several solutions that take the actuator constraints into account have been extensively studied [9–14]. More specially, in [9, 10], the authors utilized the VSC technique to design robust sliding mode controllers for global asymptotic stabilization of spacecraft in the presence of control input saturation and disturbances. The smooth attitude sta-

Q. Hu et al., *Control Allocation for Spacecraft Under Actuator Faults*,
https://doi.org/10.1007/978-981-16-0439-3_4

bilizing control containing hyperbolic tangent functions was also discussed in [11]. Ali et al. [12] presented a method to accomplish the bounded control performance for spacecraft attitude maneuver with backstepping control. Ruiter [13] proposed an adaptive attitude tracking control scheme for a rigid spacecraft with linearly parameterized disturbances. Recently, Su and Zheng [14] presented a saturated PD-like method for spacecraft attitude maneuver under the disturbance-free case. To deal with possible actuator misalignment, [15] has developed an adaptive control law to accomplish attitude maneuver in the presence of relatively small gimbals' alignment error of variable speed control moment gyros. An extended Kalman filter was used to develop methods for on-orbit actuator alignment calibration [16], but the uncertain inertia was not considered. In [17], a novel adaptive tracking controller was synthesized for Hamiltonian systems which is successfully applied to a spacecraft with both the inertia and the actuator uncertainties. Another recent paper in [18] proposed an adaptive control approach for satellite formation flying, in which backstepping technique is utilized to handle thrust magnitude error and misalignment.

In this chapter, the control objective is to design a finite-time robust attitude control strategy for spacecraft in the presence of external disturbances, actuator misalignment and even control constraints. The proposed control law is a novel combination of the nonsingular terminal sliding mode with robust least squares CA method. With the help of the terminal sliding mode, the spacecraft can achieve the desired rotation maneuvering in finite time and the ACS is designed to be robust to the external disturbances. By utilizing robust least squares CA, the spacecraft ACS can distribute the virtual control signal into the individual actuator properly. The main strategy of this proposed CA is to find the optimal control vector of actuator by minimizing the worst-case residual, under the condition of the uncertainty included in actuator configuration matrix and control constraints like actuator saturation. The key feature of the proposed scheme is that the designed controller ensures both attitude and velocity convergence in finite time with simple design procedures and inexpensive online computations. The effectiveness and fine performance of the proposed scheme for spacecraft attitude control system is illustrated by numerical simulations.

(*Remark: The main control schemes and contents of this chapter are from the published journal paper "Qinglei Hu, Bo Li and Aihua Zhang, Robust finite-time control allocation in spacecraft attitude stabilization under actuator misalignment, Nonlinear Dynamics, 2013, 73(1–2): 53–71". The authors appreciate the permission from the Springer Nature to reuse partly the results published in the relevant Journal.*)

4.2 Problem Formulation

4.2.1 Actuator Configuration with Misalignment

For orbiting spacecraft, loosely speaking, they have more than three actuators aligned with the spacecraft body axes. A common configuration includes four reaction wheels, in which three reaction wheels' (such as reaction wheels 1, 2, and 3) rotation axes are orthogonal to the spacecraft ontology shaft and the forth one is installed with the equiangular direction with the ontology three axis.

In this sense, employing the configuration of four reaction wheels, the spacecraft dynamics in Eq. (2.20) can be written as

$$\boldsymbol{J}\dot{\boldsymbol{\omega}} = -\boldsymbol{\omega}^{\times}\boldsymbol{J}\boldsymbol{\omega} + \boldsymbol{D}\boldsymbol{\tau}(t) + \boldsymbol{d}(t) \tag{4.1}$$

where $\boldsymbol{D}$ denotes reaction wheel configuration matrix, representing the influence of each wheel on the angular acceleration of the spacecraft, and $[\tau_1, \tau_2, \tau_3, \tau_4]^{\mathrm{T}}$ denotes the torque produced by the four reaction wheels. Note that the configuration matrix $\boldsymbol{D}$ is available for a given spacecraft.

However, in practice, the configuration of actuators will never be perfect, that is to say, whether due to finite manufacturing tolerances or warping of the spacecraft structure during launch, some alignment errors can always exist. In this section, actuator alignment error is mathematically modelled for the configuration misalignment. Referring to Fig. 4.1, it is assumed that the reaction wheel mounted on X axis is tilted over nominal direction with constant angles, $\Delta\alpha_1$ and $\Delta\beta_1$, also for other reaction wheels mounted left are assumed to be tilted over nominal direction with $\Delta\alpha_2$, $\Delta\beta_2$, $\Delta\alpha_3$, $\Delta\beta_3$, $\Delta\alpha_4$ and $\Delta\beta_4$ respectively. To this end, the real reaction wheel torque with misalignment is expressed as

$$\begin{aligned} \boldsymbol{u} = {} & \tau_1 \begin{bmatrix} \cos\Delta\alpha_1 \\ \sin\Delta\alpha_1 \cos\Delta\beta_1 \\ \sin\Delta\alpha_1 \sin\Delta\beta_1 \end{bmatrix} + \tau_1 \begin{bmatrix} \sin\Delta\alpha_2 \cos\Delta\beta_2 \\ \cos\Delta\alpha_2 \\ \sin\Delta\alpha_2 \sin\Delta\beta_2 \end{bmatrix} \\ & + \tau_3 \begin{bmatrix} \sin\Delta\alpha_3 \cos\Delta\alpha_3 \\ \sin\Delta\alpha_3 \sin\Delta\beta_3 \\ \cos\Delta\alpha_3 \end{bmatrix} + \tau_4 \begin{bmatrix} \cos(\alpha_4 + \Delta\alpha_4)\cos(\beta_4 + \Delta\beta_4) \\ \cos(\alpha_4 + \Delta\alpha_4)\sin(\beta_4 + \Delta\beta_4) \\ \sin(\alpha_4 + \Delta\alpha_4) \end{bmatrix} \end{aligned} \tag{4.2}$$

Generally speaking, the misalignment angle errors $(\Delta\alpha_i, \Delta\beta_i)$ are very small in practice, and the following relationships are adopted to approximate Eq. (4.2)

$$\cos\Delta\alpha_i \approx \cos\Delta\beta_i \approx 1, \ \sin\Delta\alpha_i \approx \Delta\alpha_i, \ \sin\Delta\beta_i \approx \Delta\beta_i \tag{4.3}$$

Then for the considered actuator configuration, the configuration matrix $\boldsymbol{D}$ can be represented as

$$\boldsymbol{D} = \boldsymbol{D}_0 + \Delta\boldsymbol{D} \tag{4.4}$$

with

$$\boldsymbol{D}_0 = \begin{bmatrix} 1 & 0 & 0 & \cos\alpha_4 \cos\beta_4 \\ 0 & 1 & 0 & \cos\alpha_4 \sin\beta_4 \\ 0 & 0 & 1 & \sin\alpha_4 \end{bmatrix} \tag{4.5a}$$

$$\Delta\boldsymbol{D} = \begin{bmatrix} 0 & \Delta\alpha_2 \cos\Delta\beta_2 & \Delta\alpha_3 \cos\Delta\beta_3 & -\Delta\alpha_4 \sin\alpha_4 \cos\beta_4 - \Delta\beta_4 \cos\alpha_4 \sin\beta_4 \\ \Delta\alpha_1 \cos\Delta\beta_1 & 0 & \Delta\alpha_3 \sin\Delta\beta_3 & -\Delta\alpha_4 \sin\alpha_4 \sin\beta_4 + \Delta\beta_4 \cos\alpha_4 \cos\beta_4 \\ \Delta\alpha_1 \sin\Delta\beta_1 & \Delta\alpha_2 \sin\Delta\beta_2 & 0 & \Delta\alpha_4 \cos\alpha_4 \end{bmatrix} \tag{4.5b}$$

Fig. 4.1 Configuration of four reaction wheels under misalignment

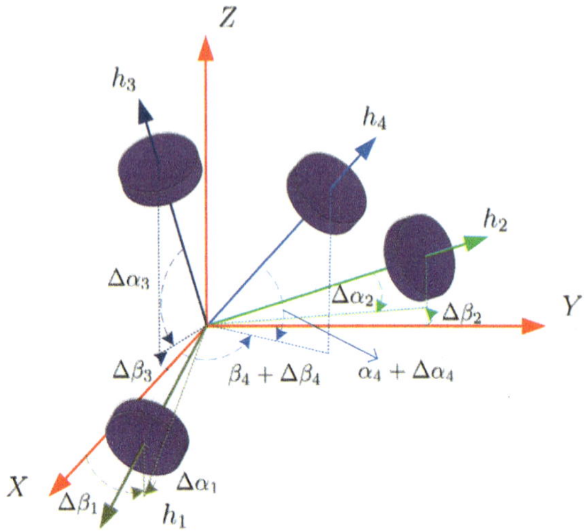

where $\boldsymbol{D}_0$ denotes the nominal value, and $\Delta \boldsymbol{D}$ denotes the uncertainty of configuration matrix. Accordingly, the spacecraft dynamics under this uncertainty can be rewritten as:

$$\boldsymbol{J}\dot{\boldsymbol{\omega}} = -\boldsymbol{\omega}^{\times}\boldsymbol{J}\boldsymbol{\omega} + (\boldsymbol{D}_0 + \Delta\boldsymbol{D})\,\boldsymbol{\tau}(t) + \boldsymbol{d}(t) \tag{4.6}$$

For synthesis of control system design, the following reasonable assumptions are made.

Assumption 4.1 The external disturbances $\boldsymbol{d}(t)$, in Eq. (4.6) include gravitational perturbations, atmospheric drag, etc., and are assumed to be bounded. Thus, it is reasonable to assume that there always exists a constant $\bar{d}$ such that

$$\|\boldsymbol{d}(t)\| \leq \bar{d}. \tag{4.7}$$

Assumption 4.2 Due to physical limitations on the reaction wheel considered, the control action generated is limited by the saturation value, i.e. $\boldsymbol{\tau}(t) \in \boldsymbol{\Omega} := \left\{\boldsymbol{\tau} \in \mathbf{R}^m \middle| \underline{\tau}_i \leq \tau_i \leq \bar{\tau}_i, i = 1, 2, 3, 4\right\}$. For simplicity, we assume that actuator output torque has the same constraint value $(\underline{\boldsymbol{\tau}}, \bar{\boldsymbol{\tau}})$, i.e.

$$\boldsymbol{\tau}(t) \in \boldsymbol{\Omega} := \left\{\boldsymbol{\tau} \in \mathbf{R}^m \middle| \underline{\tau}_i \leq \tau_i \leq \bar{\tau}_i, i = 1, 2, 3, 4\right\}. \tag{4.8}$$

Assumption 4.3 The uncertainty $\Delta\boldsymbol{D}$ due to misalignment is unknown but bounded matrix satisfying

$$\|\Delta\boldsymbol{D}\|_\infty \leq \varsigma \tag{4.9}$$

for some constant ς.

4.2.2 Control Objective

Consider the spacecraft attitude system given by Eqs. (2.14) and (4.6) with the actuator misalignment under the Assumptions 4.1–4.3. Design a control input $\boldsymbol{\tau}(t) \in \boldsymbol{\Omega}$ such that, for all physically realizable initial conditions, the states of the closed-loop system can be stabilized in finite time, which can be expressed as follows

$$\lim_{t\to T} \boldsymbol{q} = \lim_{t\to T} \boldsymbol{\omega} = \boldsymbol{0} \tag{4.10}$$

where T is the convergent time, which is function of the initial values of the system states.

4.3 Control Law Design

In this section, two steps are split for the controller design using the principle of CA [19]: (1) Design a control law specifying which total control effort to be produced (net torque, etc.), and (2) Design a control allocator that maps the total control demand onto individual actuator settings (such as reaction wheel torques, etc.).

From the principle of control allocation, an equivalent representation of the Eq. (4.6) can be written as

$$\boldsymbol{J}\dot{\boldsymbol{\omega}} = -\boldsymbol{\omega}^{\times}\boldsymbol{J}\boldsymbol{\omega} + \boldsymbol{B}_u\boldsymbol{u}(t) + \boldsymbol{d}(t) \tag{4.11a}$$

$$\boldsymbol{B}_u\boldsymbol{u}(t) = \boldsymbol{B}_\tau\boldsymbol{\tau}(t) \tag{4.11b}$$

where $\boldsymbol{u}(t)$ is the virtual control input or called combined control torque produced by the actuators, $\boldsymbol{B}_u$ is the virtual input matrix, and $\boldsymbol{B}_\tau$ is used to describe distribution of the physical actuators with $\boldsymbol{B}_\tau = \boldsymbol{D} + \Delta\boldsymbol{D}$, here, representing the influence of each actuator on the angular acceleration of the spacecraft considered. For the considered spacecraft, the virtual input matrix $\boldsymbol{B}_u$ as the identity $\boldsymbol{B}_u = \boldsymbol{I}_{3\times 3}$ will be defined, for convenience. With such a choice, the virtual input $\boldsymbol{u}(t)$ represents exactly the total torques produced by the actuators, and the following virtual equivalent plant can be given:

$$\boldsymbol{J}\dot{\boldsymbol{\omega}} = -\boldsymbol{\omega}^{\times}\boldsymbol{J}\boldsymbol{\omega} + \boldsymbol{u}(t) + \boldsymbol{d}(t) \tag{4.12a}$$

$$\boldsymbol{u}(t) = (\boldsymbol{D} + \Delta\boldsymbol{D})\,\boldsymbol{\tau}(t) \tag{4.12b}$$

In what follows, we shall develop such an attitude control law for spacecraft as given in Eqs. (2.14) and (4.9) to achieve the control target given in Sect. 4.2.2.

4.3.1 *Nonsingular Terminal Sliding Mode Based Virtual Finite Time Feedback Controller Design*

In this section, a nonsingular terminal sliding mode (NTSM) control algorithm is designed for spacecraft attitude control to achieve finite time convergence to the given attitude motion.

4.3.1.1 Design of Sliding Surface

For simplicity of expression, the following notions are introduced and used in the analysis and design of the NTSM controllers. For any vector $\boldsymbol{x} \in \mathbf{R}^n$ and $a > 0$ which is a constant, let

$$\mathrm{sig}(\boldsymbol{x})^a = \left[|x_1|^a \mathrm{sign}(x_1), \ldots, |x_n|^a \mathrm{sign}(x_n)\right]^{\mathrm{T}} \tag{4.13a}$$

$$|\boldsymbol{x}|^a = \left[|x_1|^a, \ldots, |x_n|^a\right]^{\mathrm{T}} \tag{4.13b}$$

$$\frac{1}{d\boldsymbol{x}} \mathrm{sig}(\boldsymbol{x})^a = \left[a|x_1|^{a-1}, \ldots, a|x_n|^{a-1}\right]^{\mathrm{T}} \tag{4.13c}$$

Under this definition, the following sliding surface is selected as [20]

$$\boldsymbol{s} = \mathrm{sig}\left(\dot{\boldsymbol{q}}\right)^b + \beta \boldsymbol{q} \tag{4.14}$$

where β is a constant, diagonal, positive-definite, control design matrix chosen by the designer/user, and b satisfies $1 < b < 2$.

In view of Eq. (4.13), the time derivative of the sliding surface in Eq. (4.14) yields

$$\dot{\boldsymbol{s}} = b\mathrm{diag}\left(|\dot{\boldsymbol{q}}|^{b-1}\right)\ddot{\boldsymbol{q}} + \beta\dot{\boldsymbol{q}} = b\mathrm{diag}\left(|\dot{\boldsymbol{q}}|^{b-1}\right)\left(\ddot{\boldsymbol{q}} + \frac{1}{b}\beta\mathrm{sig}\left(\dot{\boldsymbol{q}}\right)^{2-b}\right) \tag{4.15}$$

Using the Eqs. (2.14) and (4.12), and substituting these into Eq. (4.15) yield

$$\begin{aligned}\dot{\boldsymbol{s}} = b\mathrm{diag}\left(|\dot{\boldsymbol{q}}|^{b-1}\right)\Bigg[&\frac{1}{2}\left(q_0\boldsymbol{I} + \boldsymbol{q}^{\times}\right)\boldsymbol{J}^{-1}\left(-\omega^{\times}\boldsymbol{J}\omega + \boldsymbol{u} + \boldsymbol{d}\right)\\ &- \frac{1}{4}\|\boldsymbol{\omega}\|^2\boldsymbol{q} + \frac{1}{b}\beta\mathrm{sig}\left(\dot{\boldsymbol{q}}\right)^{2-b}\Bigg]\end{aligned} \tag{4.16}$$

4.3.1.2 NTSM Based Virtual Feedback Controller Design

For the derivation of the control law, let us choose a positive definite Lyapunov function

$$V = \frac{1}{2}\boldsymbol{s}^{\mathrm{T}}\boldsymbol{s} \tag{4.17}$$

Then, differentiating V along the solution of Eq. (4.12a) and using Eqs. (4.14) and (4.16) give

$$\begin{aligned}\dot{V} = \boldsymbol{s}^{\mathrm{T}} b\operatorname{diag}\left(|\dot{\boldsymbol{q}}|^{b-1}\right)\bigg[&\frac{1}{2}\left(q_0 \boldsymbol{I}+\boldsymbol{q}^{\times}\right) \boldsymbol{J}^{-1}\left(-\omega^{\times} \boldsymbol{J} \omega+\boldsymbol{u}+\boldsymbol{d}\right) \\ &-\frac{1}{4}\|\boldsymbol{\omega}\|^{2} \boldsymbol{q}+\frac{1}{b} \beta \operatorname{sig}(\dot{\boldsymbol{q}})^{2-b}\bigg]\end{aligned} \tag{4.18}$$

In view of Eq. (4.18), the control law is defined as

$$\boldsymbol{u}=\boldsymbol{\omega}^{\times} \boldsymbol{J} \omega-\boldsymbol{J}\left(q_0 \boldsymbol{I}+\boldsymbol{q}^{\times}\right)^{-1}\left(\frac{2}{b} \beta \operatorname{sig}(\dot{\boldsymbol{q}})^{2-b}-\frac{1}{2}\|\boldsymbol{\omega}\|^{2} \boldsymbol{q}+\rho \operatorname{sign}(\boldsymbol{s})\right) \tag{4.19}$$

where ρ is an appropriate constant yet to be determined.

Then, using the design control law in Eqs. (4.17), (4.16) can be rewritten as

$$\begin{aligned}\dot{V} &=\frac{1}{2} \boldsymbol{s}^{\mathrm{T}} b\operatorname{diag}\left(|\dot{\boldsymbol{q}}|^{b-1}\right)(-\rho \operatorname{sign}(\boldsymbol{s})+\boldsymbol{d}(t)) \\ &\leq-\frac{1}{2} \sum_{i=1}^{2} b|\dot{q}_i|^{b-1}\left(\rho-\bar{\boldsymbol{d}}\right)|s_i|\end{aligned} \tag{4.20}$$

Let ρ satisfies $\rho>\bar{\boldsymbol{d}}$, and then Eq. (4.18) can be further written as

$$\dot{V} \leq-\frac{1}{2} \sum_{i=1}^{2} b|\dot{q}_i|^{b-1}\left(\rho-\bar{\boldsymbol{d}}\right)|s_i|<0 \tag{4.21}$$

for $\dot{q}_i \neq 0$ $(i=1,2,3)$, and the condition for Lyapunov stability is satisfied.

While for the case $\dot{q}_i$, the following analysis will be further discussed for the stability of the overall system. Using the designed control law, the following can be given as

$$\ddot{\boldsymbol{q}}=-\frac{1}{b} \beta \operatorname{sig}(\dot{\boldsymbol{q}})^{2-b}-\frac{1}{2} \rho \operatorname{sign}(\boldsymbol{s})+\frac{1}{2} \bar{\boldsymbol{d}} \tag{4.22}$$

For $\dot{q}_i=0$, Eq. (4.22) can be rewritten as

$$\ddot{\boldsymbol{q}}=-\frac{1}{2} \rho \operatorname{sign}(\boldsymbol{s})+\frac{1}{2} \bar{\boldsymbol{d}} \tag{4.23}$$

It is easy to know that when $s_i>0$, $\ddot{\boldsymbol{q}}=-\frac{1}{2} \rho+\frac{1}{2} \bar{\boldsymbol{d}}<\mathbf{0}$, and when $s_i<0$, $\ddot{\boldsymbol{q}}=\frac{1}{2} \rho+\frac{1}{2} \bar{\boldsymbol{d}}>\mathbf{0}$, and then we can conclude that $\dot{q}_i=0$ is not an attractor. Correspondingly, it can be concluded that the sliding mode $\boldsymbol{s}=0$ can be reached at a certain time t^* from anywhere in the phase plane in finite time. Once on the sliding mode surface, i.e., $s_i=0$, one may have

$$\dot{q}_i = -\beta_i^{\frac{1}{b}} \mathrm{sig}(q_i)^b \tag{4.24}$$

From this, we can obtain that $\dot{q}_i$ and q_i can converge to the origin along the sliding mode surface at time [20, 21]

$$t_i = t^* + \frac{b}{b-1}\beta_i^{-\frac{1}{b}} |q_i(t^*)|^{\frac{b-1}{b}} \tag{4.25}$$

Therefore, $\boldsymbol{q} \to \mathbf{0}$ in finite time, and then $\boldsymbol{\omega} \to \mathbf{0}$.

Then, we have the following conclusion.

Theorem 4.1 *Consider the spacecraft dynamics governed by Eqs. (2.14) and (4.12) under the Assumption 4.1. With the nonsingular terminal sliding surface defined in Eq. (4.14) and control law in Eq. (4.19), the closed-loop system can be driven onto the sliding surface* $\boldsymbol{s} = \mathbf{0}$ *in finite time.*

Proof The proof can be easily obtained from above analysis and omitted here. □

Remark 4.1 For the design control law in Eq. (4.19), there is no singularity problem when $\dot{\boldsymbol{q}}$ crosses zero at a certain time by implementing the term sig $(\dot{\boldsymbol{q}})^{2-b}$, no like the conventional terminal sliding mode control case.

Remark 4.2 For the control law in Eq. (4.19), only three control parameters β, b and ρ, are needed to design, which means that inexpensive online computations are required from the point view of a feature of practical importance in reducing the usage of onboard resources in terms of computing power and memory size. And also this is user/designer friendly in that is does not involve a time-consuming design procedure and demands little redesigning or reprogramming during spacecraft operation.

While for the selection of these parameters, ρ is related to the bound of external disturbances. In general, this bound is difficult to measure, and therefore ρ is usually chosen large enough to ensure the robust stability. For β and b, they are related to the convergence time and system performance, and can be selected properly by the requirement of the system performance.

4.3.2 Robust Least Squares-Based Control Allocator Design Under Actuator Misalignment

Control allocation (CA) is useful for control of over-actuated spacecraft systems, and deals with suitably distributing the desired total control command among the individual actuators while considering the actuator constraints. Using CA, the actuator selection task is separated from the regulation task in the control design. Generally, the CA problem is formulated as an optimization problem systematically handling: redundant sets of actuators, actuator constraints and minimizing power consumption, etc. There is extensive literature on CA which discusses different algorithms, approaches and applications [19, 22–27]. While, for the case with uncertainty of configuration matrix due to misalignment, robust least squares-based control allocation is an effective way and well developed in recent research [28–30]. In what follows, the principle of this scheme is investigated for the considered plant.

In [20, 28, 29], the considered control allocation problem can be stated as: Given a configuration matrix $\boldsymbol{D}_0$ with its uncertain term $\Delta\boldsymbol{D}$, and virtual control $\boldsymbol{u}(t)$ designed by Sect. 4.3.1, the optimal actuator control vector can be described by

$$\boldsymbol{\tau} = \arg \min_{\underline{\boldsymbol{\tau}} \le \iota \le \bar{\boldsymbol{\tau}}} \max_{\|\Delta\boldsymbol{D}\| \le \varsigma} \|(\boldsymbol{D}_0 + \Delta\boldsymbol{D})\boldsymbol{\tau} - \boldsymbol{u}\| \tag{4.26}$$

subject to $\|\Delta\boldsymbol{D}\|_\infty \le \varsigma$, and $\boldsymbol{\tau}(t) \in \boldsymbol{\Omega} := \left\{\boldsymbol{\tau} \in \mathbf{R}^m \middle| \underline{\boldsymbol{\tau}} \le \boldsymbol{\tau}_i \le \bar{\boldsymbol{\tau}}, i = 1, 2, 3, 4\right\}$.

For a variable $\boldsymbol{\tau}$ in the interval of $(\underline{\boldsymbol{\tau}}, \bar{\boldsymbol{\tau}})$, the worst-case residual can be given

$$\gamma(\boldsymbol{\tau}) = \max_{\|\Delta\boldsymbol{D}\|_\infty \le \varsigma} \|(\boldsymbol{D}_0 + \Delta\boldsymbol{D})\boldsymbol{\tau} - \boldsymbol{u}\| \tag{4.27}$$

And further the triangle inequality

$$\begin{aligned} \gamma(\boldsymbol{\tau}) &\le \max_{\|\Delta\boldsymbol{D}\|_\infty \le \varsigma} (\|\boldsymbol{D}_0\boldsymbol{\tau} - \boldsymbol{v}\| + \|\Delta\boldsymbol{D}\boldsymbol{\tau}\|) \\ &= \|\boldsymbol{D}_0\boldsymbol{\tau} - \boldsymbol{v}\| + \max_{\|\Delta\boldsymbol{D}\|_\infty \le \varsigma} \|\Delta\boldsymbol{D}\boldsymbol{\tau}\| \end{aligned} \tag{4.28}$$

Let

$$\Delta\boldsymbol{D} = \frac{\varsigma}{\boldsymbol{\tau}}\varepsilon\boldsymbol{\tau}^{\mathrm{T}} \tag{4.29}$$

where

$$\varepsilon = \begin{cases} \frac{\boldsymbol{D}_0\boldsymbol{\tau} - \boldsymbol{u}}{\|\boldsymbol{D}_0\boldsymbol{\tau} - \boldsymbol{u}\|} & \text{if } \boldsymbol{D}_0\boldsymbol{\tau} \ne \boldsymbol{u} \\ \boldsymbol{I} & \text{otherwise} \end{cases} \tag{4.30}$$

in which $\boldsymbol{I}$ is a unite vector. In this case, the worst-case residual in the direction of ε can be rewritten as

$$\gamma(\boldsymbol{\tau}) = \|\boldsymbol{D}_0\boldsymbol{\tau} - \boldsymbol{u}\| + \varsigma\|\boldsymbol{\tau}\| \tag{4.31}$$

and satisfies

$$\|\boldsymbol{D}_0\boldsymbol{\tau} - \boldsymbol{u}\| + \varsigma\|\boldsymbol{\tau}\| \le \kappa \tag{4.32}$$

where κ is the upper bound of the residual to be minimized by finding the optimal $\boldsymbol{\tau}$ in the interval of $(\underline{\boldsymbol{\tau}}, \bar{\boldsymbol{\tau}})$. Thus, this problem can be written as a standard second-order cone programming problem as

$$\min_{\boldsymbol{\tau}, \mu, \kappa} \kappa \tag{4.33}$$

subject to

$$\|\boldsymbol{D}_0\boldsymbol{\tau} - \boldsymbol{u}\| \le \kappa - \mu \tag{4.34a}$$

$$\varsigma\|\boldsymbol{\tau}\| \le \mu \tag{4.34b}$$

$$\underline{\tau} \le \tau_i \le \bar{\tau} \tag{4.34c}$$

with the variables $\boldsymbol{\tau}$, μ, and κ.

Accordingly, the following statements can be concluded.

Theorem 4.2 *The optimal solution $\boldsymbol{\tau}$ to the robust least square control allocation problem is given by*

$$\boldsymbol{\tau} = \begin{cases} \left(\eta \boldsymbol{I} + \boldsymbol{D}_0^{\mathrm{T}} \boldsymbol{D}_0\right)^{-1} \boldsymbol{D}_0^{\mathrm{T}} \boldsymbol{u} & \text{if } \eta \triangleq \frac{(\kappa-\mu)\mu\varsigma^2}{\mu^2+\varsigma^2 z} > 0 \\ \boldsymbol{D}_0^{\dagger} \boldsymbol{u} & \text{otherwise} \end{cases} \tag{4.35}$$

where $\eta > 0$, $z = \underline{\tau}^2 + \bar{\tau}^2$, κ and μ are the optimal solution to the above problem, and $\boldsymbol{D}_0^{\dagger}$ is a right pseudo-inverse of matrix of matrix $\boldsymbol{D}_0$, with $\boldsymbol{D}_0^{\dagger} = \boldsymbol{D}_0^{\mathrm{T}} \left(\boldsymbol{D}_0 \boldsymbol{D}_0^{\mathrm{T}}\right)^{-1}$.

Proof The proof is omitted here and please see [28–30] for more details. In the next section, numerical simulation and comparison are given to verify the success of the NTSM control law in conjunction with the RLSCA technique. □

4.4 Simulation Results

To verify the effectiveness and performance of the proposed attitude stabilization control scheme, numerical simulations have been carried out using the rigid spacecraft system given in Eqs. (2.14) and (4.6). For all numerical simulation presented in this section, the parameters are provided in Table 4.1.

To show the effectiveness of the proposed NTSM incorporating with RLSCA, simulations were done under the given initial condition. And the external disturbances are assumed to be

$$d(t) = 1 \times 10^{-3} \cdot \begin{bmatrix} 3\cos(10t) + 4\sin(3t) - 1.0 \\ -1.5\sin(2t) + 3\cos(5t) + 1.5 \\ 3\sin(10t) - 8\sin(4t) + 1.0 \end{bmatrix} \text{N.m}$$

All the computations and plots are performed using the MATLAB/SIMULINK software package.

Table 4.1 Simulation parameters and initial condition

Model parameters	Inertia moments $\boldsymbol{J} = [20, 0, 0.9; 0, 17, 0; 0.9, 0, 15]\ \text{kg} \cdot \text{m}^2$, initial angular velocity $\boldsymbol{\omega}_0 = [0.1; -0.1; -0.05]\ \text{rad} \cdot \text{s}^{-1}$, initial quaternion $\boldsymbol{q}_0 = [0.9; -0.3; 0.26; 0.18]$, $\alpha_4 = 35.26\,\text{deg}$, $\beta_4 = 45\,\text{deg}$, $\Delta\alpha_i = [0.2; 0.1; 0.2; 0.3]\,\text{rad}$, $\Delta\beta_i = [0.2; 0.2; 0.3; 0.2]\,\text{rad}$
NTSM	$b = 1.1,\ \beta = 0.7,\ \rho = 0.3$
RLSCA	$\varsigma = 0.4,\ \bar{\tau} = 0.15\ \text{N.m},\ \underline{\tau} = -0.15\ \text{N.m}$

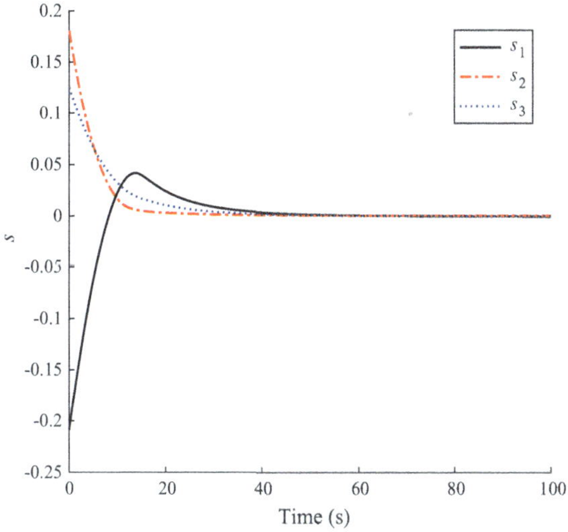

Fig. 4.2 The time responses of the sliding mode variable

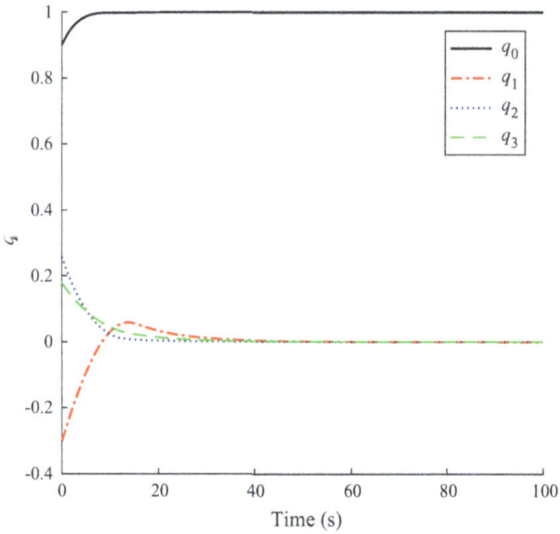

Fig. 4.3 The time responses of the quaternion

4.4.1 Simulation Results with Misalignment

Figure 4.2 shows the time response of the sliding mode variable. It can be easily obtained that the sliding mode variable will converge to the origin in finite time. The time responses of the system dynamics are shown in Figs. 4.3, 4.4, 4.5 and 4.6. From Fig. 4.3, it can be observed that the desirable orientation is achieved within 40 s, i.e., q_0 will converge to 1 and $\boldsymbol{q}_v$ will converge to the origin within 40 s. Similar conclusion can also be made from Figs. 4.4, 4.5 and 4.6, which further verifies the superiority of the proposed scheme. From Figs. 4.7 and 4.8, one can easily obtain

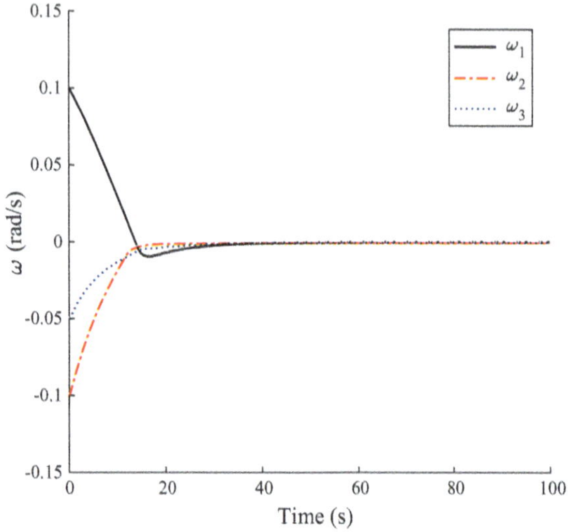

Fig. 4.4 The time responses of the angular velocity

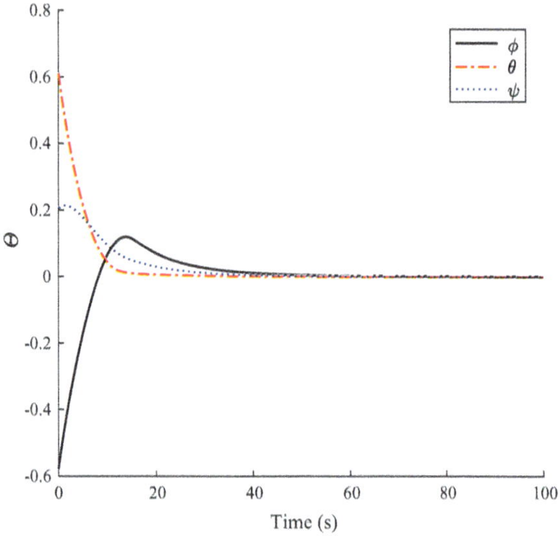

Fig. 4.5 The time responses of the Euler angle

that the actual control torques and the actuator torques is suitable for a spacecraft and the saturation limitation 0.15 N.m is never violated. In addition, from the point view of practice, the spacecraft attitude responses using Euler angles $[\phi, \theta, \psi]$ (ϕ, θ and ψ are, respectively, the roll, pitch, and yaw angles), are shown as Fig. 4.9, which further demonstrate the effectiveness and fine performance of the proposed NTSM + RLSCA scheme.

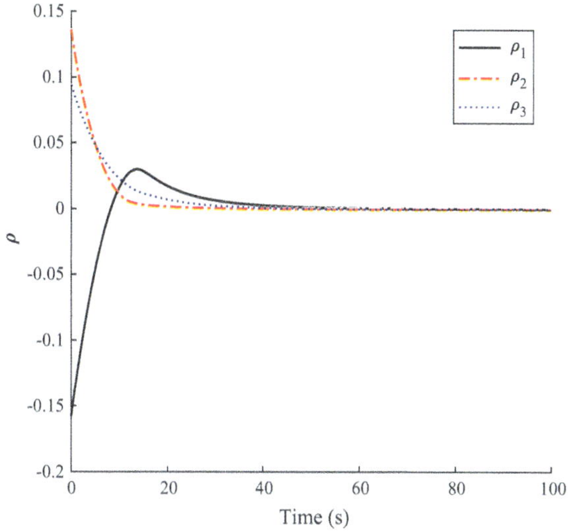

Fig. 4.6 The time responses of the MRPs

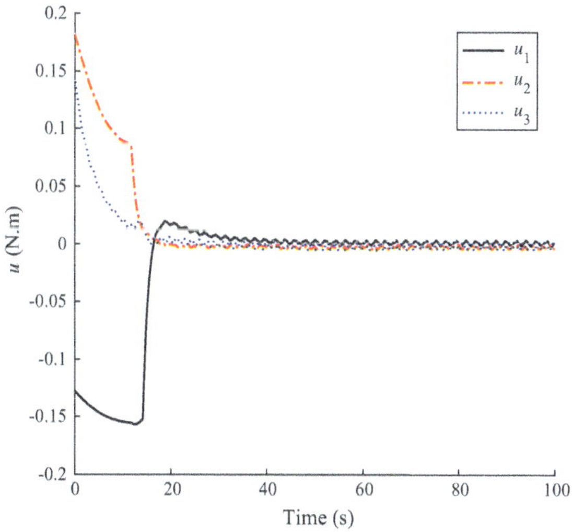

Fig. 4.7 The time responses of the actual control torques

4.4.2 Energy Consumption Analysis

Fuel or electrical energy saving for spacecraft in orbit is of crucial importance to prolong the working life and achieve the target missions, and therefore the energy consumption optimal problem should also be considered for spacecraft attitude control system. To analyze the system energy consumption, the performance function which is derived from the optimal control allocation problem is defined as

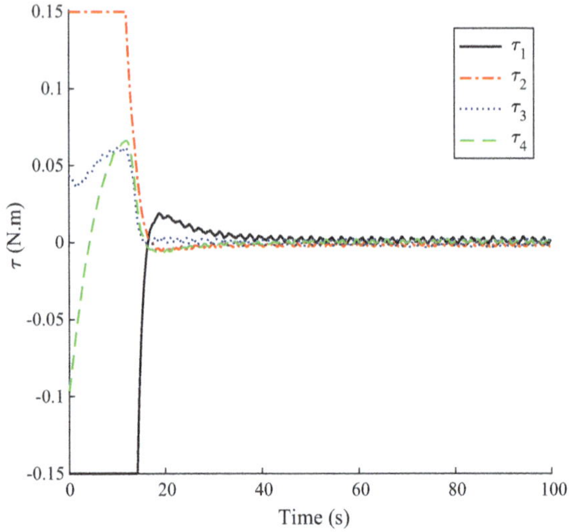

Fig. 4.8 The time responses of actuator torques

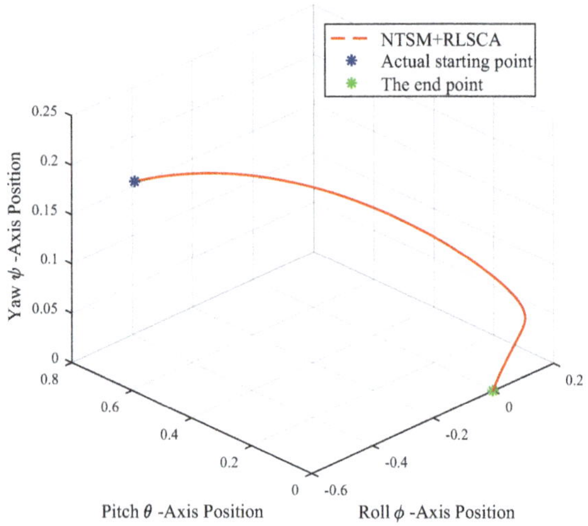

Fig. 4.9 The time responses of the Euler angles

$$E = \frac{1}{2}\int_0^T \|\tau\|^2 dt$$

where T denotes the simulation time, and $T = 100$s is chosen in the simulation From Fig. 4.10, it can be seen that the energy consumption using NTSM+RLSCA developed in this chapter is very low. This illustrates and verifies the effectiveness and fine performance of the developed theoretical results in Sect. 4.3, which can provide the theoretical basis for the practical application of the advanced control theory to spacecraft ACS.

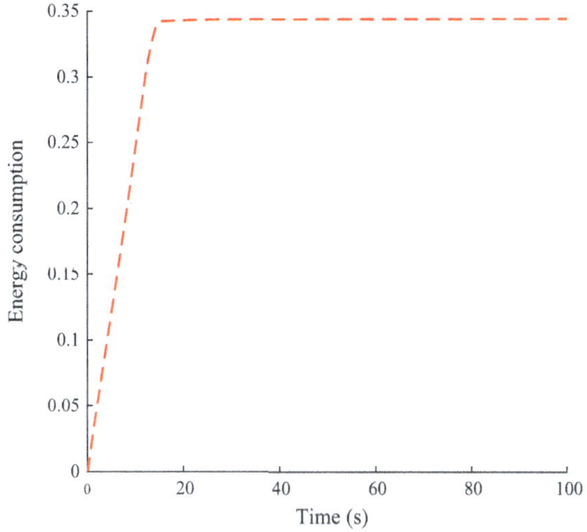

Fig. 4.10 The time responses of the energy consumption

4.5 Conclusions

In this chapter, the finite-time control method has been incorporated into a CA framework to solve the attitude control problem of spacecraft in presence of uncertain configuration matrix due to misalignment, external disturbances and actuator saturation as well. The proposed finite-time control allocation scheme, which uses nonsingular terminal sliding mode technique to design the virtual feedback control to achieve finite-time convergence to desired attitude position under undesired disturbances, and then employs the robust least squares-based CA scheme to suitably distribute the total virtual control effort into the active actuators. It should be noted that the proposed control schemes can achieve the fast attitude finite-time convergence with high accuracy under the actuator misalignment and actuator saturation. Numerical implementation of the new controller was also presented to demonstrate the effectiveness and fine performance of the proposed scheme. In practical, the aging of spacecraft's actuator components often results in the loss of actuating effectiveness or failures. The unexpected actuator faults may lead to the system stability reduction or even catastrophic consequence. Thus, high reliability and fault-tolerant ability show great importance in the design of spacecraft control system. In the next chapter, the attitude stabilization problem of a rigid spacecraft actuated by thrusters is addressed by utilizing the finite-time control and CA techniques, in the presence of external disturbance, input saturation, and inertia uncertainty. The proposed control scheme is also possesses the capacity of guaranteeing the desirable stability and system performance of the spacecraft ACS even in the event of actuator faults.

References

1. Hu, Q.L., Ma, G.F. (2005) Variable structure control and active vibration suppression of flexible spacecraft during attitude maneuver. Aerospace Science and Technology 9(4): 307–317
2. Hu, Q.L., Xiao, B. (2011) Fault-tolerant sliding mode attitude control for flexible spacecraft under loss of actuator effectiveness. Nonlinear Dynamics 64(1–2): 13–23
3. Liu, K.F., Xia, Y.Q., Zhu, Z., Basin, M.V. (2012) Sliding mode attitude tracking of rigid spacecraft with disturbance. Journal of the Franklin Institute 349(2): 413–440
4. Chen, Y.P., Lo, S.C. (1993) Sliding-mode controller design for spacecraft attitude tracking maneuvers. IEEE Transactions on Aerospace and Electronic Systems 29(4): 1328–1333
5. Mu, C., Xu, W., Sun, C. (2016) On switching manifold design for terminal sliding mode control. Journal of The Franklin Institute 353: 1553–1572
6. Song, Z.Y., Duan, C., Su, H.S., Hu, J.C. (2018) Full-order sliding mode control for finite-time attitude tracking of rigid spacecraft. IET Control Theory and Applications 12(8): 1086–1094
7. Ding, S.H., Li, S.H. (2009) Stabilization of the attitude of a rigid spacecraft with external disturbances using finite-time control techniques. Aerospace Science and Technology 13(4): 256–265
8. Zhu, Z., Xia, Y.Q., Fu, M.Y. (2011) Attitude stabilization of rigid spacecraft with finite-time convergence. International Journal of Robust and Nonlinear Control 21(6): 686–702
9. Boskovic, J.D., Li, S., Mehra, R.K. (2001) Robust adaptive variable structure control of spacecraft under control input saturation. Journal of Guidance, Control, and Dynamics 24(1): 14–22
10. Hu, Q.L. (2009) Robust adaptive sliding mode attitude maneuvering and vibration damping of three-axis-stabilized flexible spacecraft with actuator saturation limits. Nonlinear Dynamics 55(4): 301–321
11. Hu, Q.L. (2008) Sliding mode maneuvering control and active vibration damping of three-axis stabilized flexible spacecraft with actuator dynamics. Nonlinear Dynamics 52(3): 227–248
12. Ali, I., Radice, G., Kim, J. (2010) Backstepping control design with actuator torque bound for spacecraft attitude maneuver. Journal of Guidance, Control, and Dynamics 33(1): 254–259
13. De, R.A. (2010) Adaptive spacecraft attitude tracking control with actuator saturation. Journal of Guidance, Control, and Dynamics 33(5): 1692–1695
14. Su, Y.X., Zheng, C.H. (2011) Globally asymptotic stabilization of spacecraft with simple saturated proportional derivative control. Journal of Guidance, Control, and Dynamics 31(6): 1932–1935
15. Yoon, H., Tsiotras, P. (2005) Adaptive spacecraft attitude tracking control with actuator uncertainties. In: Proceeding of the AIAA Guidance Navigation and Control Conference and Exhibit, San Francisco, California
16. Fosbury, A.M., Nebelecky, C.K. (2009) Spacecraft actuator alignment estimation. In: Proceeding of the AIAA Guidance Navigation and Control Conference and Exhibit, Chicago, Illinois
17. Yoon, H., Agrawal, B.N. (2009) Adaptive control of uncertain hamiltonian multi-input multi-output systems: with application to spacecraft control. IEEE Transactions on Control Systems Technology 17(4): 900–906
18. Lima H.C., Bang, H. (2009) Adaptive control for satellite formation flying under thrust misalignment. Acta Astronautica 65(1–2): 112–122
19. Härkegård, O. (2003) Backstepping and control allocation with applications to flight control. Ph d thesis, Linkoping Univ, Linkoping, Sweden
20. Feng, Y., Yu, X.H., Man, Z.H. (2002) Nonsingular terminal sliding mode control of rigid manipulator. Automatica 38(12): 2159–2167
21. Li, S.H., Wang, Z., Fei, S.M. (2011) Comments on the paper: Robust controllers design with finite time convergence for rigid spacecraft attitude tracking control. Aerospace Science and Technology 15(3): 193–195
22. Buffington, J.M., Enns, D.F. (1996) Lyapunov stability analysis of daisy chain control allocation. Journal of Guidance, Control, and Dynamics 19(6): 1226–1230
23. Durham, W. C. (1993) Constrained control allocation. Journal of Guidance, Control, and Dynamics 16(4): 717–725

24. Durham, W.C. (1994) Constrained control allocation - three-moment problem. Journal of Guidance, Control, and Dynamics 17(2): 330–336
25. Härkegård, O. (2004) Dynamic control allocation using constrained quadratic programming. Journal of Guidance, Control, and Dynamics 27(6): 1028–1034
26. Alwi, H.and Edwards, C. (2008) Fault tolerant control using sliding modes with online control allocation. Automatica 44(7): 1859–1866
27. Casavola, A., Garone, E. (2010) Fault-tolerant adaptive control allocation schemes for overactuated systems. International Journal of Robust and Nonlinear Control 20(17): 1958–1980
28. Cui, L., Yang, Y. (2011) Disturbance rejection and robust least-squares control allocation in flight control system. Journal of Guidance, Control, and Dynamics 34(6): 1632–1643
29. Ghaoui, L.E., Lebret, H. (1997) Robust solutions to least squares problems with uncertain data. SIAM Journal on Matrix Analysis and Applications 18(4): 1035–1064
30. Ghaoui, L.E., Oustry, F., Lebret, H. (1998) Robust solutions to uncertain semi-definite programs. SIAM Journal on Optimization 9(1): 33–52

Chapter 5
Finite-Time Fault-Tolerant Spacecraft Attitude Control with Torque Saturation

5.1 Introduction

On-orbiting spacecraft are working for a specified long time under the effects of environmental disturbances, uncertainties, component failures, and input saturation. All these issues are big challenges for the design of spacecraft attitude control with the requirements of the adequate performance and stability. If these influences can't be considered during the control system design, it may deteriorate the control system performance and even drive the system to be unstable. It should be noted that the rapid, accurate and reliable response to various attitude mission commands is needed urgently for the future spacecraft, CubeSats or nanosatellites. All the above mentioned challenges bring considerable difficulty to design a spacecraft ACS with the capabilities of high-precise stability, high reliability and fault tolerance against to the possible actuator uncertainties and faults. Recently, several FTC schemes have been presented in the literature based on several inspired approaches, such as linear matrix inequalities [1], $H\infty$ [2], adaptive control [3–5], sliding-mode control [6], fuzzy logic [7, 8], and neural networks [9]. Reference [3] presented a direct model reference adaptive control scheme to compensate for actuator failures and even structure damages to ensure closed-loop system stability. In [4], an indirect robust adaptive control scheme was proposed for attitude tracking with actuator failures, external disturbance, and model uncertainties. A positive-definite gain matrix was constructed with the transpose of an actuation distribution matrix. Based on the passive method in [4], the controller in [6] was designed with the terminal sliding surface and accomplished finite-time convergence under actuator faults. In [10], a fault-tolerant control law was derived without the measurement of angular velocity. Shen et al. [11] presented a fault detection observer based on an indirect fault identification method, and using the fault information obtained, a feedback control law was proposed to accommodate system faults. In [12], multiple actuator faults were dealt with utilizing the adaptive fault compensation control scheme, in which the control gain matrix was constructed as the product of two positive definite matrices to handle the uncertainty. In [13], a fault-tolerant reconfigurable controller was pre-

Q. Hu et al., *Control Allocation for Spacecraft Under Actuator Faults*,
https://doi.org/10.1007/978-981-16-0439-3_5

sented by introducing the sliding surface into the dual-layer gain adaptive scheme, which accelerated the system convergence.

Although the mentioned control designs can deal with actuator faults, to our best knowledge, very few among them can achieve finite-time stability. As it is known, asymptotic stability theoretically implies that system converges to equilibrium when the convergence time tends to infinity, which cannot meet the requirements during highly critical mission phases. A finite-time control technique provides a faster convergence rate, higher precision control performance, and better disturbance rejection properties. The existing finite time control methods can be roughly divided into two categories: the homogeneous domination approach [14] and the method based on Lyapunov function [15–18]. However, the homogeneous approach cannot estimate or even adjust the settling time, and cannot handle parameter uncertainty. Thus, Lyapunov-based methods were proposed for employing the terminal sliding-mode surface [15–18]. In [18], model predictive control and terminal sliding mode control were combined in a compound double layer structure, in which the model predictive control technique provided the required optimal target for the lower level sliding-mode controller. In [19], the integral sliding mode controller was employed for the finite-time control problem. In [20], the adaptive fast terminal sliding mode control combined with nonlinear disturbance observer were designed for finite-time attitude tracking problem.

Another problem needs to be considered is the input saturation. Due to the physical structure and energy consumption constraints on board, the available actuation is always limited. If there exist serious discrepancies between commanded signals and actual control efforts, performance degradation or (even worse) system instability arises. This problem has attracted considerable attention recently. In [21, 22], two saturated controller formulations were proposed with two bounded functions to explicitly account for input saturation. Ali et al. [23], Zhu et al. [24], de Ruiter [25], Shen et al. [11], and Zou et al. [26] developed various methods to effectively accommodate limited control authorities. However, little consideration was given to make full use of the available control authorities on the level of a single actuator, especially with actuator faults or failures in the literature.

In this chapter, spacecraft attitude is guaranteed to converge to desired equilibrium in finite-time in the presence of external disturbance, input saturation, and inertia uncertainty. Specifically, a time-varying terminal sliding-mode control is designed with combination of a signal strictly constrained by the known maximum torque. The smoothness of the control can be adjusted by changing the controller's parameters. The advantages are that it can accomplish finite-time convergence, and eliminate the singularity problems without any switch. Another contribution is that the algorithm is adjusted such that it can be applied to the case with uncertain faults or failures by broadening the available torque envelope and loosening the conservative condition. The system is proved to be finite-time stable using the Lyapunov stability theory. In this chapter, the PICA method is utilized to distribute the optimal thrust signal. A detailed analysis of the selection of the positive-definite diagonal control parameter $\boldsymbol{A}$ is provided, and a modification $\boldsymbol{A}$ is proposed on account of the CA matrix $\boldsymbol{D}$, such that the sum of the absolute value of every row in $\boldsymbol{D}^{\mathrm{T}}\boldsymbol{A}$ is equal. Based on the modification A, it can be guaranteed that all functional actuators have the potential to reach its output limit simultaneously. The proposed control scheme shows feasibility in practical implementation with inertia-free and simple structure. Numerical simulations are conducted to show the performance of the controller, and illustrate the performance properties under different parameters.

(Remark: The main control schemes and contents of this chapter are from the published journal paper "Qinglei Hu, Xiao Tan and Maruthi R. Akella, Finite-time fault-tolerant spacecraft attitude control with torque saturation, Journal of Guidance, Control, and Dynamics, 2017, 40(10): 2524–2537". The authors appreciate the permission from the American Institute of Aeronautics and Astronautics to reuse partly the results published in the relevant Journal.)

5.2 Problem Formulation

Considering the external disturbance torques such as air drags and solar pressure, it is reasonable to assume the disturbance is bounded by a common maximum magnitude, expressed by

$$|d_i(t)| \leq d_{\mathrm{m}}, \; i = 1, 2, 3 \tag{5.1}$$

It is of great practical value to consider actuator saturation explicitly. Assume that the three components of control torques $\boldsymbol{\tau}$ are bounded by a common limit u_m and the control authorities always dominate the unknown disturbances. This assumption is expressed by

$$|\tau_i(t)| \leq u_m, i = 1, 2, 3 \text{ and } u_m = d_m + f_0 \tag{5.2}$$

for a positive scalar f_0.

The control objective is to design a saturated controller for plant (2.14) and (2.20) to achieve finite-time attitude stabilization under assumption (5.2) for all physically realizable initial conditions. Before giving the control design, some definitions and lemmas are needed.

5.2.1 Definitions and Lemmas

Definition 5.1 Consider the nonlinear system $\dot{x} = f(x, u)$, where x is a state vector, u is the input vector. The solution is finite-time stable if for all $x(t_0) = x_0$, there exists a settling time $T(x_0) < \infty$, such that $\lim_{t \to t_0 + T} \|x(t)\| = 0$.

Definition 5.2 ([27]) Consider the nonlinear system $\dot{x} = f(x, u)$, where x is a state vector, u is the input vector. The solution is practical finite-time stable if, for all $x(t_0) = x_0$, there exist $\epsilon > 0$ and $T(\epsilon, x_0) < \infty$, such that $\|x(t)\| < \epsilon$, for all $t \geq t_0 + T$.

Lemma 5.1 ([28]) *For any real numbers* $\lambda_1 > 0, \lambda_2 > 0, 0 < r < 1$, *a Lyapunov condition of finite-time stability can be given in the form as* $\dot{V}(x) + \lambda_1 V(x) + \lambda_2 V^r(x) \leq 0$ *with initial value* $V(x_0)$, *where the settling time can be estimated by*

$$T_0 \leq \frac{1}{\lambda_1(1-r)} \ln \frac{\lambda_1 V^{1-r}(x_0) + \lambda_2}{\lambda_2}.$$

Lemma 5.2 ([29]) *For any real numbers* $\lambda > 0, 0 < r < 1$, *finite-time stability can be given in the form as* $\dot{V}(x) + \lambda V^r(x) \leq 0$. *The reaching time is estimated by* $T_0 \leq \frac{V^{1-r}(x_0)}{\lambda(1-r)}$.

Lemma 5.3 *For arbitrary matrices* $A > 0$, $B > 0$, $A \in R^{n\times n}$, $B \in R^{n\times n}$, *assuming the eigenvalues of* A *and* B *are* $0 < \lambda_1 \leq \lambda_2 \leq \cdots \leq \lambda_n$ *and* $0 < \mu_1 \leq \mu_2 \leq \cdots \leq \mu_n$, *then the eigenvalue of* AB *satisfies* $\lambda(AB) \in [\lambda_1\mu_1, \lambda_n\mu_n]$.

Proof: See Appendix 1.

Lemma 5.4 *For arbitrary vector* $x \in R^3$, *positive definite diagonal matrix* $A \in R^{3\times 3}$ *and positive scalar* $\delta > 0$, *the inequality*

$$\lambda_{\min}(A)x^T \tanh\left(\frac{x}{\delta}\right) \leq x^T A \tanh\left(\frac{x}{\delta}\right) \leq \lambda_{\max}(A)x^T \tanh\left(\frac{x}{\delta}\right)$$

holds.

Proof: See Appendix 2.

5.3 Saturated Finite-Time Fault-Tolerant Control Law Design

In this section, the design procedure is shown to address the attitude stabilization problem. Firstly, a novel time-varying sliding manifold is introduced, and then applied to formulate the basic saturated finite-time controller. The Lyapunov analysis is followed to prove finite time convergence with explicitly considering actuator saturations. Then this basic controller is extended to compensate for the actuator faults and finite time stability is also ensured. Note that this paper generally assumes that perfect measurements of $\boldsymbol{q}$ and $\boldsymbol{\omega}$ are available, whereas no knowledge of $\boldsymbol{d}$ other than d_m is assumed.

5.3.1 *Novel Time-Varying Terminal Sliding Mode Design*

The fast terminal sliding mode (FTSM) manifold proposed in [16] has the form $\boldsymbol{S} = \boldsymbol{\omega} + k_1\boldsymbol{q}_v + k_2\boldsymbol{q}_v^{r_1/r_2}$, where $k_1 > 0$ and $k_2 > 0$, r_1 and r_2 are positive odd integers satisfying $0 < r_1/r_2 < 1$. The time derivative of the sliding manifold $\boldsymbol{S}$ is given by $\dot{\boldsymbol{S}} = \dot{\boldsymbol{\omega}} + k_1\dot{\boldsymbol{q}}_v + k_2 r_1 \boldsymbol{q}_v^{r_1/r_2-1}\dot{\boldsymbol{q}}_v/r_2$. It implies that if $\boldsymbol{q}_v = 0$ and $\dot{\boldsymbol{q}}_v \neq 0$, then the singularity will occur due to $r_1/r_2 - 1 < 0$. If the controller formulation contains the derivative of the sliding mode, which is very likely since many works utilize the so-called equivalent control component, then the singularity may demand very large control torques. In order to avoid the singularity problem, Zou et al. [14] proposed a modified terminal sliding mode surface

$$\boldsymbol{S}' = \boldsymbol{\omega} + k_1\boldsymbol{q}_v + k_2\beta(\boldsymbol{q}_v)$$

with $\beta(\boldsymbol{q}_v)$ defined as

$$\beta(q_i) = \begin{cases} q_i^{r_1/r_2}, & if\ \overline{S_i} = 0\ or\ \overline{S_i} \neq 0, |q_i| \geq \epsilon_0 \\ l_1 q_i + l_2 \mathrm{sign}(q_{vi}) q_i^2, & if\ \overline{S_i} \neq 0, |q_i| < \epsilon_0 \end{cases}, \text{ for } i = 1, 2, 3$$

where ϵ_0 is a small positive constant, $\overline{\boldsymbol{S}} = \boldsymbol{\omega} + k_1\boldsymbol{q}_v + k_2\boldsymbol{q}_v^{r_1/r_2}$, $l_1 = (2 - r_1/r_2)\epsilon_0^{r_1/r_2-1}$, $l_2 = (r_1/r_2 - 1)\epsilon_0^{r_1/r_2-2}$.

Theorem 5.1 *Consider the rigid spacecraft system in Eqs. (2.14) and (2.20) with the conventional FTSM surface satisfying* $\boldsymbol{S} = \boldsymbol{\omega} + \eta_1\boldsymbol{q}_v + \eta_2\boldsymbol{q}_v^r = 0$. *For arbitrary* $\eta_1 > 0$, $\eta_2 > 0$ *and* $0 < r < 1$ *with initial states* $q_0(0) \in (-1, 1]$, *the equilibrium* $\{\boldsymbol{q}_v = 0, q_0 = 1, \omega = 0\}$ *can be reached in finite time* T_s.

Proof: See Appendix 3.

As Theorem 5.1 suggests that once the conventional FTSM surface is reached and maintained, the system states will reach the desirable equilibrium in finite time. However, based on the conventional FTSM surface, the singularity problem may arise and lead to an unbounded control signal. Moreover, controllers based on the modified FTSM surface inevitably contain switch functions in their formulations and may give rise to system chattering. To this end, an alternative way to address the singularity problem is the novel time-varying FTSM surface proposed as follow

$$\boldsymbol{s} = \boldsymbol{\omega} + k_1 k^2 \boldsymbol{q}_v + k_2 k^2 \boldsymbol{q}_v^{r_1/r_2} = \boldsymbol{\omega} + \boldsymbol{x} \tag{5.3}$$

where $k(t)$ is a time-varying parameter, κ and k_2 are positive scalar constants, r_1 and r_2 are positive odd integers which satisfy $1/2 < r_1/r_2 < 1$, and $\boldsymbol{x} = \Delta k_1 k^2 \boldsymbol{q}_v + k_2 k^2 \boldsymbol{q}_v^{r_1/r_2}$. The updating law for k is naturally derived from the term cancellations in the following Lyapunov stability analysis to ensure the whole system states to be bounded and stable.

5.3.2 Basic Saturated Finite-Time Controller Design

The saturated finite-time controller is designed as

$$\boldsymbol{\tau} = -u_m \tanh\left(\frac{\boldsymbol{s}}{k^2\delta}\right) \tag{5.4}$$

And the auxiliary time-varying parameter $k(t)$ defined in Eq. (5.3) satisfies the following first-order dynamics:

$$\begin{aligned} \dot{k} =& \frac{-\gamma_k k}{1+4\gamma_k(1-q_0)}\left[\sum_{i=1}^{3}|q_i s_i| + u_m \sum_{i=1}^{3}\frac{|x_i|}{k^2}\left(\tanh\frac{|s_i|}{k^2\delta} + \tanh\frac{|x_i|}{k^2\delta}\right)\right. \\ &\left. +\delta u_m \sum_{i=1}^{3}\left(\tanh\frac{|s_i|}{k^2\delta} + \tanh\frac{|x_i|}{k^2\delta}\right)\right] \end{aligned} \tag{5.5}$$

Note that γ_k and are small positive scalars. Then another conclusion can be stated as:

Theorem 5.2 *Consider the spacecraft attitude dynamics described by Eqs. (2.14) and (2.20) in the presence of the bounded disturbance and limited control authorities described by Eq.* (5.2). *With the sliding mode surface in Eq.* (5.3) *and the control law in Eqs.* (5.4), (5.5), *the closed-loop system is globally practical finite-time stable, i.e., the novel sliding manifold* $\boldsymbol{s}$ *converges to region* o_s *in finite time, then the attitude error* $\boldsymbol{q}$ *and* $\boldsymbol{\omega}$ *finally converge into regions* o_q *and* o_ω *in finite time.*

$$\begin{aligned} o_s &= \frac{3k_0^2 u_m}{u_m - d_m - \sigma}\delta, \\ o_q &= \max\left\{\frac{o_s}{2k_1\mu^2}, \left(\frac{o_s}{2k_2\mu^2}\right)^{r_2/r_1}\right\}, \\ o_\omega &= k_1 k_0^2 o_q + k_2 k_0^2 o_q^{r_1/r_2} + o_s \end{aligned}$$

where k_0 *is a positive scalar that denotes the initial value of* k *for brevity;* σ *is a user-selected positive scalar and* μ *is the lower bound of* $k(t)$ *as discussed later.*

Proof: See Appendix 4.

Remark 5.1 The proposed scheme firstly assures the global asymptotic stability of attitude and velocity, and then local finite-time stability is attained via a TSM technique. The finite-time stabilization result established here is of great practical interest because the asymptotic convergence rate is at best exponential, while finite-time control can provide a faster convergence near the equilibrium via user specified design choices for the controller parameters. Besides the convergence rate, the finite time convergence can assure higher accuracies, better disturbance rejection properties, and better robustness against uncertainties as compared to asymptotic convergence [14]. In the Sect. 5.4, numerical simulations will illustrate this.

Remark 5.2 It should be pointed out that there are extensive spacecraft attitude control schemes available in literature. Here, to mention a few, the control schemes in [21–23] can achieve asymptotic convergence while the controllers in [14, 17, 27] guarantee finite-time stability. In this work, a simple structure controller based on time-varying FTSM surface is obtained. Generally, the proposed scheme in Eq. (5.4) can be effectively implemented under many practical constraints such as actuator saturation, external disturbance, and unknown or uncertain inertia distribution, which can hardly be simultaneously achieved by aforementioned controllers. Moreover, the smoothness of the proposed controller varies with time according to a set of user-defined parameters.

Remark 5.3 Unlike the previous nonsingular sliding mode controller [30] where a hard saturation is enforced on the unbounded singular term, the control law formulation in Eq. (5.4) contains no item related to the derivative of the sliding surface. Thus, the singularity problem of FTSM is avoided in this work by the time-varying sliding mode surface technique. To the best of authors' knowledge, though the time-varying FTSM surface has been studied in [31], this paper presents the first attempt to utilize the time-varying adaptive gain technique to address singularity problem and system chattering with a uniform finite-time control formulation.

Remark 5.4 The controller formulation is mainly inspired by [22], where the sharpness function therein approaches zero when time goes infinitely. In the proposed controller, with a proper selection of parameters, the sharpness function has a positive lower bound and thus possible chattering is eliminated. Unlike previous time-varying analog of the variable structure controllers, such as in [22], the analog of sliding mode surface can only be reached as time goes infinitely. In such cases, it is better to define s as an auxiliary variable or a filter variable rather than a sliding mode surface.

5.3.3 *Saturated Fault-Tolerant Finite-Time Controller Design*

When considering the actuation degradation (partial power loss or total failure) of onboard thrusters, the relation between the actual control torques and command signals is modeled as $\boldsymbol{\tau} = DF$ and $F = E(t)F_c$, where $F = [F_1, F_2, F_3, \ldots, F_n]^{\mathrm{T}}$ is the actual control forces exerted by n thrusters, the notation $\boldsymbol{F}_c$ denotes the commanded force signals, $D \in R^{3\times n}$ is the thruster distribution matrix and $E = \mathrm{diag}(e_1(t), e_2(t), \ldots, e_n(t)) \in R^{n\times n}$ denotes the effectiveness matrix. $E(t)$ is an identity matrix only if all actuators are healthy. Otherwise, the effectiveness matrix is a diagonal matrix with its diagonal elements ranging from 0 to 1. The case of $e_i(t) = 0$ indicates that the ith thruster has either totally failed or been shut down purposely. Note that the effectiveness matrix is time-varying and unknown to designers for the lack of fault detection and diagnosis (FDD) scheme.

In this subsection, the attitude dynamics with faulty thrusters is governed by

$$\boldsymbol{J}\dot{\boldsymbol{\omega}} = -\boldsymbol{\omega} \times \boldsymbol{J}\boldsymbol{\omega} + DE(t)F_c + \boldsymbol{d} \tag{5.6}$$

Assuming the nominal output of each thruster has an identical upper bound, expressed by

$$|F_{ci}| \leq F_m,\ i = 1, 2 \ldots n \tag{5.7}$$

If a spacecraft with only three actuators suffers from a failure of one or more thrusters, the vehicle becomes under-actuated system. In this case, attitude stabilization from arbitrary initial states becomes physically challenging [4]. In the following analysis, we assume that there are n thrusters (n > 3) properly mounted along the vehicle, which means the distribution matrix is made full row rank, and functional (not necessarily healthy) thrusters can provide sufficient torques to perform given maneuvers. Obviously, $\boldsymbol{D}\boldsymbol{E}\boldsymbol{D}^{\mathrm{T}}$ remains positive definite. Though the effectiveness matrix is unknown and uncertain, it is necessary to assume that the available control authorities under faulty actuation are able to produce combined torques such that

$$\frac{F_m \lambda_{\min}(\boldsymbol{D}\boldsymbol{E}\boldsymbol{D}^{\mathrm{T}})\lambda_{\min}(\boldsymbol{A})}{\|\boldsymbol{D}^T\boldsymbol{A}\|_\infty} = d_{\mathrm{m}} + f_1,\ f_1 > 0 \tag{5.8}$$

where $\boldsymbol{A}$ is an arbitrary diagonal matrix that satisfies

$$\boldsymbol{A} > 0, \boldsymbol{D}\boldsymbol{E}\boldsymbol{D}^{\mathrm{T}}\boldsymbol{A} > 0,$$

$\lambda_{\min}(\boldsymbol{D}\boldsymbol{E}\boldsymbol{D}^{\mathrm{T}})$ and $\lambda_{\max}(\boldsymbol{D}\boldsymbol{E}\boldsymbol{D}^{\mathrm{T}})$ denote the minimum and maximum eigenvalues of $\boldsymbol{D}\boldsymbol{E}\boldsymbol{D}^{\mathrm{T}}$. For convenience, two positive constants are notated as

$$u_{m1} \triangleq \Delta F_m \frac{\lambda_{\max}(\boldsymbol{D}\boldsymbol{D}^{\mathrm{T}})\lambda_{\max}(\boldsymbol{A})}{\|\boldsymbol{D}^{\mathrm{T}}\boldsymbol{A}\|_\infty}$$

and

$$u_{m2} \triangleq \Delta F_m \frac{\lambda_{\min}(\boldsymbol{D}\boldsymbol{E}\boldsymbol{D}^{\mathrm{T}})\lambda_{\min}(\boldsymbol{A})}{\|\boldsymbol{D}^{\mathrm{T}}\boldsymbol{A}\|_\infty} = d_{\mathrm{m}} + f_1.$$

There exists a diagonal matrix

$$\boldsymbol{\Delta} = \mathrm{diag}(\Delta_1, \Delta_2, \ldots, \Delta_{\mathrm{n}}) \in \mathrm{R}^{\mathrm{n}\times\mathrm{n}}$$

where $\Delta_i = \sqrt{e_i}$, then $\|\boldsymbol{\Delta}\|^2 = \|\boldsymbol{E}\| \leq 1$ and

$$\lambda_{\max}(\boldsymbol{D}\boldsymbol{E}\boldsymbol{D}^{\mathrm{T}}) = \lambda_{\max}(\boldsymbol{D}\boldsymbol{\Delta}(\boldsymbol{D}\boldsymbol{\Delta})^{\mathrm{T}}) = \|\boldsymbol{D}\boldsymbol{\Delta}\|^2 \leq \|\boldsymbol{D}\|^2\|\boldsymbol{\Delta}\|^2 \leq \|\boldsymbol{D}\|^2 = \lambda_{\max}(\boldsymbol{D}\boldsymbol{D}^{\mathrm{T}}).$$

Thus, we have

$$u_{m2} \leq \frac{F_m}{\|\boldsymbol{D}^{\mathrm{T}}\boldsymbol{A}\|_\infty}\lambda(\boldsymbol{D}\boldsymbol{E}\boldsymbol{D}^{\mathrm{T}}\boldsymbol{A}) \leq u_{m1}$$

from Lemma 5.3.

Then the simple saturated finite-time controller is extended to a faulty scenario as

$$F_c = -\frac{F_m \boldsymbol{D}^{\mathrm{T}} \boldsymbol{A}}{||\boldsymbol{D}^{\mathrm{T}} \boldsymbol{A}||_\infty} \tanh\left(\frac{\boldsymbol{s}}{k^2\delta}\right) \tag{5.9}$$

And the auxiliary time-varying parameter $k(t)$ satisfies the following first-order dynamics

$$\begin{aligned} \dot{k} =& \frac{-\gamma_k k}{1+4\gamma_k(1-q_0)}\left[\sum_{i=1}^{3}|q_i s_i| + u_{m1}\sum_{i=1}^{3}\frac{|x_i|}{k^2}\tanh\frac{|s_i|}{k^2\delta} + u_{m2}\sum_{i=1}^{3}\frac{x_i}{k^2}\tanh\frac{x_i}{k^2\delta}\right. \\ &\left. + \delta u_{m2}\sum_{i=1}^{3}\left(\tanh\frac{|s_i|}{k^2\delta} + \tanh\frac{|x_i|}{k^2\delta}\right)\right] \end{aligned} \tag{5.10}$$

Then the following statements can be given as:

Theorem 5.3 *Consider the spacecraft attitude dynamics in Eq.* (5.6) *with faulty actuators whose output magnitude satisfies conditions in Eqs.* (5.7), (5.8) . *With the proposed control law in Eqs.* (5.9), (5.10) , *the closed-loop system is globally stabilized in finite time.*

Proof The proof of Theorem 5.3 is similar to the previous analysis, and divided into three main steps: closed-loop asymptotically stabilization, reaching phase of the novel FTSM surface and dynamics along the surface.

Step 1: Consider a Lyapunov function candidate

$$V = \frac{1}{2}\boldsymbol{\omega}^{\mathrm{T}} \boldsymbol{J}\boldsymbol{\omega} + k^2 \boldsymbol{q}^{\mathrm{T}}\boldsymbol{q} + k^2(1-q_0)^2 + \frac{k^2}{2\gamma_k} \tag{5.11}$$

Following straightforward algebra, the time derivative of V with the updating law in Eq. (5.10) is established as

$$\begin{aligned} \dot{V} =& (\boldsymbol{s}-\boldsymbol{x})^{\mathrm{T}}\left[-\boldsymbol{\omega}\times\boldsymbol{J}\boldsymbol{\omega} - \frac{\boldsymbol{D}\boldsymbol{E}\boldsymbol{D}^{\mathrm{T}}\boldsymbol{A}}{\|\boldsymbol{D}^{\mathrm{T}}\boldsymbol{A}\|_\infty}F_m\tanh\left(\frac{\boldsymbol{s}}{k^2\delta}\right) + \boldsymbol{d}\right] + k^2\boldsymbol{q}_v^{\mathrm{T}}\boldsymbol{\omega} + 4k\dot{k}(1-q_0) + \frac{k\dot{k}}{\gamma_k} \\ =& (\boldsymbol{s}-\boldsymbol{x})^{\mathrm{T}}\left[-\boldsymbol{\omega}\times\boldsymbol{J}\boldsymbol{\omega} - \frac{\boldsymbol{D}\boldsymbol{E}\boldsymbol{D}^{\mathrm{T}}\boldsymbol{A}}{\|\boldsymbol{D}^{\mathrm{T}}\boldsymbol{A}\|_\infty}F_m\tanh\left(\frac{\boldsymbol{s}}{k^2\delta}\right)\right] + (\boldsymbol{s}-\boldsymbol{x})^{\mathrm{T}}\boldsymbol{d} + k^2\boldsymbol{q}_v^{\mathrm{T}}\boldsymbol{\omega} \\ & - k^2\sum_{i=1}^{3}|q_i s_i| - u_{m1}\sum_{i=1}^{3}|x_i|\tanh\frac{|s_i|}{k^2\delta} + u_{m2}\sum_{i=1}^{3}\frac{x_i}{k^2}\tanh\frac{x_i}{k^2\delta} \\ & - u_{m2}k^2\delta\sum_{i=1}^{3}\left(\tanh\frac{|s_i|}{k^2\delta} + \tanh\frac{|x_i|}{k^2\delta}\right) \end{aligned}$$

Considering Lemmas 5.3, 5.4 and the assumption in Eq. (5.8), one obtains

$$
\begin{aligned}
\dot{V} \leq & -\frac{F_m \lambda_{\min}(\boldsymbol{D}\boldsymbol{E}\boldsymbol{D}^{\mathrm{T}}\boldsymbol{A})}{\|\boldsymbol{D}^{\mathrm{T}}\boldsymbol{A}\|_{\infty}} \boldsymbol{s}^{\mathrm{T}} \tanh\left(\frac{\boldsymbol{s}}{k^2\delta}\right) - u_{m2}\boldsymbol{x}^{\mathrm{T}} \tanh\left(\frac{\boldsymbol{x}}{k^2\delta}\right) + \boldsymbol{s}^{\mathrm{T}}\boldsymbol{d} - \boldsymbol{x}^{\mathrm{T}}\boldsymbol{d} \\
& - u_{m2}k^2\delta \sum_{i=1}^{3} \left(\tanh\frac{|s_i|}{k^2\delta} + \tanh\frac{|x_i|}{k^2\delta}\right) - k^2\boldsymbol{q}_v^{\mathrm{T}}\boldsymbol{x} \\
\leq & - u_{m2}\boldsymbol{s}^{\mathrm{T}} \tanh\left(\frac{\boldsymbol{s}}{k^2\delta}\right) - u_{m2}\boldsymbol{x}^{\mathrm{T}} \tanh\left(\frac{\boldsymbol{x}}{k^2\delta}\right) + \boldsymbol{s}^{\mathrm{T}}\boldsymbol{d} - \boldsymbol{x}^{\mathrm{T}}\boldsymbol{d} \\
& - u_{m2}k^2\delta \sum_{i=1}^{3} \left(\tanh\frac{|s_i|}{k^2\delta} + \tanh\frac{|x_i|}{k^2\delta}\right) - k^2\boldsymbol{q}_v^{\mathrm{T}}\boldsymbol{x} \\
\leq & - f_1\|\boldsymbol{s}\|_1 - f_1\|\boldsymbol{x}\|_1 - k^2\boldsymbol{q}_v^{\mathrm{T}}x \leq -f_1\|\boldsymbol{\omega}\|_1 - k^2\boldsymbol{q}_v^{\mathrm{T}}x
\end{aligned}
\tag{5.12}
$$

With a similar analysis to Theorem 5.2, the lower bound on k is assured and asymptotical attitude stabilization is then guaranteed. Moreover, the analysis procedures of Step 2 and Step 3 are almost the same as previous and thus are omitted here.

Remark 5.5 Note that in control law formulation in Eq. (5.9), the signals are directly assigned to command the nominal force output. Though the real force that actuators exert on the vehicle is unknown to designers because of the existence of faults, the finite-time stabilization of the overall system is achieved.

Remark 5.6 The designed controller in Eq. (5.9) rigorously obeys the actuator saturation limits, that is

$$
\begin{aligned}
\|F_c\|_{\infty} &= \frac{F_m}{\|\boldsymbol{D}^{\mathrm{T}}\boldsymbol{A}\|_{\infty}} \|\boldsymbol{D}^{\mathrm{T}}\boldsymbol{A} \tanh\left(\frac{s}{k^2\delta}\right)\|_{\infty} \\
&\leq \frac{F_m}{\|\boldsymbol{D}^{\mathrm{T}}\boldsymbol{A}\|_{\infty}} \|\boldsymbol{D}^{\mathrm{T}}\boldsymbol{A}\|_{\infty} \|\tanh\left(\frac{s}{k^2\delta}\right)\|_{\infty} \\
&= F_m
\end{aligned}
$$

5.3.4 *The Advantage of Modification Matrix A*

As the positive-definite matrix $\boldsymbol{A}$ has almost no constraint in the proof process of Theorem 5.3, there are many alternative matrices to choose from. In fact, the modification matrix $\boldsymbol{A}$ is introduced to ensure that all functional actuators have the potential to reach its output limit simultaneously. In a plant with very limited control authority, every available actuation counts, and thus this modification is much more favorable in severe faulty occasions.

Considering a scenario when 6 two-sided thrusters (A two-sided thruster has exhausts in two opposite directions) with the same nominal maximum thrust capability are symmetrically mounted on a vehicle in the configuration shown in Fig. 5.1. Then the distribution matrix can be simply determined by the distance between the center of mass to the mounting position, that is,

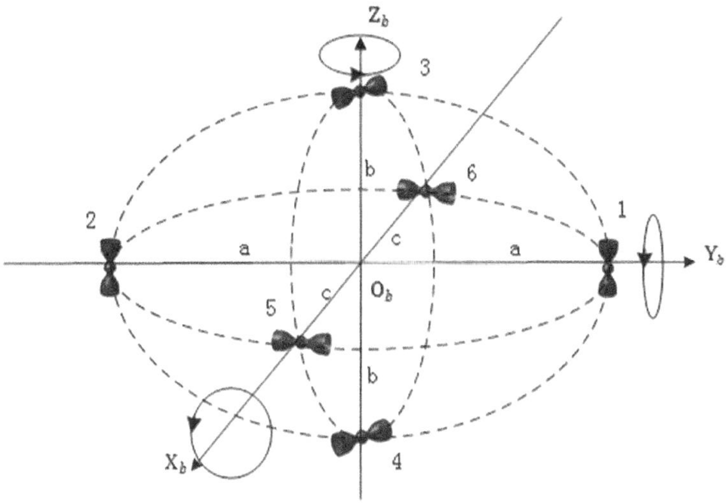

Fig. 5.1 A typical symmetric configuration

$$D = \begin{bmatrix} a & -a & 0 & 0 & 0 & 0 \\ 0 & 0 & b & -b & 0 & 0 \\ 0 & 0 & 0 & 0 & c & -c \end{bmatrix}, \text{ where } a, b, c > 0$$

The modification matrix A is then chosen such that the sum of the absolute value of every row in $D^T A$ is equal. In this condition, due to the properties of saturated function tanh, thrusters onboard have the potential to reach their output limits simultaneously and form the largest torque envelope. One simple example for selecting the modification matrix is $A = \text{diag}(\frac{1}{a}, \frac{1}{b}, \frac{1}{c})$. In the following is a visual demonstration on the comparison of the available control torques between the proposed scheme and a scheme proposed in [4]; note that both schemes rigorously enforce the actuator saturation limit.

For a fair comparison, consider a specific configuration in [4]

$$D = \begin{bmatrix} 0.8 & -0.8 & 0 & 0 & 0 & 0 \\ 0 & 0 & 0.7 & -0.7 & 0 & 0 \\ 0 & 0 & 0 & 0 & 0.7 & -0.7 \end{bmatrix} \text{m}$$

Here, the torque envelope manifolds of $A = \text{diag}(\frac{1}{8}, \frac{1}{7}, \frac{1}{7})$, $A = I$, and the one in [4] with a common thrust limit $F_m = 500\,\text{mN}$ are compared. Note that the control torque $\tau = DEF_c$ is calculated in a healthy scenario, that is, $E = I$ in three cases. The available torque envelopes of three schemes are shown in Fig. 5.2.

With $A = \text{diag}(\frac{1}{8}, \frac{1}{7}, \frac{1}{7})$ applied, the six thrusters can output their thrust limit simultaneously as shown in Fig. 5.2a. Indeed, the volume of the envelope is largest and only constrained by the thrust limit and distribution configuration. Obviously, the

Fig. 5.2 Torque envelopes of three schemes

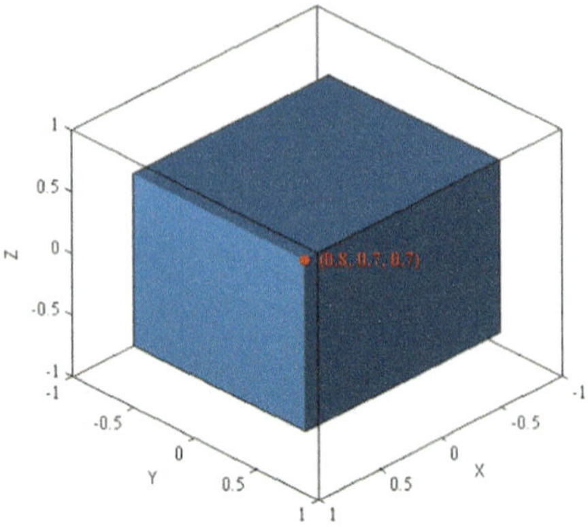

(a) available torque envelope when $\boldsymbol{A} = \text{diag}(\frac{1}{8}, \frac{1}{7}, \frac{1}{7})$

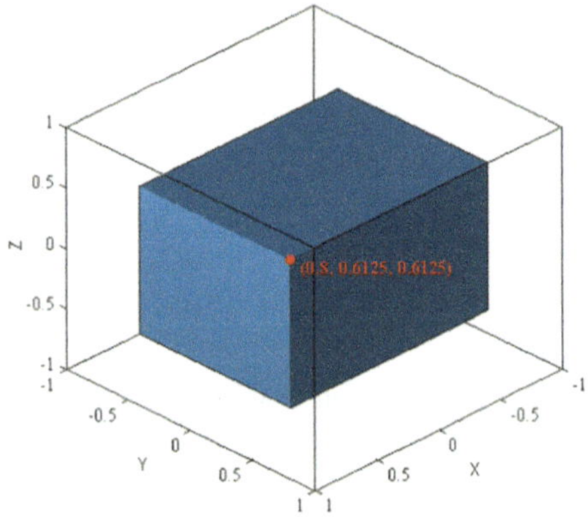

(b) available torque envelope when $\boldsymbol{A} = \boldsymbol{I}$

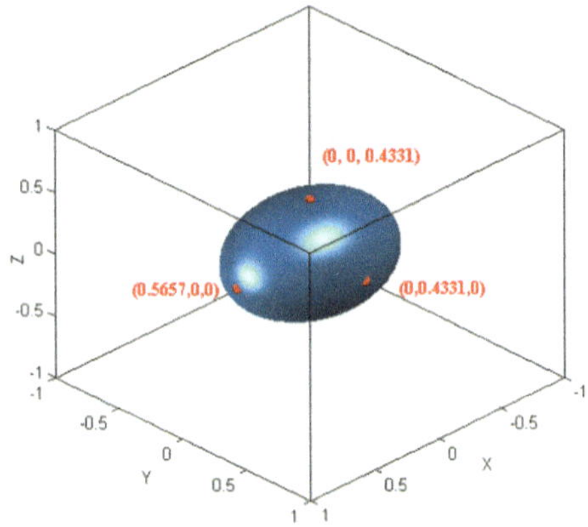

(c) available torque envelope in [4]

available torque envelope in [4] shapes an ellipsoid and is much smaller. Though it is just a straightforward modification, it helps the controller to take full use of all the available thrust and achieve better performance in a severe situation as demonstrated in simulations.

Remark 5.7 The necessary condition described in Eq. (5.8) is less conservative than the case in [4]. In [4], the functional thrusters are able to produce sufficient forces that are assumed to satisfy

$$\frac{\lambda_{\min}(\boldsymbol{D}\boldsymbol{E}\boldsymbol{D}^{\mathrm{T}})F_m}{\|\boldsymbol{D}\|} = \|\boldsymbol{L}_d\| + f_2,\ f_2 > 0 \tag{5.13}$$

$\boldsymbol{L}_d$ is lumped term of system nonlinearities and external disturbances. Presume that vector $\boldsymbol{L}_d$ and $\boldsymbol{d}$ in the present paper are identical, then, $d_{\mathrm{m}} = \|\boldsymbol{d}\|_\infty \leq \|\boldsymbol{L}_d\|$. Note that once select $\boldsymbol{A} = \mathrm{diag}(\frac{1}{\mathrm{a}}, \frac{1}{\mathrm{b}}, \frac{1}{\mathrm{c}})$, then $\|\boldsymbol{D}^{\mathrm{T}}\boldsymbol{A}\|_\infty = 1$, $\lambda_{\min}(\boldsymbol{A})\|\boldsymbol{D}\| = \frac{\|\boldsymbol{D}\|}{\max(a,b,c)} = \sqrt{2} > 1$. It is then obtained

$$\frac{\lambda_{\min}(\boldsymbol{D}\boldsymbol{E}\boldsymbol{D}^{\mathrm{T}})\lambda_{\min}(\boldsymbol{A})F_m}{\|\boldsymbol{D}^{\mathrm{T}}\boldsymbol{A}\|_\infty} \geq \frac{\lambda_{\min}(\boldsymbol{D}\boldsymbol{E}\boldsymbol{D}^{\mathrm{T}})}{\|\boldsymbol{D}\|}F_m$$

which shows the proposed controller is more favorable in faulty occasions and this property is demonstrated in Sect. 5.4

Remark 5.8 As for the occasions when the maximum thrust capability varies among the six thrusters, one way to address the problem is to let F_m be the common lower bound and that leads to a little conservation for that the potential control authority is not fully exploited. Another way is to extend the thrust limit scalar to a diagonal matrix. That is to say,

$$F_c = -F_m\frac{\boldsymbol{D}^{\mathrm{T}}\boldsymbol{A}}{\|\boldsymbol{D}^{\mathrm{T}}\boldsymbol{A}\|_\infty}\tanh(\frac{s}{k^2\delta})$$

where $F_m = \mathrm{diag}(F_{mi}), i = 1, 2, \ldots 6$ and F_{mi} denotes the nominal maximum thrust capability of ith thruster.

Remark 5.9 Controller in Eq. (5.9) is also valid and applicable in a healthy scenario. In this case, the relation between the combined control torque and the thrust command signal is a typical least squares control allocation by the pseudoinverse method.

Proof The optimal thrust signal via pseudo-inverse method is

$$F_{\mathrm{optimal}} = \boldsymbol{D}^{\dagger}\boldsymbol{\tau} = \boldsymbol{D}^{\dagger}\boldsymbol{D}F_c.$$

With Eq. (5.9) in mind, one obtains

$$F_{\mathrm{optimal}} = -\frac{F_m\boldsymbol{D}^{+}\boldsymbol{D}\boldsymbol{D}^{\mathrm{T}}\boldsymbol{A}}{||\boldsymbol{D}^{\mathrm{T}}\boldsymbol{A}\|_\infty}\tanh(\frac{s}{k^2\delta}) = -\frac{F_m\boldsymbol{D}^{\mathrm{T}}\boldsymbol{A}}{||\boldsymbol{D}^{\mathrm{T}}\boldsymbol{A}||_\infty}\tanh(\frac{s}{k^2\delta}) = F_c.$$

Table 5.1 Parameters of a rigid spacecraft

Mission	Attitude control
Initial quaternion	$\left[q_0, \boldsymbol{q}_v^{\mathrm{T}}\right]^{\mathrm{T}} = [0.9, -0.3, 0.26, 0.18]^{\mathrm{T}}$
Initial angular Velocity	$\boldsymbol{\omega} = [0.1, -0.1, -0.05]^{\mathrm{T}}$ rad/s
Attitude control type	Three-axis control by six thrusters
Maximum thrust	0.5 N
Actuator fault type	Efficiency loss and failure

5.4 Simulation Results

To analyze the favorable properties of the proposed control schemes, numerical simulations have been carried out and results are presented in this section. Throughout the simulations, the inertia matrix is given by

$$\boldsymbol{J} = \begin{bmatrix} 20 & 0 & 0.9 \\ 0 & 17 & 0 \\ 0.9 & 0 & 15 \end{bmatrix} \mathrm{kg} \cdot \mathrm{m}^2,$$

the distribution matrix is

$$\boldsymbol{D} = \begin{bmatrix} 0.8 & -0.8 & 0 & 0 & 0 & 0 \\ 0 & 0 & 0.7 & -0.7 & 0 & 0 \\ 0 & 0 & 0 & 0 & 0.7 & -0.7 \end{bmatrix} \mathrm{m},$$

and external disturbance is $\boldsymbol{d}(t) = d_m[\sin(0.8t), \cos(0.5t), \cos(0.3t)]^{\mathrm{T}}$ N.m. Note that the disturbance generating function is adopted from [4] among its alternatives to illustrate the disturbance rejection property of the designed controller. And no knowledge of disturbance other than d_m is assumed to be known to designers. Other simulation parameters of the stabilization mission are shown in the Table 5.1.

In this section, three different cases are numerically studied: (1) healthy thrusters, (2) thrusters with partial efficiency (with a mean efficiency value 70%), and (3) thrusters with a more severer fading actuation (mean value 30%). All these simulations are done under the thrust capability limit constraints. The following simulations verify the theoretical analysis declared in Sect. 5.3.

Case 1. Healthy actuators

With all healthy thrusters, the controller in Eq. (5.4) is utilized. Here, with a specific performance requirement, the guidelines to choose proper parameters to implement the controller are demonstrated. The available control torques u_m is calculated as 0. 7 N.m and external disturbance bound is set as $d_m = 0.06$ N.m. The attitude stabilization target, parameterized as steady-state Pointing Precision Index, is preset as $\|\boldsymbol{q}_v\| \leq 1 \times 10^{-3}$. From the previous analysis, γ_k should be sufficiently small and then μ is within a near neighborhood of k_0. To simplify the tuning process,

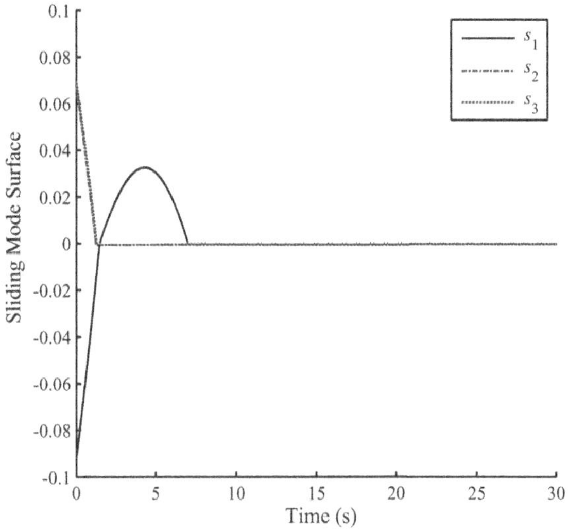

Fig. 5.3 Time response of the sliding mode surface

we arbitrarily choose $\gamma_k = 0.01$, and $k_0 = 1$, and $\mu \geq k_0/2 = 0.5$ is assumed (and validated by later numerical simulation).

Considering the phase when the system states slide along the chosen surface, the transient response is governed by the sliding surface and the kinematic equation in Eq. (2.14). With this in mind, we can set some criteria such as the state convergence rate and choose proper parameters. In the aforementioned case, the sliding surface parameters $k_1 = 1, k_2 = 0.1, r_1 = 5, r_2 = 7$ are chosen. With the explicit upper bound on the convergence zone in Theorem 5.2, when $\zeta = 0.1$ is chosen, $\delta = 1 \times 10^{-4}$ is obtained via a straightforward calculation.

The time response of the sliding mode surface is given in Fig. 5.3. Figures 5.4, 5.5, 5.6 and 5.7 show the time response of the system states. Obviously, the motion of state parameters is consistent with previous theoretical analysis, that is, the sliding surface is reached at 8s and the state parameters are maintained on it afterwards, then attitude error and velocity converge to origin within finite time. The time response of adaptive parameter in Fig. 5.8 converges to a positive scalar larger than $k_0/2$. Moreover, the control torque and thrust shown in Figs. 5.9, 5.10. Another favorable property of the proposed finite-time controller is that it has fine disturbance rejection capability and achieves higher-precision control performance. This property is depicted in Fig. 5.11.

Case 2. Fading and failed actuators

To model a thruster with unknown efficiency loss and total failure (or shut down purposely), the effectiveness of each actuator is generated by following equation

$$e_i(t) = \varepsilon + 0.1\text{rand}(t_i) + 0.1\sin(0.5t + i\pi/3),\ t_i = \text{mod}(t + \Delta t_i, T),\ i = 1, 2, ..., 6 \quad (5.14)$$

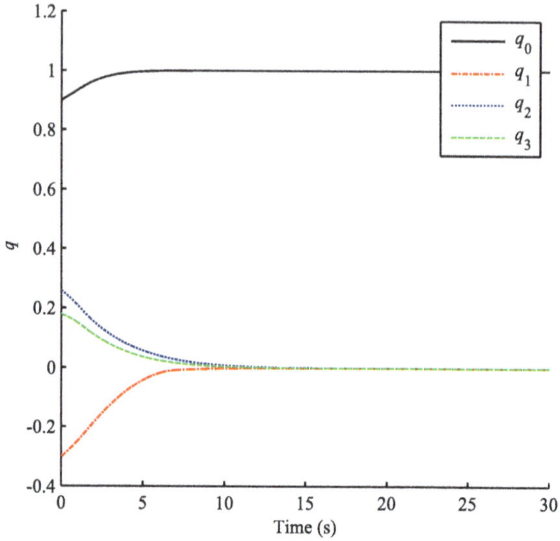

Fig. 5.4 Time response of the quaternion

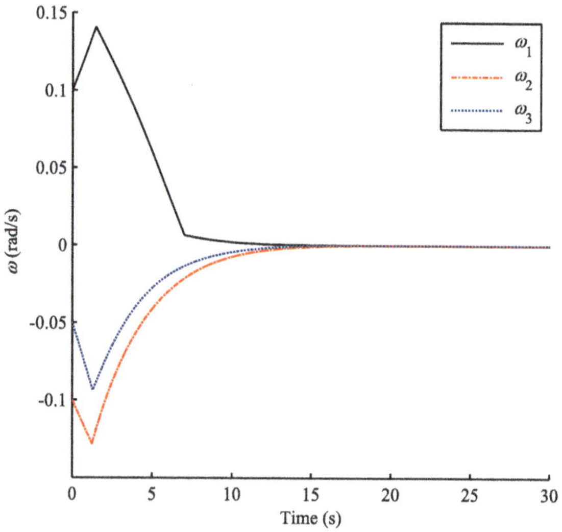

Fig. 5.5 Time response of the angular velocity

ε is a positive constant, $\text{rand}(\cdot) \in (-1, 1)$ is a random number generator, $\Delta t_i = 2i - 2$ and sampling time $T = 12$ s. The three terms in Eq. (5.14) simulate the average part, the 'jump' part and the continuous part in the efficiency matrix respectively. For failed actuation simulation, thrusters 2, 4, 6 are purposely shut down at 8 s, 12 s, 16 s, separately.

This subsection represents the fault-tolerant ability of proposed controller under fading and completely failed actuation. In this case, the average functional efficiency parameter is chosen as $\varepsilon = 0.7$. Note that effectiveness matrix is unknown

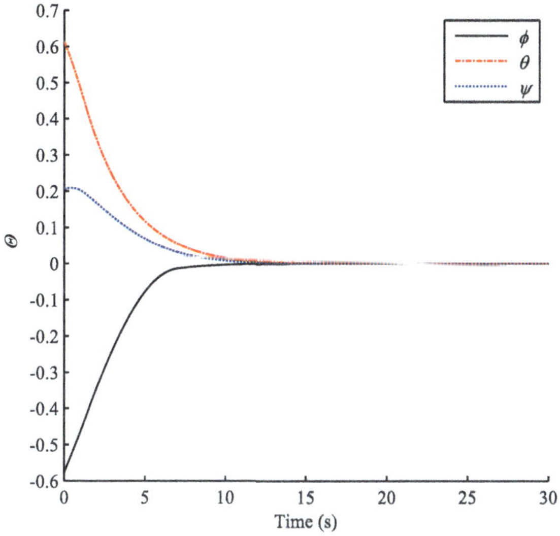

Fig. 5.6 Time response of the Euler angle

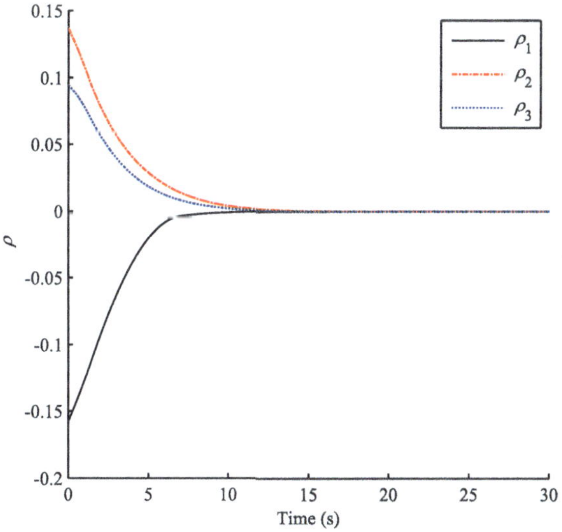

Fig. 5.7 Time response of the MRPs

and uncertain to designers, and both assumptions on $\lambda_{\min}(\boldsymbol{DED}^{\mathrm{T}})$ are assured here ($d_m = 0.06\,\mathrm{N.m}$). The effectiveness factor is shown in Fig. 5.12.

The parameters of the controller in Eq. (5.9) remain the same as in Case 1, and $\boldsymbol{A} = \mathrm{diag}(1/0.8, 1/0.7, 1/0.7)$, $u_{m2} = 0.1$, $u_{m1} = F_m \frac{\lambda_{\max}(\boldsymbol{DD}^{\mathrm{T}})\lambda_{\max}(\boldsymbol{A})}{\|\boldsymbol{D}^{\mathrm{T}}\boldsymbol{A}\|_{\infty}} = 0.9143$.

Figure 5.13 shows the time response of the sliding mode surface. The attitude responses are presented in Figs. 5.14, 5.15, 5.16 and 5.17. It can be seen that the system states will convergent to the origin within 20 s and the proposed controller can

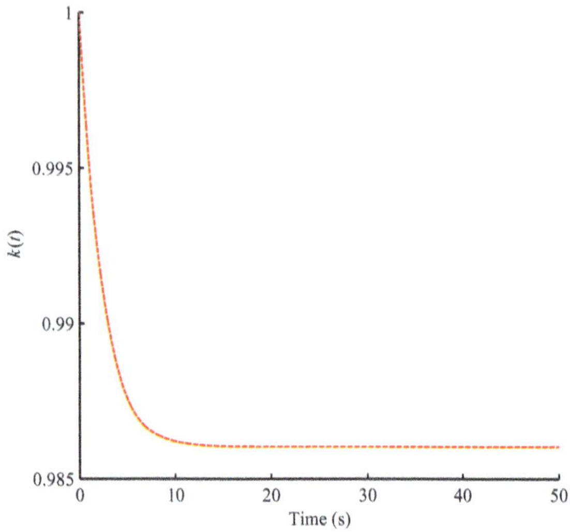

Fig. 5.8 Time response of the adaptive parameter

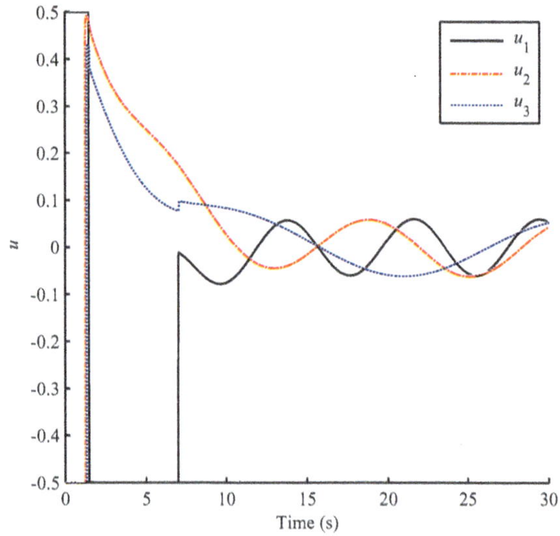

Fig. 5.9 Time response of the control torque

achieve a high precision stabilization within limited thrusts. The employment of the novel time-varying FTSM surface is the main reason for these favorable properties.

The time response of the control torque and actual thrusts are shown in Figs. 5.18 and 5.19. As for the actual thrust exerted by faulty actuators, it is shown in Fig. 5.19 that the proposed controllers rigorously enforce the thrust saturation constraint. The robustness against external disturbance is further demonstrated with different disturbance bounds in Case 3.

Fig. 5.10 Time response of the thrusters

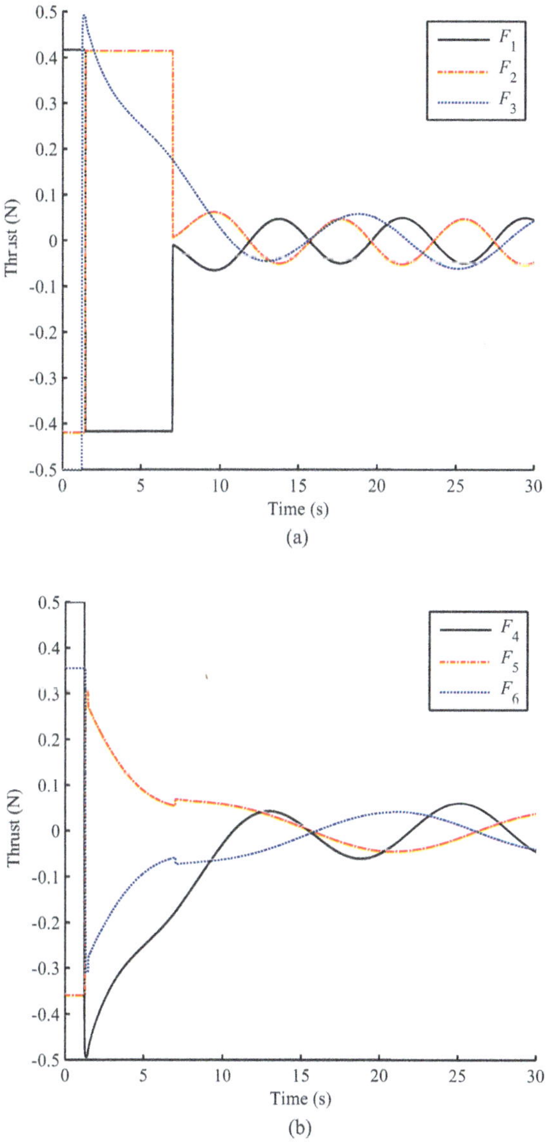

Fig. 5.11 Pointing Precision Index

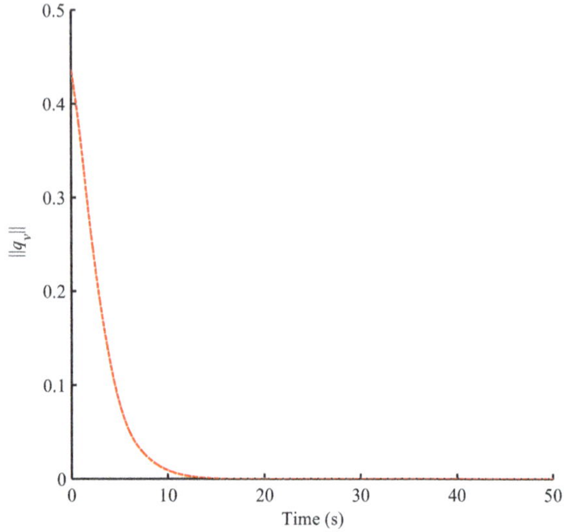

Case 3. Severe faulty scenario

In this section, two severe faulty scenarios are considered: (1) $\epsilon = 0.3$, $d_m = 0.03$ N.m, (2) $\varepsilon = 0.3$, $d_m = 0.06$ N.m. Note that in this case, assumptions on $\lambda_{\min}(\boldsymbol{DED}^{\mathrm{T}})$ for controller in Eq. (5.9) may not be satisfied. The control parameters are well tuned, most of which remain the same but $k_1 = 0.2$, $k_2 = 0.05$ in the controller in Eq. (5.9).

From Fig. 5.20, the proposed controller can handle the faults or failures and assure a certain degree of stabilization. With controller in Eq. (5.9) applied, it can be observed that it performs a fast convergence and fine steady-state motions in both scenarios. Moreover, the controller in Eq. (5.9) almost maintains its favorable properties despite the increase of external disturbance, which shows good robustness against disturbances.

To sum up, simulation results show that the proposed control methods enable to achieve the control target even in the presence of external disturbance, input saturation and inertia uncertainty. Furthermore, the proposed control methods also have good robustness to actuator faults. In addition, extensive simulations have also been performed with various initial conditions, control parameters, and fault situations. These results confirm that the stabilization of the closed-loop system is accomplished.

5.5 Conclusions

This chapter presents a structurally simple finite-time controller of spacecraft attitude with a novel time-varying fast terminal sliding mode surface. The algorithm is then extended to accommodate unknown actuator faults. In the basic controller design,

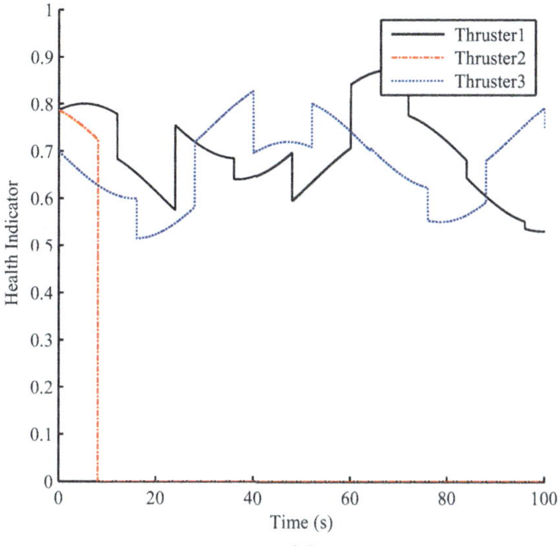

(a)

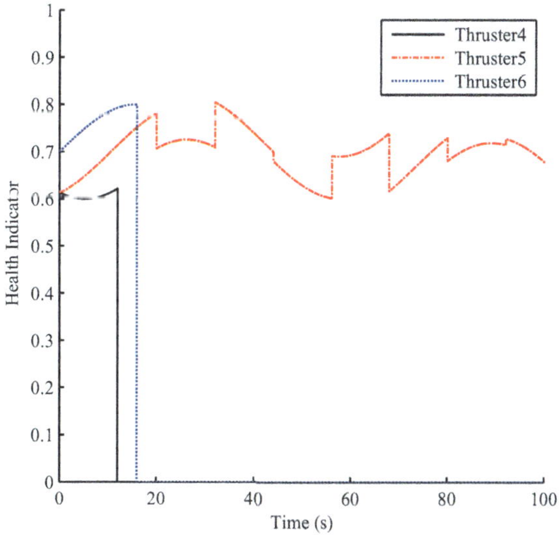

(b)

Fig. 5.12 Effectiveness factor versus time

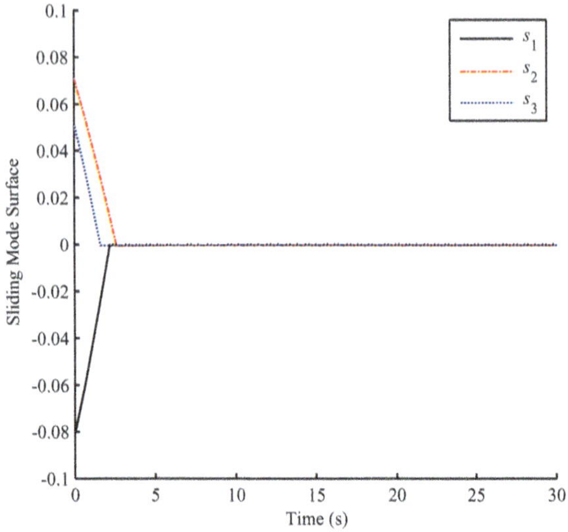

Fig. 5.13 Time response of the sliding mode surface

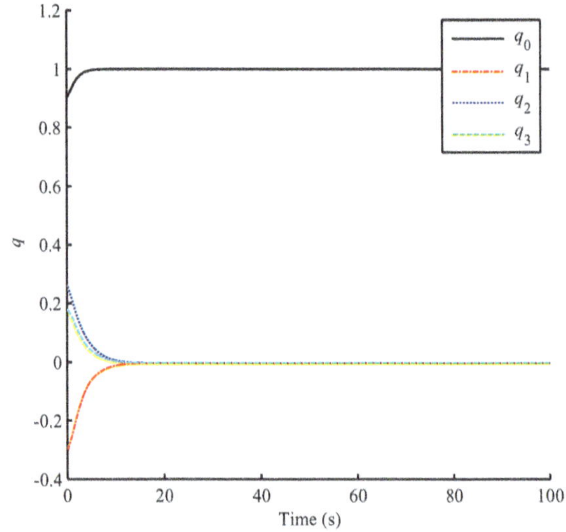

Fig. 5.14 Time response of the quaternion

an attempt is made to combine the time-varying sliding mode control with FTSM surface to accomplish finite-time convergence. Theoretical analysis shows that the novel sliding surface eliminates the conventional singularity problems without any modified or switched terms. External disturbance and actuator saturation are well addressed as well within the proposed novel framework. This simple algorithm is then developed to the faulty scenarios without fault detection and diagnosis scheme. Once an actuator fault or failure occurs, in order to make full use of the remaining functional thrusts, an extension version of the common method to passively deal with uncertain faults is designed. This modification helps to broaden the available torque

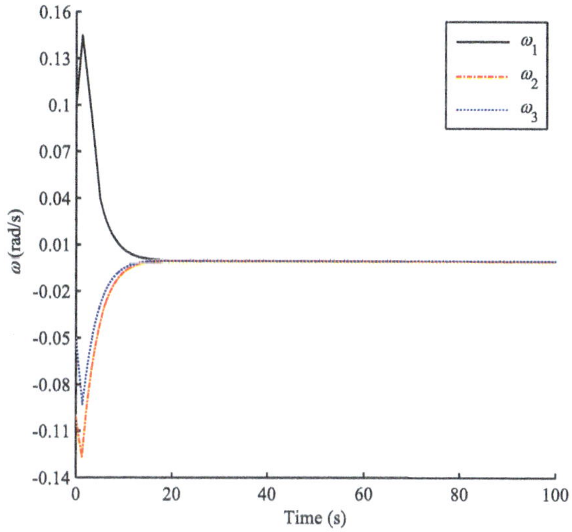

Fig. 5.15 Time response of the angular velocity

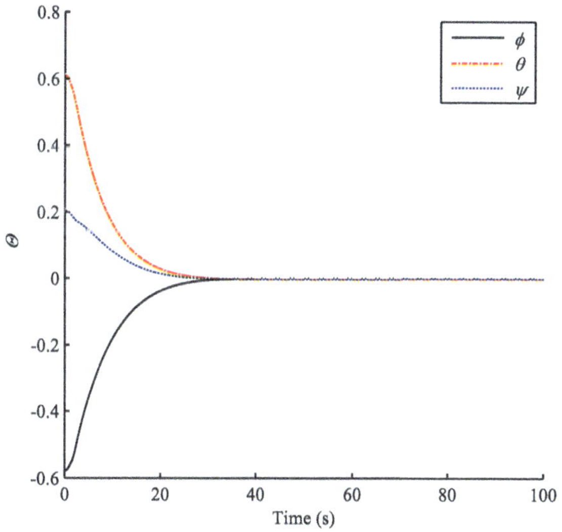

Fig. 5.16 Time response of the Euler angle

envelopes and loosen the conservative condition. Numerical simulations validate several desirable performance features of the simple structure control scheme.

However, the FTC scheme presented in this chapter is a CA base PFTC method to make the control system to be robust against the actuator faults and external disturbances. In the next chapters, the authors attempt to utilize some active approaches, such as extended state observer and iterative learning observer based FDD to deal with the unknown uncertainties and actuator faults respectively. And the CA problem is also investigated by taking the uncertainties and actuator faults into consideration, using the optimal CA and robust CA methods.

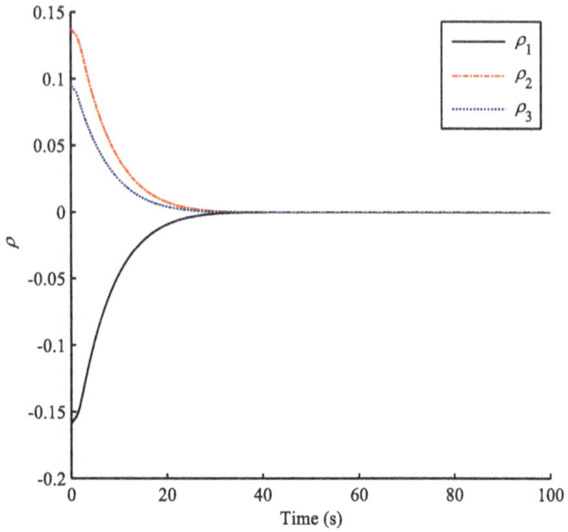

Fig. 5.17 Time response of the MRPs

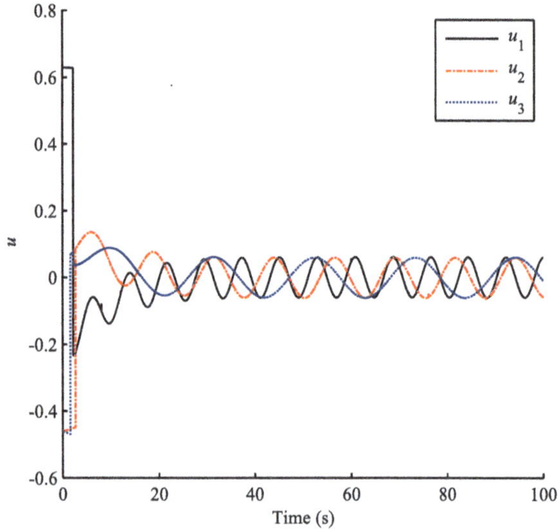

Fig. 5.18 Time response of the control torque

Appendices

Appendix 1

Proof of Lemma 5.3

First prove $\lambda(\boldsymbol{AB}) > 0$. Since real-valued $\boldsymbol{A}$ and $\boldsymbol{B}$ are symmetric and positive definite, there exist invertible matrices $\boldsymbol{P}$ and $\boldsymbol{Q}$ such that $\boldsymbol{A} = \boldsymbol{P}^{\mathrm{T}}\boldsymbol{P}$ and $\boldsymbol{B} = \boldsymbol{Q}^{\mathrm{T}}\boldsymbol{Q}$. The polynomial of eigenvalues is

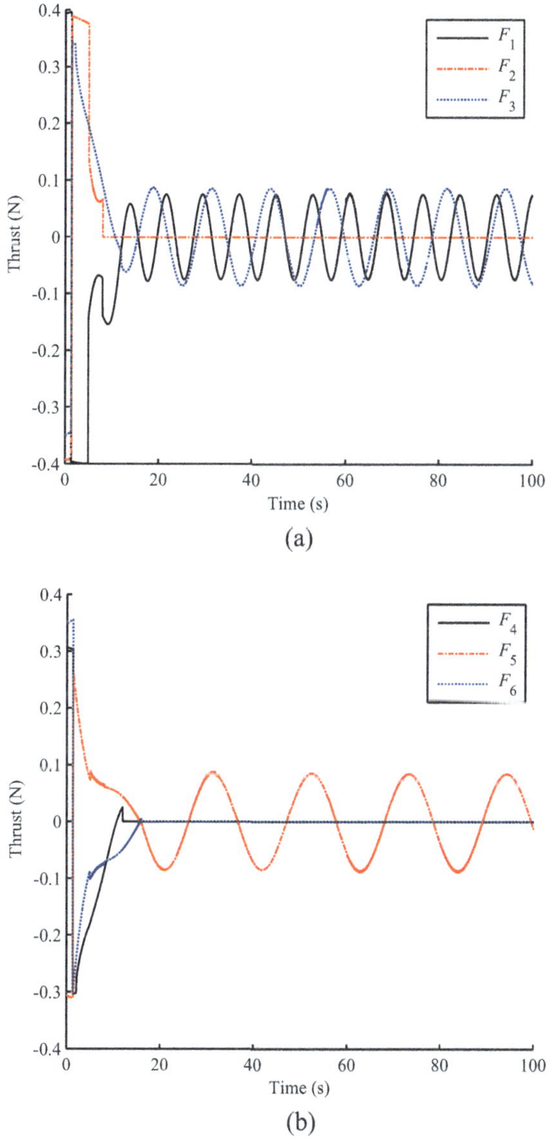

Fig. 5.19 Actual thrust of the proposed controller

Fig. 5.20 Pointing Precision Index in severe situation

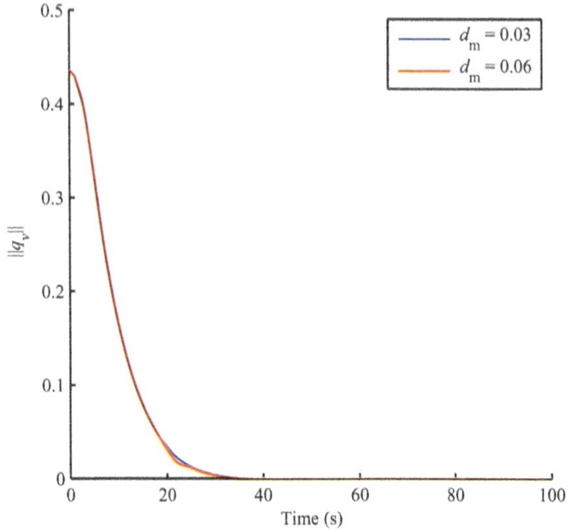

$$\begin{aligned}\det(\lambda \boldsymbol{I} - \boldsymbol{AB}) &= \det(\boldsymbol{P}^{\mathrm{T}}(\lambda \boldsymbol{I} - \boldsymbol{P}\boldsymbol{Q}^{\mathrm{T}}\boldsymbol{Q}\boldsymbol{P}^{\mathrm{T}})(\boldsymbol{P}^{\mathrm{T}})^{-1}) \\ &= \det(\lambda \boldsymbol{I} - (\boldsymbol{Q}\boldsymbol{P}^{\mathrm{T}})^{\mathrm{T}}\boldsymbol{Q}\boldsymbol{P}^{\mathrm{T}}) \\ &= 0.\end{aligned}$$

Since $(\boldsymbol{Q}\boldsymbol{P}^{\mathrm{T}})^{\mathrm{T}}\boldsymbol{Q}\boldsymbol{P}^{\mathrm{T}} > 0$, $\lambda(\boldsymbol{AB}) > 0$ is guaranteed. Note that, if $\boldsymbol{C} \geq 0$, $\boldsymbol{C} \in R^{n\times n}$, $\lambda(\boldsymbol{AC}) \geq 0$ is assured via the same procedures.

Then, there exists an orthogonal matrix $\boldsymbol{U}$ so that $\boldsymbol{U}^{\mathrm{T}}\boldsymbol{AU} = \mathrm{diag}(\lambda_1, \lambda_2, \ldots, \lambda_{\mathrm{n}})$. Consider matrix $\boldsymbol{B} - \lambda_1\mu_1\boldsymbol{A}^{-1}$, one has

$$\begin{aligned}&\boldsymbol{U}^{\mathrm{T}}(\boldsymbol{B} - \boldsymbol{A}_1\mu_1\boldsymbol{A}^{-1})\boldsymbol{U} \\ &= \boldsymbol{U}^{\mathrm{T}}\boldsymbol{BU} - \mu_1\mathrm{diag}(1, \lambda_1\lambda_2{}^{-1}, \ldots, \lambda_1\lambda_{\mathrm{n}}^{-1}) \\ &= (\boldsymbol{U}^{\mathrm{T}}\boldsymbol{BU} - \mu_1\boldsymbol{I}) + \mathrm{diag}(0, (1 - \lambda_1\lambda_2^{-1})\mu_1, \ldots., (1 - \lambda_1\lambda_{\mathrm{n}}^{-1})\mu_1)\end{aligned}$$

Note that $\boldsymbol{U}^{\mathrm{T}}\boldsymbol{BU} - \mu_1\boldsymbol{I} \geq 0$ and

$$\mathrm{diag}(0, (1 - \lambda_1\lambda_2^{-1})\mu_1, \ldots., (1 - \lambda_1\lambda_{\mathrm{n}}^{-1})\mu_1) \geq 0$$

Thus, $\boldsymbol{B} - \lambda_1\mu_1\boldsymbol{A}^{-1} \geq 0$ and

$$\lambda_{\min}(\boldsymbol{A}(\boldsymbol{B} - \lambda_1\mu_1\boldsymbol{A}^{-1})) = \lambda_{\min}(\boldsymbol{AB} - \lambda_1\mu_1\boldsymbol{I}) \geq 0$$

is derived, i.e., $\lambda_{\min}(\boldsymbol{AB}) \geq \lambda_1\mu_1$.

With a similar derivation, conclusion can be drawn that $\lambda_{\max}(\boldsymbol{A}(\boldsymbol{B} - \lambda_n\mu_n\boldsymbol{A}^{-1})) = \lambda_{\max}(\boldsymbol{AB} - \lambda_n\mu_n\boldsymbol{I}) \leq 0$. Thus, Lemma 5.3 is obtained.

Appendix 2

Proof of Lemma 5.4

First note that vector $\tanh(\frac{x}{\delta})$ can be reformed as

$$\tanh\left(\frac{\boldsymbol{x}}{\delta}\right) = \begin{bmatrix} b_1 & & \\ & b_2 & \\ & & b_3 \end{bmatrix} \boldsymbol{x}$$

where

$$b_i = \begin{cases} \tanh\left(x_i/\delta\right)/x_i & \text{for } x_i \neq 0 \\ 1/\delta & \text{otherwise} \end{cases} \quad \text{for } i = 1, 2, 3$$

Obviously, $0 < b_i \leq 1/\delta$ holds. That is to say, $\tanh(\frac{x}{\delta}) = \boldsymbol{B}\boldsymbol{x}$, where the diagonal matrix $\boldsymbol{B} > 0$. Then,

$$\begin{aligned} &\boldsymbol{x}^{\mathrm{T}}\boldsymbol{A}\tanh\left(\frac{\boldsymbol{x}}{\delta}\right) - \lambda_{\min}(\boldsymbol{A})\boldsymbol{x}^{\mathrm{T}}\tanh\left(\frac{\boldsymbol{x}}{\delta}\right) \\ &= \boldsymbol{x}^{\mathrm{T}}(\boldsymbol{A} - \lambda_{\min}(\boldsymbol{A})\boldsymbol{I}\tanh\left(\frac{\boldsymbol{x}}{\delta}\right) \\ &= \boldsymbol{x}^{\mathrm{T}}(\boldsymbol{A} - \lambda_{\min}(\boldsymbol{A})\boldsymbol{I})\boldsymbol{B}\boldsymbol{x} \end{aligned}$$

Due to $\boldsymbol{A}$ being a diagonal matrix, $(\boldsymbol{A} - \lambda_{\min}(\boldsymbol{A})\boldsymbol{I})\boldsymbol{B}$ is diagonal as well. From Lemma 5.3, $\lambda((\boldsymbol{A} - \lambda_{\min}(\boldsymbol{A})\boldsymbol{I})\boldsymbol{B}) \geq 0$, thus $(\boldsymbol{A} - \lambda_{\min}(\boldsymbol{A})\boldsymbol{I})\boldsymbol{B} \geq 0$ holds, i.e.,

$$\boldsymbol{x}^{\mathrm{T}}\boldsymbol{A}\tanh\left(\frac{\boldsymbol{x}}{\delta}\right) \geq \lambda_{\min}(\boldsymbol{A})\boldsymbol{x}^{\mathrm{T}}\tanh\left(\frac{\boldsymbol{x}}{\delta}\right)$$

holds.

The inequality $(\boldsymbol{A} - \lambda_{\max}(\boldsymbol{A})\boldsymbol{I})\boldsymbol{B} \leq 0$ can be shown via a similar analysis. Thus

$$\boldsymbol{x}^{\mathrm{T}}\boldsymbol{A}\tanh\left(\frac{\boldsymbol{x}}{\delta}\right) \leq \lambda_{\max}(\boldsymbol{A})\boldsymbol{x}^{\mathrm{T}}\tanh\left(\frac{\boldsymbol{x}}{\delta}\right)$$

holds and Lemma 5.4 is obtained.

Appendix 3

Proof of Lemma 5.1

Note $\omega = -\eta_1\boldsymbol{q}_v - \eta_2\boldsymbol{q}_v^r$ holds. Consider the candidate Lyapunov function on equilibrium $\{q_0 = 1, \boldsymbol{q}_v = 0\}$

$$\begin{aligned} V_1 &= \frac{1}{2}[q_1^2 + q_2^2 + q_3^2 + (1 - q_0)^2] \\ &= 1 - q_0 \end{aligned}$$

and its derivative can be obtained as

$$\begin{aligned} \dot{V}_1 &= \frac{1}{2}\boldsymbol{q}_v^{\mathrm{T}}\omega \\ &= -\frac{1}{2}\eta_1 \boldsymbol{q}^{\mathrm{T}}\boldsymbol{q}_v - \frac{1}{2}\eta_2 \boldsymbol{q}_v^{\mathrm{T}}\boldsymbol{q}_v^r \\ &\leq 0 \end{aligned}$$

then the equilibrium $q_0 = 1$ is shown to be attractive.

Further consider another semi-positive-definite scalar function on equilibrium $\{q_0 = -1, \boldsymbol{q}_v = 0\}$

$$V_2 = \frac{1}{2}[q_1^2 + q_2^2 + q_3^2 + (1 + q_0)^2] = 1 + q_0$$

The time derivative of V_2 can be derived as

$$\begin{aligned} \dot{V}_2 &= -\frac{1}{2}\boldsymbol{q}_v^{\mathrm{T}}\omega \\ &= \frac{1}{2}\eta_1 \boldsymbol{q}_v^{\mathrm{T}}\boldsymbol{q}_v + \frac{1}{2}\eta_2 \boldsymbol{q}_v^{\mathrm{T}}\boldsymbol{q}_v^r \\ &\geq 0 \end{aligned}$$

Since the state trajectory departs from the equilibrium, $q_0 = -1$ is shown to be a saddle point. Note $q_0(0) \neq -1$, and

$$\begin{aligned} q_0 &= \dot{V}_2 \\ &\geq \frac{1}{2}\eta_1(1 - q_0^2) + \frac{1}{2}\eta_2(1 - q_0^2)^{(r+1)/2} \\ &> 0 \end{aligned}$$

for $q_0 \neq 1$, which means $q_0(t)$ is an increasing function. For the case $q_0(0) < 0$, after a finite time $T_{s1} = q_0(0)/c$, where

$$c = \frac{1}{2}\eta_1(1 - q_0^2(0)) + \frac{1}{2}\eta_2(1 - q_0^2(0))^{(r+1)/2}$$

Then, the inequality

$$(1 - q_0)^2 \leq (1 - q_0)(1 + q_0) = q_1^2 + q_2^2 + q_3^2$$

holds. For the case $q_0(0) \geq 0$, the inequality holds as well.

Next, define scalar function $V_q = q_1^2 + q_2^2 + q_3^2$, and thus $V_1 \leq V_q$ holds after T_{s1}. It further follows that

$$\dot{V}_1 \leq -\frac{1}{2}\eta_1 V_1 - \frac{1}{2}\eta_2 V_1^{(r+1)/2}$$

Noted that $\frac{1}{2} < \frac{1+r}{2} < 1$. By Lemma 5.1, $\boldsymbol{q}_v = 0$ and $q_0 = 1$ can be reached in another finite time T_{s2}, and accordingly, $\boldsymbol{\omega} = 0$ is guaranteed as well. Therefore $\{\boldsymbol{q}_v = 0, q_0 = 1, \boldsymbol{\omega} = 0\}$ can be reached in finite time $T_s = T_{s1} + T_{s2}$.

Appendix 4

Proof of Lemma 5.2

The proof of Theorem 5.2 can be divided into three steps: firstly the closed-loop asymptotic stabilization is shown; and then the reaching phase of sliding mode surface is proved to be within finite time; finally, dynamics along the sliding surface nearby is further discussed.

Step 1: The systems stabilization analysis

Consider the following Lyapunov function candidate

$$\begin{aligned} V &= \frac{1}{2}\boldsymbol{\omega}^{\mathrm{T}}\boldsymbol{J}\boldsymbol{\omega} + k^2\boldsymbol{q}_v^{\mathrm{T}}\boldsymbol{q}_v + k^2(1-q_0)^2 + \frac{k^2}{2\gamma_k} \\ &= \frac{1}{2}\boldsymbol{\omega}^{\mathrm{T}}\boldsymbol{J}\boldsymbol{\omega} + 2k^2(1-q_0) + \frac{k^2}{2\gamma_k} \end{aligned} \tag{5.15}$$

Following straightforward algebra, the time derivative of V evaluated along the closed-loop trajectories can be established as

$$\begin{aligned} \dot{V} &= \boldsymbol{\omega}^{\mathrm{T}}\boldsymbol{J}\dot{\boldsymbol{\omega}} + k^2\boldsymbol{q}_v^{\mathrm{T}}\boldsymbol{\omega} + 4k\dot{k}(1-q_0) + \frac{k\dot{k}}{\gamma_k} \\ &= \boldsymbol{\omega}^{\mathrm{T}}(-\boldsymbol{\omega}\times\boldsymbol{J}\boldsymbol{\omega} + \boldsymbol{\tau} + \boldsymbol{d}) + k^2\boldsymbol{q}_v^{\mathrm{T}}\boldsymbol{\omega} + 4k\dot{k}(1-q_0) + \frac{k\dot{k}}{\gamma_k} \\ &= (\boldsymbol{s}-\boldsymbol{x})^{\mathrm{T}}\left(-\boldsymbol{\omega}\times\boldsymbol{J}\boldsymbol{\omega} - u_m\tanh\frac{\boldsymbol{s}}{k^2\delta} + \boldsymbol{d}\right) + k^2\boldsymbol{q}_v^{\mathrm{T}}\boldsymbol{\omega} \\ &\quad + 4k\dot{k}(1-q_0) + \frac{k\dot{k}}{\gamma_k} \\ &= -\boldsymbol{s}^{\mathrm{T}}\left[u_m\tanh\frac{\boldsymbol{s}}{k^2\delta} - \boldsymbol{d}\right] + \boldsymbol{x}^{\mathrm{T}}\left[u_m\tanh\frac{\boldsymbol{s}}{k^2\delta} - \boldsymbol{d}\right] \\ &\quad + k^2\boldsymbol{q}_v^{\mathrm{T}}\boldsymbol{\omega} + 4k\dot{k}(1-q_0) + \frac{k\dot{k}}{\gamma_k} \end{aligned}$$

In view of the time-varying function in Eq. (5.5), it yields

$$\begin{aligned}
\dot{V} \leq & -\boldsymbol{s}^{\mathrm{T}}\left[u_m \tanh\frac{\boldsymbol{s}}{k^2\delta} - \boldsymbol{d}\right] + \boldsymbol{x}^{\mathrm{T}}\left[u_m \tanh\frac{\boldsymbol{s}}{k^2\delta} - \boldsymbol{d}\right] + k^2\boldsymbol{q}_v^{\mathrm{T}}\boldsymbol{\omega} - k^2\sum_{i=1}^{3}|q_i s_i| \\
& - u_m\sum_{i=1}^{3}|x_i|\left(\tanh\frac{|s_i|}{k^2\delta} + \tanh\frac{|x_i|}{k^2\delta}\right) - u_m k^2\delta\sum_{i=1}^{3}\left(\tanh\frac{|s_i|}{k^2\delta} + \tanh\frac{|x_i|}{k^2\delta}\right) \\
\leq & -u_m\boldsymbol{s}^{\mathrm{T}}\tanh\frac{\boldsymbol{s}}{k^2\delta} - u_m\boldsymbol{x}^{\mathrm{T}}\tanh\frac{\boldsymbol{x}}{k^2\delta} - u_m k^2\delta\sum_{i=1}^{3}\left(\tanh\frac{|s_i|}{k^2\delta} + \tanh\frac{|x_i|}{k^2\delta}\right) \\
& + \boldsymbol{s}^{\mathrm{T}}\boldsymbol{d} - \boldsymbol{x}^{\mathrm{T}}\boldsymbol{d} - k^2\boldsymbol{q}_v^{\mathrm{T}}\boldsymbol{x}
\end{aligned}$$

For any $a > 0$ and all real scalar x, the inequality holds

$$x\tanh(x/a) \geq |x| - a\tanh(|x|/a)$$

Then,

$$\begin{aligned}
\dot{V} & \leq -(u_m - d_m)\|\boldsymbol{s}\|_1 - (u_m - d_m)\|\boldsymbol{x}\|_1 - k^2\boldsymbol{q}_v^{\mathrm{T}}\boldsymbol{x} \\
& = -f_0\|\boldsymbol{s}\|_1 - f_0\|\boldsymbol{x}\|_1 - k^2\boldsymbol{q}_v^{\mathrm{T}}x \\
& \leq -f_0\|\boldsymbol{\omega}\|_1 - k^2\boldsymbol{q}_v^{\mathrm{T}}\boldsymbol{x} \\
& \leq 0
\end{aligned} \tag{5.16}$$

Therefore, $\boldsymbol{s}, \boldsymbol{x}, \boldsymbol{\omega}$ and $k^2\boldsymbol{q}_v^{\mathrm{T}}\boldsymbol{x}$ are L_1 integral, and $\boldsymbol{\omega}, k$ are bounded. Furthermore, $\dot{\boldsymbol{\omega}}$ from Eq. (2.20) and $\dot{k}$ from Eq. (5.5) are bounded. Note that

$$\begin{aligned}
\frac{\mathrm{d}}{\mathrm{d}t}(k^2\boldsymbol{q}_v^{\mathrm{T}}\boldsymbol{x}) = & 4k^3\dot{k}(k_1\boldsymbol{q}_v^{\mathrm{T}}\boldsymbol{q}_v + k_2\boldsymbol{q}_v^{\mathrm{T}}\boldsymbol{q}_v^{r_1/r_2}) \\
& + k^4\left(2k_1\sum_{i=1}^{3}q_i\dot{q}_i + \frac{r_1 + r_2}{r_2}k_2\sum_{i=1}^{3}q_i^{r_1/r_2}\dot{q}_i\right)
\end{aligned}$$

are bounded as well, which implies that $\boldsymbol{\omega}$ and $k^2\boldsymbol{q}_v^{\mathrm{T}}\boldsymbol{x}$ are uniformly continuous functions with a finite integral over $[0, +\infty)$. Then Barbalat's lemma applies, and

$$\left.\begin{aligned}
\lim_{t\to\infty}\boldsymbol{\omega}(t) = 0 \\
\lim_{t\to\infty}k^2\boldsymbol{q}_v^{\mathrm{T}}\boldsymbol{x} = 0
\end{aligned}\right\} \Rightarrow \begin{aligned}
\lim_{t\to\infty}s(t) = 0 \\
\lim_{t\to\infty}k^2\boldsymbol{q}_v = 0
\end{aligned}$$

is derived. To show the asymptotic convergence of $\boldsymbol{q}_v$, k^2 should not converge to zero unless $\boldsymbol{q}_v$ converges to origin. A deeper discussion on the dynamics of k is stated below in Assertion 1.

For a given γ_k, from Eq. (5.16), there exist positive functions such that

$$\int_0^{+\infty} \|s(t)\|_1 \mathrm{d}t \le c_1(\gamma_k),$$
$$\int_0^{+\infty} \|x(t)\|_1 \mathrm{d}t \le c_2(\gamma_k),$$
$$\int_0^{+\infty} k^2 \sum_{i=1}^{3} (|q_i c v_i| \mathrm{d}t \le c_3(\gamma_k),$$
$$\int_0^{+\infty} k^2 \boldsymbol{q}_v^{\mathrm{T}} x \mathrm{d}t \le c_4(\gamma_k)$$

for $\forall t \ge 0$. Assertion 1: For the updating law Eq. (5.5), with any given $k(0) = k_0 > 0$ and $\mu > 0$, where $k_0 > \mu$, if there exists a small $\gamma_k > 0$ such that

$$k_0^2 - \mu^2 \ge 2\gamma_k [u_m c_1 + 3u_m c_2 + c_3 + c_4] \tag{5.17}$$

then $k_0 \ge k(t) \ge \mu > 0$ for all $t \ge 0$.

Integrating both sides of Eq. (5.16) on time interval $[0, +\infty)$, one obtains

$$\begin{aligned} & f_0 \int_0^{+\infty} \|\boldsymbol{\omega}(t)\|_1 \mathrm{d}t + \int_0^{+\infty} k^2 \boldsymbol{q}_v^{\mathrm{T}} \boldsymbol{x} \mathrm{d}t \\ & \le f_0 \int_0^{+\infty} \|\boldsymbol{s}(t)\|_1 \mathrm{d}t + f_0 \int_0^{+\infty} \|\boldsymbol{x}(t)\|_1 \mathrm{d}t + \int_0^{+\infty} k^2 \boldsymbol{q}_v^{\mathrm{T}} \boldsymbol{x} \mathrm{d}t \\ & \le V(0) - V(\infty) \end{aligned}$$

Note where $V(\infty)$ denotes the value of $V(t)$ when $t \to \infty$.

Note that $V(t)$ is semi-positive and

$$\int_0^{+\infty} k^2 \sum_{i=1}^{3} (|q_i \omega_i|) \mathrm{d}t \le k_0^2 \int_0^{+\infty} \|\boldsymbol{\omega}(t)\|_1 \mathrm{d}t$$

It establishes a rough upper-bound estimate on the functions $c_i(\gamma_k), i = 1, 2, 3, 4$ in terms of k_0 and γ_k. That is,

$$\frac{f_0}{k_0^2} c_3 + c_4 \le V(0)$$

$$f_0 c_1 + f_0 c_2 + c_4 \le V(0)$$

where

$$V(0) \le \frac{1}{2} \lambda_{\max}(\boldsymbol{J}) \|\boldsymbol{\omega}(0)\|^2 + 4k_0^2 + k_0^2 / 2\gamma_k.$$

In Eq. ((5.5)), note that the following inequalities hold:

$$1 + 4\gamma_k(1 - q_0) \geq 1, \ \sum_{i=1}^{3} |q_i s_i| \leq \|\boldsymbol{s}\|_1,$$

$$\delta u_m \sum_{i=1}^{3} \left(\tanh \frac{|s_i|}{k^2 \delta} + \tanh \frac{|x_i|}{k^2 \delta} \right) \leq 6 \delta u_m$$

$$\begin{aligned}
\sum_{i=1}^{3} \frac{|x_i|}{k^2} \left(\tanh \frac{|s_i|}{k^2 \delta} + \tanh \frac{|x_i|}{k^2 \delta} \right) &\leq 2 \sum_{i=1}^{3} \frac{|x_i|}{k^2} \\
&\leq 2 \sum_{i=1}^{3} |k_1 q_i + k_2 q_i^{r_1/r_2}| \\
&\leq 6(k_1 + k_2)
\end{aligned}$$

Then

$$\dot{k} \geq -\gamma_k k[\|\boldsymbol{s}\|_1 + 6u_m\delta + 6u_m(k_1 + k_2)] \triangleq -\gamma_k \chi k$$

is derived, where the notation χ denotes the upper bound of $[\|\boldsymbol{s}\|_1 + 6u_m\delta + 6u_m(k_1 + k_2)]$. Because the lower bound on $k(t)$ is described simply by a first-order dynamics, it is shown that $k \geq 0$ for all the time. Since $\dot{k}(t) \leq 0$, then $k(t)$ is decreasing with upper bound k_0.

Taking a closer inspection of the update law Eq. (5.5) for $k(t)$, one can further derive that

$$\begin{aligned}
k\dot{k} \geq &-\gamma_k \left[k^2 \sum_{i=1}^{3} |q_i s_i| + u_m \sum_{i=1}^{3} |x_i| \left(\tanh \frac{|s_i|}{k^2 \delta} + \tanh \frac{|x_i|}{k^2 \delta} \right) \right. \\
&\left. + u_m \sum_{i=1}^{3} k^2 \delta \left(\tanh \frac{|s_i|}{k^2 \delta} + \tanh \frac{|x_i|}{k^2 \delta} \right) \right] \\
\geq &-\gamma_k \left[k^2 \sum_{i=1}^{3} (|q_i x_i| + |q_i \omega_i|) + 2u_m \sum_{i=1}^{3} |x_i| + u_m \sum_{i=1}^{3} (|s_i| + |x_i|) \right] \\
\geq &-\gamma_k \left[u_m ||\boldsymbol{s}||_1 + 3u_m \|\boldsymbol{x}\|_1 + k^2 \sum_{i=1}^{3} (|q_i \omega_i|) + k^2 \boldsymbol{q}_v^{\mathrm{T}} \boldsymbol{x} \right].
\end{aligned}$$

Upon integrating this equation on time interval $[0, +\infty)$, one obtains

$$k^2(\infty) \geq k_0^2 - 2\gamma_k \int_0^{+\infty} [u_m \|\boldsymbol{s}\|_1 + 3u_m \|\boldsymbol{x}\|_1 + k^2 \sum_{i=1}^{3} (|q_i \omega_i|) + k^2 \boldsymbol{q}_v^{\mathrm{T}} \boldsymbol{x}] \mathrm{d}t$$

Hence, if a sufficiently small γ_k that satisfies the Eq. (5.17) is found, then $k(t) \geq \mu$ is guaranteed for all the time.

Assumption Eq. (5.17) imposes an implicit condition that γ_k needs to satisfy in order to assure the positive lower bound of $k(t)$. Though an explicit condition on γ_k is favorable, finding such a solution is the main technical difficulty with the approach outlined here. However, simulation studies show that, when γ_k is chosen to be small ($\gamma_k << 1$), convergence of $k(t)$ to some nonzero value is usually attainable. In fact, the lower bound of $k(t)$ is typically observed to be within a near neighborhood of the initial value k_0. A similar analysis of the adaptive parameter behavior is presented in [21].

Assuming that a sufficiently small γ_k is founded that satisfies Eq. (5.17), recalling that $\lim_{t\to\infty} \boldsymbol{x}(t) = \lim_{t\to\infty} k_1 k^2 \boldsymbol{q}_v + k_2 k^2 \boldsymbol{q}_v^{r_1/r_2} = 0$, it follows $\lim_{t\to\infty} \boldsymbol{q}_v = 0$.

Step 2: Reaching phase analysis

After pre-multiplying Eq. (5.3) by $\boldsymbol{J}$, and differentiating it with respect to time, one obtains

$$\boldsymbol{J}\dot{\boldsymbol{s}} = \boldsymbol{N} + \boldsymbol{\tau} + \boldsymbol{d} \tag{5.18}$$

where

$$\boldsymbol{N} = -\boldsymbol{\omega} \times \boldsymbol{J}\boldsymbol{\omega} + k^2 \boldsymbol{J}\left(k_1 \dot{\boldsymbol{q}}_v + \frac{r_1}{r_2} k_2 \frac{\dot{\boldsymbol{q}}_v}{\boldsymbol{q}_v^{1-r_1/r_2}}\right) + 2k\dot{k}\boldsymbol{J}\left(k_1 \boldsymbol{q}_v + k_2 \boldsymbol{q}_v^{r_1/r_2}\right)$$

To streamline analysis, let $\bar{N}(t)$ denote the maximum value of $\{|N_1|, |N_2|, |N_3|\}$ after time t.

The conventional singularity problems may arise in the derivative of $\boldsymbol{s}$ when $\dot{\boldsymbol{q}}_v \neq 0$ and $\boldsymbol{q}_v = 0$. Note that $\lim_{t\to\infty} \boldsymbol{q} = 0$ and

$$\lim_{t\to\infty} \boldsymbol{q}_v = \lim_{t\to\infty} \frac{1}{2}(q_0 I + \boldsymbol{q}_v \times)\boldsymbol{\omega} = 0$$

is guaranteed. With L'Hôpital's Rule applied, one obtains

$$\lim_{t\to\infty} \frac{(\dot{q}_i)^2}{(q_i^{1-r_1/r_2})^2} = \lim_{t\to\infty} \frac{2\dot{q}_i \ddot{q}_i q_i^{2(r_1/r_2)-1}}{2(1 - r_1/r_2)\dot{q}_i} = \lim_{t\to\infty} \frac{\ddot{q}_i q_i^{2(r_1/r_2)-1}}{(1 - r_1/r_2)} = 0$$

From previous analysis, $\lim_{t\to\infty} \frac{\dot{q}_i}{q_i^{1-r_1/r_2}} = 0$ for $i = 1,2,3$, which indicates $\lim_{t\to\infty} N(t) = 0$. Hence, after an existing though unknown finite time T_{11}, $\bar{N}(t) \leq \zeta < f_0$ holds for $t > T_{11}$.

Consider a scalar function $V_s = \frac{1}{2}\boldsymbol{s}^{\mathrm{T}} \boldsymbol{J}\boldsymbol{s}$ for $t > T_{11}$, the derivative along the state trajectory is

$$\begin{aligned}\dot{V}_s &= \boldsymbol{s}^{\mathrm{T}} \boldsymbol{J} \dot{\boldsymbol{s}} = \boldsymbol{s}^{\mathrm{T}} \left(\boldsymbol{N} + \boldsymbol{d} - u_m \tanh\left(\frac{\boldsymbol{s}}{k^2\delta}\right)\right) \\ &\le \left(\bar{N}(t) + d_m - u_m\right) \|\boldsymbol{s}\|_1 + k^2 \delta u_m \sum_{i=1}^{3} \tanh\left(\frac{s}{k^2\delta}\right) \\ &\le -\left(f_0 - \zeta - 3k^2\delta u_m / \|\boldsymbol{s}\|_1\right) \|\boldsymbol{s}\|_1\end{aligned}$$

for $t \ge T_{11}$.

Note that $V(s) \le \frac{1}{2}\bar{\lambda}_{\boldsymbol{J}} \|\boldsymbol{s}\|^2$ and $\|\boldsymbol{s}\| \le \|\boldsymbol{s}\|_1$ (where $\bar{\lambda}_{\boldsymbol{J}}$ denotes the maximum eigenvalue of the inertia matrix). Then, it is derived that

$$\dot{V}_s \le -(f_0 - \zeta - 3k^2\delta u_m / \|\boldsymbol{s}\|_1) \left(\sqrt{2}/\sqrt{\bar{\lambda}_{\boldsymbol{J}}}\right) \sqrt{V_s}, \ t \ge T_{11}$$

It follows from the above inequality that $\dot{V}_s + \eta_1 V_s^{1/2} \le 0 (t \ge T_{11}\ ,\ \|\boldsymbol{s}\|_1 > o_s)$ where $\eta_1 = \left(\sqrt{2}/\sqrt{\bar{\lambda}_{\boldsymbol{J}}}\right) (f_0 - \sigma - 3k^2\delta u_m / o_s)$. o_s is a small region containing origin with sufficient small parameters.

Actually, due to the negative impact by the inertia uncertainty and sampled computation, etc., the motion cannot precisely reach $s = 0$ and moves along the sliding surface nearby [32].

According to Lemma 5.2, the neighborhood of the proposed sliding mode surface is approached in another finite time T_{12}, and the systems states remain in region o_s after $T_1 = T_{11} + T_{12}$.

Step 3: Dynamics along the sliding surface

In this subsection, we prove that $\boldsymbol{q}_v, \boldsymbol{\omega}$ converge into small regions o_q, o_ω respectively in finite time. Note that $\|\boldsymbol{s}\|_1 \le o_s$, one obtains

$$\omega_i + k_1 k^2 q_i + k_2 k^2 q_i^{r_1/r_2} = \xi,\ |\xi| \le 0_s,\ i = 1,2,3 \tag{5.19}$$

Equation (5.19) can be rewritten as

$$\omega_i + \left(k_1 k^2 - \frac{\xi}{2q_i}\right) q_i + \left(k_2 k^2 - \frac{\xi}{2q_i^{r_1/r_2}}\right) q_i^{r_1/r_2} = 0,$$

Based on Theorem 5.1, as long as $k_1\mu^2 - \frac{\xi}{2q_i} > 0$ and $k_2\mu^2 - \frac{\xi}{2q_i^{r_1/r_2}} > 0$, the system states approach to $\{\boldsymbol{q}_v = 0, q_0 = 1, \omega = 0\}$ in finite time $T = T_1 + T_s$. Thus, the attitude error $\boldsymbol{q}_v$ will converge to the region $|q_i| \le o_q$ in finite time, and so is the velocity tracking error converging to the region $|\omega_i| \le o_\omega$. The proof is completed.

References

1. Haj Brahim, I., Chaabane, M., Mehdi,D. (2017) Fault-tolerant control for T-S fuzzy descriptor systems with sensor faults: an lmi approach. International Journal of Fuzzy Systems volume 19: 516–527
2. Yang, G.H., Ye, D. (2008) Adaptive fault-tolerant H∞ control against sensor failures. IET Control Theory and Applications 2(2): 95–107
3. Guo,J.X., Tao,G., Liu, Y. (2014) Adaptive actuator failure and structural damage compensation of nasa generic transport model. Journal of Dynamic Systems, Measurement, and Control 136(3): 031009
4. Cai, W.C., Liao, X.H., Song, D.Y. (2008) Indirect robust adaptive fault-tolerant control for attitude tracking of spacecraft. Journal of Guidance, Control, and Dynamics 31(5): 1456–1463
5. Sun, L. (2020) Adaptive fault-tolerant constrained control of cooperative spacecraft rendezvous and docking. IEEE Transactions on Industrial Electronics 67(4): 3107–3115
6. Shen, Q., Wang, D.W., Zhu, S.Q., Poh, E.K. (2015) Finite-time fault-tolerant attitude stabilization for spacecraft with actuator saturation. IEEE Transactions on Aerospace and Electronic Systems 51(3): 2390–2405
7. Jing, Y.H., Yang, G.H. (2019) Fuzzy adaptive fault-tolerant control for uncertain nonlinear systems with unknown dead-zone and unmodeled dynamics. IEEE Transactions on Fuzzy Systems 27(12): 2265–2278
8. Huo, B.Y., Xia, Y.Q., Yin, L.J., Fu, M.Y. (2017) Fuzzy adaptive fault-tolerant output feedback attitude-tracking control of rigid spacecraft. IEEE Transactions on Systems, Man, and Cybernetics: Systems 47(8): 1898–1908
9. Talebi, H.A., Khorasani, K., Tafazoli, S. (2009) A recurrent neural-network-based sensor and actuator fault detection and isolation for nonlinear systems with application to the satellite's attitude control subsystem. IEEE Transactions on Neural Networks 20(1): 45–60
10. Xiao, B., Hu, Q.L., Zhang, Y.M., Huo, X. (2014) Fault-tolerant tracking control of spacecraft with attitude-only measurement under actuator failures. Journal of Guidance, Control, and Dynamics 37(3): 838–849
11. Shen, Q., Yue, C.F., Goh, C.H., Wang, D.W. (2019) Active fault-tolerant control system design for spacecraft attitude maneuvers with actuator saturation and faults. IEEE Transactions on Industrial Electronics 66(5): 3763–3772
12. Ma, Y.J., Jiang, B., Tao, G., Yang, H. (2019) Adaptive direct compensation control scheme for spacecraft with multiple actuator faults. Journal of Guidance, Control, and Dynamics 42(4): 923–930
13. Hu, H., Liu, L., Wang,Y.J., Cheng, Z.T., Luo, Q.Q. (2020) Active fault-tolerant attitude tracking control with adaptive gain for spacecrafts. Aerospace Science and Technology 98: 105706
14. Du, H.B., Li, S.H. (2012) Finite-time attitude stabilization for a spacecraft using homogeneous method. Journal of Guidance, Control, and Dynamics 35(3): 740–748
15. Venkataraman, S.T., Gulati, S. (1991) Terminal sliding modes: A new approach to nonlinear control synthesis. In: Fifth International Conference on Advanced Robotics' Robots in Unstructured Environments, IEEE, pp 443–448
16. Yu, X.H., Man, Z.H. (2002) Fast terminal sliding-mode control design for nonlinear dynamical systems. IEEE Transactions on Circuits and Systems I: Fundamental Theory and Applications 49(2): 261–264
17. Zou, A.M., Kumar, K.D., Hou, Z.G., Liu, X. (2011) Finite-time attitude tracking control for spacecraft using terminal sliding mode and chebyshev neural network. IEEE Transactions on Systems, Man, and Cybernetics, Part B (Cybernetics) 41(4): 950–963
18. Bayat, F. (2019) Model predictive sliding control for finite-time three-axis spacecraft attitude tracking. IEEE Transactions on Industrial Electronics 66(10): 7986–7996
19. Guo, Y., Huang, B., Song, S.M., Li, A.J., Wang, C.Q. (2019) Robust saturated finite-time attitude control for spacecraft using integral sliding mode. Journal of Guidance, Control, and Dynamics 42(2): 440–446

20. Amrr, S.M., Nabi, M. (2020) Finite-time fault tolerant attitude tracking control of spacecraft using robust nonlinear disturbance observer with anti-unwinding approach. Advances in Space Research 66(7): 1659–1671
21. Boskovic, J.D., Li, S. M., Mehra, R. K. (2004) Robust tracking control design for spacecraft under control input saturation. Journal of Guidance, Control, and Dynamics 27(4): 627–633
22. Wallsgrove, R.J., Akella, M.R. (2005) Globally stabilizing saturated attitude control in the presence of bounded unknown disturbances. Journal of Guidance, Control, and Dynamics 28(5): 957–963
23. Ali, I., Radice, G., Kim, J. (2010) Backstepping control design with actuator torque bound for spacecraft attitude maneuver. Journal of Guidance, Control, and Dynamics 33(1): 254–259
24. Zhu, Z., Xia, Y.Q., Fu, M.Y. (2011) Adaptive sliding mode control for attitude stabilization with actuator saturation. IEEE Transactions on Industrial Electronics 58(10): 4898–4907
25. De Ruiter, A.H.J. (2010) Adaptive spacecraft attitude control with actuator saturation. Journal of Guidance, Control, and Dynamics 33(5): 1692–1696
26. Zou, A.M., De Ruiter, A.H.J., Kumar, K.D. (2019) Disturbance observer-based attitude control for spacecraft with input MRS. IEEE Transactions on Aerospace and Electronic Systems 55(1): 384–396
27. Zhu, Z., Xia, Y.Q., Fu, M.Y. (2011) Attitude stabilization of rigid spacecraft with finite-time convergence. International Journal of Robust and Nonlinear Control 21(6): 686–702
28. Yu, S.H., Yu, X.H., Shirinzadeh, B., Man, Z.H. (2005) Continuous finite-time control for robotic manipulators with terminal sliding mode. Automatica 41(11): 1957–1964
29. Bhat, S.P., Bernstein, D.S. (2000) Finite-time stability of continuous autonomous systems. SIAM Journal on Control and Optimization 38(3): 751–766
30. Hou, Z.W., Wang, Y.J., Liu, L. (2014) Improved nonsingular fast terminal sliding mode controller for nonlinear systems. In: 2014 International Conference on Mechatronics and Control (ICMC), IEEE, pp 678–683
31. Zhao, L., Jia, Y.M. (2015) Finite-time attitude tracking control for a rigid spacecraft using time-varying terminal sliding mode techniques. International Journal of Control 88(6): 1150–1162
32. Shtessel, Y., Taleb, M., Plestan, F. (2012) A novel adaptive-gain supertwisting sliding mode controller: Methodology and application. Automatica 48(5): 759–769

Chapter 6
Extended State Observer Based Optimal Attitude Robust Control of Spacecraft

6.1 Introduction

In practical applications, the external disturbances and uncertainties are always exist for the spacecraft attitude control system. They are difficult to be dealt with during the theoretical analyses, but crucial for the successful implementation of theoretical results. These problems become more challenging in the sense of optimal control especially taking time optimal, fuel/energy optimal/minimum, or some other optimal performance indexes at the same time. Due to its inherent robustness and optimality with respect to nonlinearity and uncertainty, the nonlinear optimal control is a potential approach to solve the nonlinear attitude control problem. Motivated by these difficulties, this chapter addresses the problem of robust optimal control of spacecraft attitude stabilization in the presence of parametric uncertainties, external disturbances and actuator saturation simultaneously.

To actively compensate for the so-called uncertainty and/or disturbance, Extended State Observer (ESO) has been considered as an effective way [1]. The key idea of ESO is that system uncertainties and disturbances are considered as an added or extended state of the system, and then all the states including the extended one will be estimated accurately and quickly by ESO. In [2], the problem of attitude control for a spacecraft model with inertia uncertainty and external disturbance has been investigated by utilizing ESO and further combined with an adaptive updating law. For the antenna pointing control of a large flexible satellite system, ESO was developed to estimate the total uncertain of the system in [3]. Additionally, Reference [4] proposed a new robust optimal control strategy for flexible attitude tracking maneuvers in the presence of external disturbances, through the method of Control Lyapunov Function (CLF) incorporating with ESO. While, in the sense of convergence of ESO, there are few rigorous proofs for the stabilization of the closed-loop system. Reference [5] tried to address the convergence of a nonlinear high-gain extended state observer even in the presence of uncertainties, but the asymptotical stability was achieved under some restrictive assumptions. And the stability analysis when the ESO is incorporated with CLF is still lacking. In this chapter, an ESO is developed

Q. Hu et al., *Control Allocation for Spacecraft Under Actuator Faults*,
https://doi.org/10.1007/978-981-16-0439-3_6

to estimate and compensate the specified uncertainties including actuators misalignments and parametric uncertainties while ensuring uniformly ultimately bounded estimation error in the sense of finite-time stability.

Based on the reconstructed information from the ESO or other mechanisms, the attitude control is also significant for spacecraft. Although there exists many research results to deal with the attitude control under uncertainties or disturbances, but some optimization problems such as time optimal, fuel/energy optimal/minimum, or some other optimal performance indexes, cannot be considered at the same time. Inverse optimal control technique is an alternative approach to derive optimal feedback control laws for spacecraft attitude control problem. Robust inverse optimal-control approach [6] circumvents the task of solving the Hamilton–Jacobi–Bellman (HJB) partial differential equation and results in an optimal controller with respect to a set of cost functions. For robust inverse optimal approach, the knowledge of CLF is very important for the designer, and it can solve HJB equations associated with a meaningful cost and a stable control law for an auxiliary nonlinear system. In [7], the attitude tracking control problem for spacecraft with external disturbances was addressed using the robust inverse optimal control method, associated with CLF. Further consider the spacecraft attitude tracking control problem with external disturbances and an uncertain inertia matrix, Luo [8] proposed an inverse optimal adaptive control law by means of adaptive CLF method. Recently, Ames [9] demonstrated that a class of modified rapid exponentially stabilizing CLF (RES-CLF) that enforce rapid exponential convergence to the zero dynamic surface can be used to achieve the stability of the periodic orbit in the full-order dynamics, thereby extending the stabilizing controllers by utilizing point-wise minimum norm method [10], which is optimal with respect to a cost function.

Another important problem encountered in practice is that of actuator saturation. The occurrence of actuator saturation will subsequently reduce performances and even destabilize the closed-loop system, if the system is not equipped with an appropriate control scheme to reject the issue of saturation. As such, several solutions/methods have been proposed to deal with actuator saturation constraints, such as inverse tangent-based function, explicit saturation function, anti-windup scheme and some other ways [11–14] as a direct way. That is to say, the functions/methods can be imposed directly on the controllers such that the control signals are produced within the specified limitations. Additionally, the actuator saturation issue also can be dealt with in the process of control allocation as an indirect way. For instance, some practical actuators constraints were added in the quadratic programming scheme, in order to obtain an optimal control effort with the rejection of actuator saturation [15] utilizing control allocator.

This chapter will address the optimal control problem of the spacecraft ACS under the external disturbances and parametric uncertainties, and control input saturation is taken into consideration explicitly. An ESO is first investigated to estimate all the states of the system, which is used to compensate for specified total uncertainty. And also the uniformly ultimately bounded stability in finite time of ESO is guaranteed via a rigorous Lyapunov analysis. And then, an inverse optimal CLF approach is presented to guarantee the asymptotic stability of the closed-loop system, such that an optimal/minimum performance index can be achieved simultaneously. On the basis of input-output linearization of the dynamics transverse to the zero dynamic manifold, a rapid exponential stabilization CLF based optimal control scheme is developed, and the actuator saturation is dumped successfully by utilizing the quadratic programming technique. It should be noted that the key feature of the proposed strategies is that they stabilize asymptotically the closed-loop system with respect to meaningful cost functions, even in the presence of uncertainties and actuator constraints.

(*Remark: The main control schemes and contents of this chapter are from the published journal paper "Bo Li, Qinglei Hu and Guangfu Ma, Extended State Observer based robust attitude control of spacecraft with input saturation, Aerospace Science and Technology, 2016, 50: 173–182". The authors appreciate the permission from the Elsevier Paris to reuse partly the results published in the relevant Journal.*)

6.2 Problem Formulation

Consider the spacecraft attitude control system given by Eqs. (2.17) and (2.20) in the presence of system uncertainties (actuator deviations that denoted as $\Delta \boldsymbol{D}$, and inertia matrix uncertainty denoted as $\Delta \boldsymbol{J}$), design a control input $\boldsymbol{\tau}(t)$ such that, the states of the closed-loop system are ensured to be asymptotically stable, i.e. $\lim_{t\to\infty} \boldsymbol{\rho}(t) = \lim_{t\to\infty} \boldsymbol{\omega}(t) = 0$. In what follows, attitude control schemes will be developed for spacecraft system to achieve the targets.

Remark 6.1 It should be noted that, for orbiting spacecraft, there are more than three actuators aligned with the three body-fixed axes of the spacecraft, therefore, a typical application case with an overactuated dynamic system. A common configuration with four reaction wheels is chosen as the spacecraft actuators in this paper. In such a configuration, the rotational axes of three reaction wheels (such as reaction wheels 1, 2, and 3) are orthogonal to the spacecraft's orthogonal shaft and the fourth one is installed with the equiangular direction with the orthogonal to each other three axes.

6.3 Robust Optimal Attitude Controller Design

6.3.1 Extended State Observer Design

First of all, define a coordinate transformation as $z_1 = \boldsymbol{\omega} + k\boldsymbol{\rho}$. In view of Eqs. (2.17) and (2.20), it yields to

$$\dot{z}_1 = \boldsymbol{A} + \boldsymbol{B}\boldsymbol{\tau} + \boldsymbol{W} \tag{6.1}$$

with $\boldsymbol{A} = -\boldsymbol{J}_0^{-1}\boldsymbol{\omega}^{\times}\boldsymbol{J}_0\boldsymbol{\omega} + k\boldsymbol{G}(\boldsymbol{\rho})\boldsymbol{\omega}$, $\boldsymbol{B} = \boldsymbol{J}_0^{-1}\boldsymbol{D}_0$, $\boldsymbol{W} = (\boldsymbol{J}_0^{-1} + \Delta\bar{\boldsymbol{J}})\boldsymbol{d} + \boldsymbol{E}$, $\boldsymbol{E} = -\boldsymbol{J}_0^{-1}\boldsymbol{\omega}^{\times}\Delta\boldsymbol{J}\boldsymbol{\omega} + \boldsymbol{J}_0^{-1}\Delta\boldsymbol{D}\boldsymbol{\tau} - \Delta\bar{\boldsymbol{J}}(-\boldsymbol{\omega}^{\times}(\boldsymbol{J}_0 + \Delta\boldsymbol{J})\boldsymbol{\omega} + (\boldsymbol{D}_0 + \Delta\boldsymbol{D})\boldsymbol{\tau})$.

Utilizing the ESO technique [1], an extended state variable $z_2 = \boldsymbol{W}$ is defined with $\dot{z}_2 = \dot{\boldsymbol{W}} = \boldsymbol{g}(t)$, and here $\boldsymbol{g}(t)$ is assumed to be unknown but bounded, that is to say $|g_i(t)| \leq \bar{g},\ \ i = 1,\ 2,\ 3$. Afterwards, the system plant governed by Eq. (6.1) is extended as

$$\begin{cases} \dot{z}_1 = \boldsymbol{A} + \boldsymbol{B}\boldsymbol{\tau} + z_2 \\ \dot{z}_2 = \boldsymbol{g}(t) \end{cases} \tag{6.2}$$

Now, denoting $\boldsymbol{Z}_1$ and $\boldsymbol{Z}_2$ as the outputs of the observer, $\boldsymbol{e} = \boldsymbol{Z}_1 - z_1$ as the observer error of coordinate transformation with $\boldsymbol{e}_v = [e_{v1},\ e_{v2},\ e_{v3}]^{\mathrm{T}} = \mathrm{sign}(\boldsymbol{e})$. And in accordance with [1, 16], an extended state observer is conducted in the form of

$$\begin{cases} \dot{\boldsymbol{Z}}_1 = \boldsymbol{Z}_2 + \boldsymbol{A} + \boldsymbol{B}\boldsymbol{\tau} - \beta_0 \boldsymbol{e}_{\mathrm{v}} - \beta_1 \boldsymbol{e} \\ \dot{\boldsymbol{Z}}_2 = -\beta_2 \boldsymbol{e} - \beta_3 \mathrm{han}(\boldsymbol{e}_{\mathrm{v}},\ \alpha,) - \beta_4 \mathrm{sign}(\boldsymbol{e}_{\mathrm{v}}) \end{cases} \tag{6.3}$$

where $\mathrm{han}(\boldsymbol{e}_{\mathrm{v}},\ \alpha)$ is a nonlinear function defined as $\mathrm{han}(\boldsymbol{e}_{\mathrm{v}},\ \alpha) = \mathrm{sig}^{\alpha}(\boldsymbol{e}_{\mathrm{v}}) = |\boldsymbol{e}_{\mathrm{v}}|^{\alpha}\mathrm{sign}(\boldsymbol{e}_{\mathrm{v}})$, and $\beta_i > 0,\ i = 0,\ 1, \ldots,\ 4$ are observer gains. Therefore, the following statements can be concluded.

Theorem 6.1 *Consider the system plant in Eq. (6.2) with the ESO in Eq. (6.3), there exist appropriate observer gains* $\beta_i > 0,\ i = 0,\ 1, \ldots,\ 4$ *and* α *such that the observed values* $\boldsymbol{Z}_1$ *and* $\boldsymbol{Z}_2$ *converge into actual states* z_1 *and extended state* z_2 *in finite time, respectively.*

Proof **(1) Stability of the ESO**
Define the auxiliary variables $\boldsymbol{E}_1 = \boldsymbol{e} = \boldsymbol{Z}_1 - z_1,\ \boldsymbol{E}_2 = \boldsymbol{Z}_2 - z_2$, and $\boldsymbol{e}_v = [e_{v1}\ e_{v2}\ e_{v3}]^{\mathrm{T}}$, then the error dynamics can be established as

$$\begin{cases} \dot{\boldsymbol{E}}_1 = \boldsymbol{E}_2 - \beta_0 \boldsymbol{e}_v - \beta_1 \boldsymbol{E}_1 \\ \dot{\boldsymbol{E}}_2 = -\boldsymbol{g} - \beta_2 \boldsymbol{E}_1 - \beta_3 \mathrm{han}(\boldsymbol{e}_v,\ \alpha) - \beta_4 \mathrm{sign}(\boldsymbol{e}_v) \end{cases} \tag{6.4}$$

Further define $\boldsymbol{\Phi} = [\boldsymbol{E}_1\ \boldsymbol{E}_2]^{\mathrm{T}}$, and consider a Lyapunov function candidate as $\boldsymbol{V}_1 = \frac{1}{2}\boldsymbol{\Phi}^{\mathrm{T}}\boldsymbol{\lambda}\boldsymbol{\Phi}$, with $\boldsymbol{\lambda} = \mathrm{diag}(\beta_1, 1)$ is a positive defined matrix. Under this, differentiation of V_1 with respect to time yields to

$$\dot{V}_1 \leq -\boldsymbol{\Phi}^{\mathrm{T}}\boldsymbol{\Omega}\boldsymbol{\Phi} + l\,\|\boldsymbol{\Phi}\| \tag{6.5}$$

with $\boldsymbol{\Omega} = \begin{bmatrix} \beta_1^2 & 0 \\ 0 & \beta_2 \end{bmatrix}$, $l = \left(3^{\frac{\alpha}{2}}\beta_3 + \sqrt{3}\beta_4 + \sqrt{3}\bar{g}\right)$.

Because the variables $\beta_1, \beta_2, \beta_3, \beta_4$ and $\bar{g}$ are positive, $\boldsymbol{\Omega}$ is a positive define matrix and then $l > 0$. Denoting $\sigma_{\min}(\cdot),\ \ \sigma_{\max}(\cdot)$ as the minimum and maximum singular value of the specified matrix, respectively, one can get $\sigma_{\min}(\boldsymbol{\Omega})\,\|\boldsymbol{\Phi}\| \leq \boldsymbol{\Phi}^{\mathrm{T}}\boldsymbol{\Omega}\boldsymbol{\Phi} \leq \sigma_{\max}(\boldsymbol{\Omega})\,\|\boldsymbol{\Phi}\|$. To this end, it can further be obtained from Eq. (6.5) that

$$\dot{V}_1 \leq -\left(\sigma_{\min}(\boldsymbol{\Omega})\,\|\boldsymbol{\Phi}\| - l\right)\frac{V_1{}^{\frac{1}{2}}}{\sigma_{\max}(\boldsymbol{\Omega})} \tag{6.6}$$

If the observer gains are chosen to satisfy $\sigma_{\min}(\boldsymbol{\Omega})\|\boldsymbol{\Phi}\| > l$, then the error system Eq. (6.4) can converge to the small region $\|\boldsymbol{\Phi}\| \le \frac{l}{\sigma_{\min}(\boldsymbol{\Omega})} \triangleq l_R$ in finite time. This indicates that $\dot{V}_1$ is negative when $\boldsymbol{\Phi}$ enters into the compact set $\mathcal{D}_1 \triangleq \{\boldsymbol{\Phi}|\ \|\boldsymbol{\Phi}\| \le l_R\}$, and also V_1 decreases monotonically. In other words, a decreasing value of V_1 eventually drives the observer error $\boldsymbol{\Phi}$ into the above set $\mathcal{D}_1$ and then it cannot go out of $\mathcal{D}_1$. With the given initial estimates, $\|\boldsymbol{\Phi}(0)\| = 0$ is ensured. According to Lyapunov theory and Lasalle extension [17], all the states or observer errors are confined into $\mathcal{D}_1$ in the finite time T_0, which also demonstrates the uniformly ultimately bounded (UUB) stability of $\boldsymbol{\Phi}$. Furthermore, the analysis of the observer errors is presented as follows.

(2) Finite-time convergence of $\boldsymbol{E}_1$

In terms of Eq. (6.4), select another candidate Lyapunov function as $V_2 = \frac{1}{2}\boldsymbol{E}_1^{\mathrm{T}}\boldsymbol{E}_1$. By proceeding with differentiation of V_2, it yields to $\dot{V}_2 \le -(\beta_0 - \|\boldsymbol{E}_2\|)\|\boldsymbol{E}_1\|$. If selecting the observer gain $\beta_0 > \|\boldsymbol{E}_2\| + \lambda_1$ with $\lambda_1 > 0$, one can obtain that $\dot{V}_2 \le -\lambda_1\|\boldsymbol{E}_1\|$, which ensures $\dot{V}_2 < 0$ for all the $\|\boldsymbol{E}_1\| \ne 0$, and it further implies that $\boldsymbol{E}_1$ can converge to the origin in finite time. Furthermore, defining a continuous positive function $W_1 = \sqrt{2V_2} = \|\boldsymbol{E}_1\|$, one can obtain

$$\dot{W}_1 = \frac{\dot{V}_2}{\sqrt{2V_2}} \le \frac{-\lambda_1\|\boldsymbol{E}_1\|}{\|\boldsymbol{E}_1\|} = -\lambda_1 \tag{6.7}$$

Involving the comparison lemma in [17], the solution of Eq. (6.7) is established as

$$W_1(t) \le W_1(0) - \lambda_1 t \tag{6.8}$$

which can guarantee that $\|\boldsymbol{E}_1\| \equiv 0$ for all $t > \frac{\|\boldsymbol{E}_1(0)\|}{\lambda_1} \triangleq T_1$, in other words, $\boldsymbol{E}_1 = \dot{\boldsymbol{E}}_1 = 0$ can be achieved by the finite time $t = T_1$. Accordingly, solving the equivalent output injection $\boldsymbol{E}_2 - \beta_0\boldsymbol{e}_{\mathrm{v}} = 0$ results in $(\boldsymbol{e}_{\upsilon})_{\mathrm{eq}} = (\mathrm{sign}(\boldsymbol{E}_1))_{\mathrm{eq}} = \frac{1}{\beta_0}\boldsymbol{E}_2$, where the subscript "eq" denotes the equivalent value along with/under $\boldsymbol{E}_1 = \dot{\boldsymbol{E}}_1 = 0$.

(3) Finite-time convergence of $\boldsymbol{E}_2$

We now proceed to show the finite-time convergence of E_2. After a finite time $t > T_1$, the error dynamics expressed in Eq. (6.4) have the following form

$$\begin{aligned} &\dot{\boldsymbol{E}}_1 = 0,\quad \boldsymbol{E}_1 = 0,\ \text{and} \\ &\dot{\boldsymbol{E}}_2 = -\boldsymbol{g} - \beta_3\left|\frac{1}{\beta_0}\boldsymbol{E}_2\right|^{\alpha}\mathrm{sign}\left(\frac{1}{\beta_0}\boldsymbol{E}_2\right) - \beta_4\mathrm{sign}\left(\frac{1}{\beta_0}\boldsymbol{E}_2\right) \end{aligned} \tag{6.9}$$

Choose another Lyapunov function $V_3 = \frac{1}{2}\boldsymbol{E}_2^{\mathrm{T}}\boldsymbol{E}_2$, with $\dot{V}_3 \le -\frac{\beta_3}{\beta_0}\|\boldsymbol{E}_2\|^{1+\alpha} - \|\boldsymbol{E}_2\|\left(\beta_4 - \sqrt{3}\bar{g}\right)$. As a consequence, if the observer gain $\beta_4 \ge \sqrt{3}\bar{g}$ is chosen, it follows that

$$\dot{V}_3 \le -\frac{\beta_3}{\beta_0}2^{\frac{1+\alpha}{2}}(V_3)^{\frac{1+\alpha}{2}} \tag{6.10}$$

where $\alpha \in (0, 1)$ is selected. Then with Eq. (6.10) under $V_3 \equiv 0$, we also have

$$t > \frac{V_3^{\left(1-\frac{1+\alpha}{2}\right)}(0)}{\frac{\beta_3}{\beta_0} 2^{\frac{1+\alpha}{2}} \left(1 - \frac{1+\alpha}{2}\right)} = \frac{\beta_0 V_3^{\frac{1-\alpha}{2}}(0)}{\beta_3 (1-\alpha) 2^{\frac{1+\alpha}{2}}} \triangleq T_2$$

Hence, the observer error $\boldsymbol{E}_2$ can converge to zero in the finite time T_2 by the choice of $\beta_4 \geq \sqrt{3}\bar{g}$. Thereby, this is the end of the proof of Theorem 6.1.

6.3.2 *Inverse Optimal Controller Design with Uncertainties*

For the purpose of the attitude control system design, some definitions and remarks are presented as follows [8, 9].

Definition 6.1 Consider an affine system $\dot{x} = f(x) + g(x) u$, where $x \in X$ is a state vector, $u \in U$ is the input vector, and $f(x)$ is a continuous function with $f(0) = 0$. Let X be a domain of the origin, a positive definite and C^1 function $V(x) : X \rightarrow \mathbf{R}^+$ is called as a CLF of the above system, so that $\inf_u \left\{L_f V(x) + L_g V(x) u\right\} < 0$ for all $x \in X \backslash \{0\}$, where $L_f V(x)$ and $L_g V(x)$ are the Lie derivatives of the system. Obviously, the sufficient condition of the function $V(x)$ to be a CLF is

$$L_g V(x) u = 0 \Rightarrow \inf_u \{L_f V(x)\} < 0, \quad x \in X \backslash \{0\}$$

Definition 6.2 Consider an affine system of the form $\dot{x} = f(x, z) + g(x, z) u$, $\dot{z} = q(x, z)$, $x \in X$ is the controlled state, $z \in Z$ is the uncontrolled state, and $u \in U$ is the input vector. The vector fields f, g and q are assumed to be locally Lipschitz continuous with $f(0, z) = 0$, so that the surface Z defined by $x = 0$ with dynamics $\dot{z} = q(0, z)$ is invariant. Then, a continuously differentiable function $V : X \rightarrow \mathbf{R}^+$ is called as an exponentially stabilization Control Lyapunov Function (ES-CLF) if there exists positive constants $c_1, c_2, c_3 > 0$ such that

$$c_1 \|x\|^2 \leq V(x) \leq c_2 \|x\|^2,$$

$$\inf_{u \in U} \left[L_f V(x, z) + L_g V(x, z) u + c_3 V(x) \leq 0\right]$$

for all $(x, z) \in X \times Z$.

Then, Definition 6.2 can be extended to provide a means of adjusting the rate of convergence to be "sufficient rapid", and the extension is presented in Definition 6.3 as follows.

Definition 6.3 For the system given in Definition 6.3, a one-parameter family of continuously differentiable function $V_\epsilon : X \rightarrow \mathbf{R}^+$ is referred to be a rapidly exponentially stabilization Control Lyapunov Function (RES-CLF) if there exists positive constants $c_1,\ c_2,\ c_3 > 0$ such that

$$c_1\|x\|^2 \le V_\epsilon(x) \le \frac{c_2}{\epsilon}\|x\|^2$$

$$\inf_{u\in U}\left[L_f V_\epsilon(x,z) + L_g V_\epsilon(x,z)u + \frac{c_3}{\epsilon}V_\epsilon(x) \le 0\right]$$

for all $0 < \epsilon < 1$ and $(x,\ z) \in X \times Z$.

Remark 6.2 Definitions 6.2 and 6.3 are introduced by [9], in which an effective set for the control values is defined as $U_u = \{u \in U \mid L_f V(x,\ z) + L_g V(x,\ z)u + c_3 V(x) \le 0\}$ consisting of the control signals that result in $\dot{V} \le -c_3 V$. When V is an ES-CLF, it is easy to obtain that for any locally Lipschitz continuous feedback control laws $u \in U_u$, the solution of the closed-loop system satisfies $\|x(t)\| \le \sqrt{\frac{c_2}{c_1}} \cdot e^{\frac{-c_3 t}{2}}\|x(0)\|$. And $\|x(t)\| \le \frac{1}{\epsilon}\sqrt{\frac{c_2}{c_1}} \cdot e^{\frac{-c_3 t}{2\epsilon}}\|x(0)\|$ can be achieved, if V is a RES-CLF. Therefore, the rate of the exponential convergence of the states can be directly controlled with the constant $\epsilon > 0$, particularly, when a sufficient small ϵ is chosen.

6.3.2.1 The Design of the CLF for Spacecraft System

From Eqs. (2.17) and (2.20), one has

$$\begin{cases} \dot{\omega} = -J_0^{-1}\omega^\times J_0\omega + J_0^{-1}D_0\tau + W \\ \dot{\rho} = G(\rho)\omega \end{cases} \tag{6.11}$$

Then, with the definition $x = [\omega\ \rho]^T$, the spacecraft system can be rewritten as

$$\dot{x} = f(x) + g_1(x)\left(\tau + D_0^{-1}J_0 W\right) \tag{6.12}$$

with $f(x) = \begin{bmatrix} -J_0^{-1}\omega^\times J_0\omega \\ G(\rho)\omega \end{bmatrix}$, $g_1(x) = \begin{bmatrix} J_0^{-1}D_0 \\ 0 \end{bmatrix}$.

Under this condition, choosing another Lyapunov function as

$$V_4 = \frac{1}{2}a\omega^T J_0\omega + b\omega^T J_0\rho + \frac{1}{2}c\rho^T J_0\rho \tag{6.13}$$

where a, b and c are positive constants. Then the following proposition is given:

Proposition 6.1 ([18, 19]) *The Lyapunov function candidate in Eq.* (6.13) *is a CLF for the system in Eq.* (6.12).

Proof The candidate Lyapunov function can be obtained from Eq. (6.13) as $V_4 = \frac{1}{2}x^T M x$, with $M = \begin{bmatrix} aJ_0 & bJ_0 \\ bJ_0 & c \end{bmatrix}$ such that $acJ_0 > b^2 J_0^2$ and $a > 0$. Differentiation of V_4 with respect to x, yields to

$$\left(\frac{\partial V_4}{\partial \boldsymbol{x}}\right)^{\mathrm{T}} = [a\boldsymbol{J}_0\boldsymbol{\omega} + b\boldsymbol{J}_0\boldsymbol{\rho} \quad b\boldsymbol{J}_0\boldsymbol{\omega} + c\boldsymbol{\rho}]^{\mathrm{T}} \tag{6.14}$$

This results in $L_{g_1} V_4 = \left(\frac{\partial V_4}{\partial \boldsymbol{x}}\right)^{\mathrm{T}} \boldsymbol{g}_1 = (a\boldsymbol{\omega} + b\boldsymbol{\rho})^{\mathrm{T}} \boldsymbol{J}_0^{\mathrm{T}} \boldsymbol{J}_0^{-1} \boldsymbol{D}_0$. And if $L_{g_1} V_4 = 0$, we are able to obtain that $\boldsymbol{\omega} = -\frac{b}{a}\boldsymbol{\rho}$. In addition, as for $L_{\mathrm{f}} V_4 = \left(\frac{\partial V_4}{\partial \boldsymbol{x}}\right)^{\mathrm{T}} f$, it can be expressed as

$$L_{\mathrm{f}} V_4 \le -\frac{b}{a}\left(c\boldsymbol{I}_3 - \frac{b^2}{a}\boldsymbol{J}_0\right)\left(\frac{1 + \|\boldsymbol{\rho}\|^2}{4}\right)\|\boldsymbol{\rho}\|^2 \tag{6.15}$$

□

Thus, one can further obtain that $L_{\mathrm{f}} V_4 < 0$ for all $\rho \neq 0$ when $L_{g_1} V_4 = 0$ is set up, which guarantees that the candidate Lyapunov function V_4 is a CLF for the spacecraft control system in view of Eq. (6.12) according to Definition 6.1.

6.3.2.2 The Controller Design for Spacecraft System with Uncertainties

A. *The CLF based inverse optimal control law*

In view of the CLF defined in Eq. (6.13), another main result of this paper is summarized and presented as the following theorem.

Theorem 6.2 *Consider the spacecraft system governed by Eq.* (6.12) *and the CLF* $V_4(\boldsymbol{x})$ *established in Eq.* (6.13), *there exist controller gains* $\gamma > 0$ *and* $k_1 > 0$, *such that the system is asymptotically stable under the following robust inverse optimal control laws*

$$\begin{cases} \boldsymbol{\tau} = \boldsymbol{\tau}^* - \boldsymbol{D}_0^{-1}\boldsymbol{J}_0\boldsymbol{Z}_2 \\ \boldsymbol{\tau}^*(x) = -\frac{1}{2}R_{\mathrm{u}}^{-1}(\boldsymbol{x})\,\varphi(\boldsymbol{x})^{\mathrm{T}} \end{cases} \tag{6.16}$$

$$R_{\mathrm{u}}^{-1}(\boldsymbol{x}) = \begin{cases} 2\gamma + 2\frac{\phi(\boldsymbol{x}) + \sqrt{\|\phi(\boldsymbol{x})\|^2 + k_1^2\|\varphi(\boldsymbol{x})\|^2}}{\|\varphi(\boldsymbol{x})\|^2}, & \|\varphi(\boldsymbol{x})\| \neq 0 \\ 2\gamma, & \|\varphi(\boldsymbol{x})\| = 0 \end{cases} \tag{6.17}$$

with $\phi(\boldsymbol{x}) = L_{\mathrm{f}} V_4(\boldsymbol{x}) = (b\boldsymbol{J}_0\boldsymbol{\omega} + c\boldsymbol{\rho})^{\mathrm{T}}\boldsymbol{G}(\boldsymbol{\rho})\boldsymbol{\omega}$ *and* $\varphi(\boldsymbol{x}) = L_{g_1} V_4(\boldsymbol{x}) = (b\boldsymbol{J}_0\boldsymbol{\omega} + c\boldsymbol{\rho})^{\mathrm{T}} \boldsymbol{J}_0^{\mathrm{T}}\boldsymbol{J}_0^{-1}\boldsymbol{D}_0$ *respectively.*

Proof The proof of Theorem 6.2 is organized in two cases as follows.

Case 6.1 In the case of $\|\varphi(x)\| \neq 0$, the control law has the form of

$$\boldsymbol{\tau}^*(x) = -\gamma\varphi(\boldsymbol{x})^{\mathrm{T}} - \varphi(\boldsymbol{x})^{\mathrm{T}}\frac{\phi(\boldsymbol{x}) + \sqrt{\|\phi(\boldsymbol{x})\|^2 + k_1^2\|\varphi(\boldsymbol{x})\|^2}}{\|\varphi(\boldsymbol{x})\|^2} \tag{6.18}$$

With the inverse optimal control law Eqs. (6.16) and (6.17), the time derivative of the CLF V_4 evaluated along the closed-loop attitude stabilization system can be established as $\dot{V}_4 \leq -\gamma \|\varphi(\boldsymbol{x})\|^2 - \left(k_1 - \sqrt{3} l_{\mathrm{R}} \left\|\boldsymbol{D}_0^{-1}\right\| \|\boldsymbol{J}_0\|\right) \|\varphi(\boldsymbol{x})\|$. If $k_1 \geq \sqrt{3} l_{\mathrm{R}} \left\|\boldsymbol{D}_0^{-1}\right\| \|\boldsymbol{J}_0\|$ is chosen, then $\dot{V}_4 < 0$ can be guaranteed. □

Case 6.2 If $\|\varphi(\boldsymbol{x})\| = 0$, then $\boldsymbol{\tau}^*(\boldsymbol{x}) = -\gamma \varphi(\boldsymbol{x})^{\mathrm{T}} = 0$ and one can have $\dot{V}_4 = \phi(\boldsymbol{x})$. As such, for all $x \neq 0$. Therefore, for all $x \neq 0$ and $k_1 \geq \sqrt{3} l_{\mathrm{R}} \left\|\boldsymbol{D}_0^{-1}\right\| \|\boldsymbol{J}_0\|$, one can obtain $\dot{V}_4 < 0$. According to the Lyapunov theory and Lasalle lemma [17], the closed-loop spacecraft system is asymptotically stable. □

B. *The optimal performance analysis*

Consider the following performance index

$$\Psi = \int_0^{T(x)} \left(\zeta(\boldsymbol{x}) + \boldsymbol{\tau}^{\mathrm{T}} R_{\mathrm{u}}(\boldsymbol{x}) \boldsymbol{\tau}\right) dt \tag{6.19}$$

where $\zeta(\boldsymbol{x})$ and $R_{\mathrm{u}}(\boldsymbol{x})$ are two nonnegative functions for all $\boldsymbol{x} \neq 0$. Since the stable feedback control law has been already designed, this subsection shows that the proposed inverse optimal feedback control scheme is to find $\zeta(\boldsymbol{x}) \geq 0$ and $R_{\mathrm{u}}(\boldsymbol{x}) \geq 0$ such that the control law Eq. (6.16) minimizes the performance index Eq. (6.19).

To this end, let $\zeta(\boldsymbol{x}) = -\phi(\boldsymbol{x}) + \frac{1}{4} \varphi(\boldsymbol{x}) R_u^{-1}(\boldsymbol{x}) \varphi(\boldsymbol{x})^{\mathrm{T}}$; under this definition, if $\|\varphi(\boldsymbol{x})\| = 0$, then $\zeta(\boldsymbol{x}) = -\phi(\boldsymbol{x}) \geq 0$. Otherwise, if $\|\varphi(\boldsymbol{x})\| \neq 0$, one has

$$\zeta(\boldsymbol{x}) \geq -\phi(\boldsymbol{x}) + \gamma \frac{1}{2} \|\varphi(\boldsymbol{x})\|^2 + \frac{1}{2} \left(\phi(\boldsymbol{x}) + \|\phi(\boldsymbol{x})\|\right) \geq 0 \tag{6.20}$$

Let $\boldsymbol{\tau} = \boldsymbol{\tau}^* + \boldsymbol{\tau}'$, Eq. (6.19) can be further established as

$$\Psi = V_3(\boldsymbol{x}_0) + \int_0^{T(x)} \left(\boldsymbol{\tau}'^{\mathrm{T}} R_{\mathrm{u}}(\boldsymbol{x}) \boldsymbol{\tau}'\right) dt \tag{6.21}$$

Therefore, the inverse optimal control law Eq. (6.16) can minimize the performance index Eq. (6.19). More specifically, $\boldsymbol{\tau}^*$ can minimize the performance index $\Psi \leq \Psi^*$ if taking $\boldsymbol{\tau}' = 0$ for any x_0 with $V_3(x_0) = \Psi^*$.

6.3.3 CLF and Zero Dynamics Based Optimal Controller Design

6.3.3.1 The Design of the RES-CLF for Spacecraft System

With Eq. (6.12), the human-inspired output [20] consists of vector relative 2 degree associated output function $\boldsymbol{y}$ of the form

$$\boldsymbol{y} = \boldsymbol{\rho} \tag{6.22}$$

where y is the attitude based output on the region of interest. The method of Zero Dynamic (ZD) aims to drive the output defined in Eq. (6.22) (and/or its first derivatives) to zero [21]. More formally, the objective is to impose "virtual constraints" such that the system dynamics evolve on the zero dynamic surface given by $\boldsymbol{Z} = \{(\rho,\ \dot{\rho}) \mid y = 0,\ \dot{y} = 0\}$.

Differentiating output function Eq. (6.22) twice yields to

$$\ddot{y} = L_{\mathrm{f}}^2 y + L_{\mathrm{g}} L_{\mathrm{f}} y \boldsymbol{\tau}^* \triangleq \bar{\boldsymbol{L}}_{\mathrm{f}} + \boldsymbol{H}\boldsymbol{\tau}^* \tag{6.23}$$

By proceeding with application of an input-output linearizing controller of the form

$$\boldsymbol{\tau}^* = -\boldsymbol{H}^{-1}\left(\bar{\boldsymbol{L}}_f - \boldsymbol{\mu}\right) \tag{6.24}$$

for some $\boldsymbol{\mu}$, results in $\ddot{y} = \boldsymbol{\mu}$. Note that, $\boldsymbol{\mu}$ is a feed-forward item representing the torque required to remain on the manifold $\boldsymbol{Z}$. In view of above analyses, the dynamics Eqs. (6.12) and (6.22) can be decomposed into zero dynamics states $z \in \boldsymbol{Z}$ and transverse variables $\eta = [\eta_1,\ \eta_2]^{\mathrm{T}} = [y,\ \dot{y}]^{\mathrm{T}}$, and then the closed-loop dynamics in terms of $(\eta,\ z)$ take the form

$$\begin{cases} \dot{\eta} = \begin{bmatrix} 0 & \boldsymbol{I} \\ 0 & 0 \end{bmatrix} \eta + \begin{bmatrix} 0 \\ \boldsymbol{I} \end{bmatrix} \boldsymbol{\mu} \triangleq \boldsymbol{F}\eta + \boldsymbol{G}\boldsymbol{\mu} \\ \dot{z} = \boldsymbol{p}(\eta,\ z) \end{cases} \tag{6.25}$$

Denote $\tilde{\boldsymbol{f}}(\eta,\ z) = \boldsymbol{F}\eta$ and $\tilde{\boldsymbol{g}}(\eta,\ z) = \boldsymbol{G}$, one can obtain the standard formulation for a partially linearized system

$$\begin{cases} \dot{\eta} = \boldsymbol{F}\eta + \boldsymbol{G}\boldsymbol{\mu} = \tilde{\boldsymbol{f}}(\eta,\ z)\,\eta + \tilde{\boldsymbol{g}}(\eta,\ z)\,\boldsymbol{\mu}(\eta,\ z) \\ \dot{z} = \boldsymbol{p}(\eta,\ z) \end{cases} \tag{6.26}$$

Note that here $(\boldsymbol{F},\ \boldsymbol{G})$ is controllable.

Due to Hurwitz assumption, for any positive definite symmetrical matrix $\boldsymbol{Q}$ there exists a positive definite symmetrical matrix $\boldsymbol{P}$ which is the solution to the Riccati equation

$$\boldsymbol{F}^{\mathrm{T}}\boldsymbol{P} + \boldsymbol{P}\boldsymbol{F} - \boldsymbol{P}\boldsymbol{G}\boldsymbol{G}^{\mathrm{T}}\boldsymbol{P} = -\boldsymbol{Q} \tag{6.27}$$

Application of the Rayleigh–Ritz inequality [22], we have $\boldsymbol{Q} > c_3\boldsymbol{P}$, $\boldsymbol{F}^{\mathrm{T}}\boldsymbol{P} + \boldsymbol{P}\boldsymbol{F} - \boldsymbol{P}\boldsymbol{G}\boldsymbol{G}^{\mathrm{T}}\boldsymbol{P} + c_3\boldsymbol{P} \leq 0$ with $c_3 = \frac{\sigma_{\min}(\boldsymbol{Q})}{\sigma_{\max}(\boldsymbol{P})}$.

We now proceed to design the control scheme, beginning with considering a new Lyapunov function candidate as

$$V = \eta^{\mathrm{T}} \begin{bmatrix} \frac{1}{\epsilon}\boldsymbol{I} & 0 \\ 0 & \boldsymbol{I} \end{bmatrix} \boldsymbol{P} \begin{bmatrix} \frac{1}{\epsilon}\boldsymbol{I} & 0 \\ 0 & \boldsymbol{I} \end{bmatrix} \eta \triangleq \eta^{\mathrm{T}}\boldsymbol{M}\boldsymbol{P}\boldsymbol{M}\eta \triangleq \eta^{\mathrm{T}}\boldsymbol{P}\eta \tag{6.28}$$

with $c_1\|\eta\|^2 \le V(\eta) \le \frac{c_2}{\epsilon^2}\|\eta\|^2, 0<\epsilon<1, c_1=\sigma_{\min}(\boldsymbol{P}_\epsilon)$ and $c_2=\sigma_{\max}(\boldsymbol{P}_\epsilon)$. Then, another proposition in this paper is given as follows:

Proposition 6.2 ([9]) *The Lyapunov function candidate in Eq.* (6.28) *is a RES-CLF for the system in Eq.* (6.26).

Proof In view of Eq. (6.27), it follows that

$$\begin{aligned} \boldsymbol{F}^{\mathrm{T}}\boldsymbol{P}_\epsilon+\boldsymbol{P}_\epsilon\boldsymbol{F} &\le \frac{1}{c}\boldsymbol{P}_\epsilon\boldsymbol{G}\boldsymbol{G}^{\mathrm{T}}\boldsymbol{P}_\epsilon-\frac{1}{\epsilon}\boldsymbol{M}_\epsilon\boldsymbol{Q}\boldsymbol{M}_\epsilon \\ &\le \frac{1}{\epsilon}\boldsymbol{P}_\epsilon\boldsymbol{G}\boldsymbol{G}^{\mathrm{T}}\boldsymbol{P}_\epsilon-\frac{c_3}{\epsilon}\boldsymbol{P}_\epsilon \end{aligned} \tag{6.29}$$

Then, differentiation of V_ϵ with respect to time yields to

$$\begin{cases} \frac{\partial V_\epsilon}{\partial t}=L_{\bar{f}}V_\epsilon+L_{\bar{g}}V_\epsilon\boldsymbol{\mu} \\ L_{\bar{f}}V_\epsilon=\eta^{\mathrm{T}}\left(\boldsymbol{F}^{\mathrm{T}}\boldsymbol{P}_\epsilon+\boldsymbol{P}_\epsilon\boldsymbol{F}\right)\eta,\quad L_{\bar{g}}V_\epsilon=2\eta^{\mathrm{T}}\boldsymbol{P}_\epsilon\boldsymbol{G} \end{cases} \tag{6.30}$$

results in $\dot{V}_\epsilon \le -\frac{c_3}{\epsilon}\eta^{\mathrm{T}}\boldsymbol{P}_\epsilon\eta+\eta^{\mathrm{T}}\boldsymbol{P}_\epsilon\boldsymbol{G}\left(\frac{1}{\epsilon}\boldsymbol{G}^{\mathrm{T}}\boldsymbol{P}_\epsilon-2\boldsymbol{K}_\epsilon\right)\eta$ by employing an appropriate virtual control law $\boldsymbol{\mu}=\boldsymbol{K}_\epsilon\eta$. More generally, it shows that

$$\inf_{\mu\in U}\left\{\dot{V}_\epsilon+\frac{c_3}{\epsilon}V_\epsilon\right\} \le \inf_{\mu\in U}\left\{\eta^{\mathrm{T}}\boldsymbol{P}_\epsilon\boldsymbol{G}\left(\frac{1}{\epsilon}\boldsymbol{G}^{\mathrm{T}}\boldsymbol{P}_\epsilon-2\boldsymbol{K}_\epsilon\right)\eta\right\} \le 0 \tag{6.31}$$

which implies that V_ϵ defined in Eq. (6.28) is a RES-CLF.

From [9, 23], the following set is defined

$$\bar{\boldsymbol{U}}_{\text{-}}=\left\{\boldsymbol{\mu}\in\boldsymbol{U}\text{-}\,|\,L_{\bar{f}}V_\epsilon+L_{\bar{g}}V_\epsilon\boldsymbol{\mu}+\frac{c_3}{\epsilon}V_\epsilon\le 0\right\} \tag{6.32}$$

Since V_ϵ is a RES-CLF, one can obtain from the Eq. (6.31)

$$V_\epsilon(\eta(t)) \le \boldsymbol{e}^{\frac{-c_3 t}{\epsilon}}V_\epsilon(\eta(0)) \tag{6.33}$$

Once again employing the Rayleigh–Ritz inequality, Eq. (6.33) yields to $\|\eta(t)\| \le \frac{1}{\epsilon}\sqrt{\frac{c_2}{c_1}}\cdot\boldsymbol{e}^{\frac{-c_3 t}{2\epsilon}}\|\eta(0)\|$, which also indicates the norm of the output (zero dynamics manifold) $(\rho,\ \dot{\rho})$ converges to zero at an exponential rate inversely proportional to ϵ, and the rate of exponential convergence can be arbitrarily rapid by choosing a small ϵ sufficiently. □

6.3.3.2 The Quadratic Programming Based RES-CLF Min-Norm Controller Design

For RES-CLF formulated in the above subsection, there are various methods to find a feedback control law satisfying $\boldsymbol{\mu}\in\bar{\boldsymbol{U}}_\mu$. While, taking account of some practical problems, such as energy saving (minimizing energy consumption), capability for actuator saturation avoidance against actuator amplitude saturation limitations, the quadratic programming based RES-CLF optimal controller will be developed in this subsection.

Recalling the point-wise min-norm optimal control problem [23]

$$m(\eta) = \arg\min\left\{ \|\boldsymbol{\mu}\| \mid \bar{\phi}(\eta) + \bar{\varphi}(\eta)\boldsymbol{\mu} \leq 0 \right\} \tag{6.34}$$

with $\bar{\phi}(\eta) = L_{\bar{f}} V_\epsilon + \frac{c_3}{\epsilon} V_\epsilon$, and $\bar{\varphi}(\eta) = L_{\bar{g}} V_\epsilon^{\mathrm{T}}$. Under this, the solution of the above optimal control issue is

$$m(\eta) = \begin{cases} -\dfrac{\bar{\phi}(\eta) \cdot \bar{\varphi}(\eta)}{\bar{\varphi}(\eta)^{\mathrm{T}} \bar{\varphi}(\eta)}, & \bar{\phi}(\eta) > 0 \\ 0, & \bar{\phi}(\eta) \leq 0 \end{cases} \tag{6.35}$$

and it is locally Lipschitz continuous.

Given the solution of min-norm controller $m(\eta)$ has been computed in closed form as shown in Eq. (6.35), a convex quadratic programming based optimization problem is formulated as

$$\begin{aligned} m(\eta) &= \arg\min_{\mu} \left\{ \|\boldsymbol{\mu}\|_2^2 \right\} \\ \text{s.t.} \ & \bar{\phi}(\eta) + \bar{\varphi}(\eta)\boldsymbol{\mu} \leq 0 \end{aligned} \tag{6.36}$$

Note that the constraint in the above optimization problem determines the effective set of the control values given in Eq. (6.32), which can be described as $\dot{V}_\epsilon + \frac{c_3}{\epsilon} V_\epsilon \leq 0$ equivalently. Accordingly, the solution of the quadratic programming problem (6.36) is equal to the control law presented in Eq. (6.35). In this case, the final control law for the closed-loop attitude control system for a rigid spacecraft with uncertainties is

$$\begin{aligned} \boldsymbol{\tau} &= \boldsymbol{\tau}^* - \boldsymbol{D}_0^{-1} \boldsymbol{J}_0 \boldsymbol{Z}_2 \\ \boldsymbol{\tau}^* &= -\boldsymbol{H}^{-1} \left(\bar{\boldsymbol{L}}_{\mathrm{f}} - m(\eta) \right) \end{aligned} \tag{6.37}$$

Now that the point-wise min-norm problem has been formulated as an optimization problem, the actuator input saturation can directly and explicitly be introduced and implemented in the optimization process. Particularly, a quadratic programming based RES-CLF optimal issue can be constructed as

$$\begin{aligned} & \arg\min_{\mu,\, d_1} \left\{ \|\boldsymbol{\mu}\|_2^2 + p_1 d_1^2 \right\} \\ & \text{s.t.} \ \bar{\phi}(\eta) + \bar{\varphi}(\eta)\boldsymbol{\mu} \leq d_1, \ \boldsymbol{\tau} \leq \boldsymbol{\tau}_{\max}, \ -\boldsymbol{\tau} \leq \boldsymbol{\tau}_{\max} \end{aligned} \tag{6.38}$$

where $d_1 > 0$ is small variable, and $p_1 > 0$ is a large constant that penalizes the offset/bound of the RES-CLF constraints. Then, the final control law for the closed-loop attitude control system can be described by

Table 6.1 Control parameters chosen for numerical analysis

Control schemes	Control gains
ESO	$k = \text{diag}\,(6.1, 2.14, 1.2)\,, \alpha = 0.5$ $\beta_0 = \text{diag}\left(0.002\ 0.002\ 0.002\right)$ $\beta_1 = \text{diag}\left(5.0\ 10.3\ 5.0\right)$ $\beta_2 = \text{diag}\left(10\ 20\ 40\right)$ $\beta_3 = \text{diag}\left(1.8\ 3.0\ 10.2\right)$
CLF_IOC	$a = 2.0, b = 1.72, c = 1.92$ $\gamma = 0.2, k_1 = 3.72$
RESCLF_QPC	$\epsilon = 0.5, p_1 = 1.0, \tau_{\max} = 0.5$

$$
\begin{aligned}
\boldsymbol{\tau} &= \boldsymbol{\tau}^* - \boldsymbol{D}_0^{-1}\boldsymbol{J}_0\boldsymbol{Z}_2 \\
\boldsymbol{\tau}^* &= -\boldsymbol{H}^{-1}\left(\bar{\boldsymbol{L}}_{\mathrm{f}} - \boldsymbol{\mu}\right)
\end{aligned}
\tag{6.39}
$$

6.4 Simulation Results

To demonstrate the effectiveness and performance of the proposed attitude stabilization control schemes incorporated with ESO, a spacecraft model governed by Eqs. (2.17) and (2.20) is numerically simulated, in conjunction with the ESO in Eq. (6.3) and the CLF based inverse optimal control law Eq. (6.16) (noted as CLF_IOC), or RES-CLF and quadratic programing based optimal controller Eq. (6.39) (noted as RESCLF_QPC). The attitude control type is chosen as three-axis control by four reaction fly-wheels. Throughout the simulation, the inertia matrix is set as

$$
\boldsymbol{J} = \begin{bmatrix} 20 & 0 & 0.9 \\ 0 & 17 & 0 \\ 0.9 & 0 & 15 \end{bmatrix} \mathrm{kg} \cdot \mathrm{m}^2
$$

The inertia matrix uncertainty is given by

$$
\Delta \boldsymbol{J} = \begin{bmatrix} 2.7 & 0.5 & -1.0 \\ 0.2 & 2.5 & 0.2 \\ -0.3 & 0.1 & 1.8 \end{bmatrix} \mathrm{kg} \cdot \mathrm{m}^2
$$

The initial attitude MRPs and angular velocity is set as $\boldsymbol{\rho} = [-0.1579, 0.1368, 0.0947]^{\mathrm{T}}$ and $\boldsymbol{w}_0 = [0.1, \ -0.1, \ -0.05]^{\mathrm{T}}$ (rad/s). In this case, considering the actuator saturation, the available control torque of each reaction wheel is assumed to be 0.5 (N.m). According to the practical situation, the assembling angles $\alpha_4 = 35.26$

Fig. 6.1 Time response of the observation error $\boldsymbol{E}_1$ in CLF_IOC

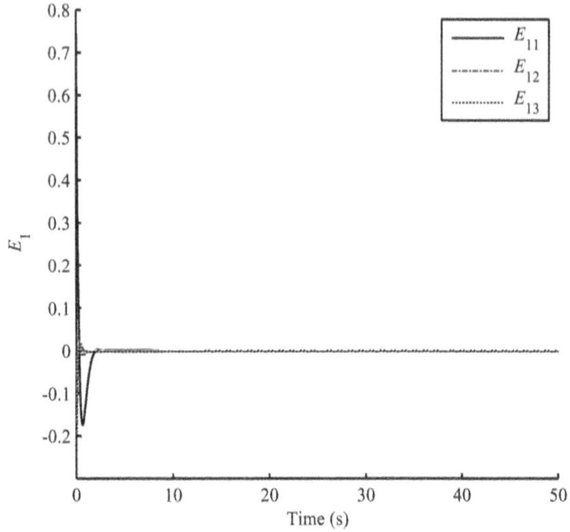

Fig. 6.2 Time response of the observation error $\boldsymbol{E}_1$ in RESCLF_QPC

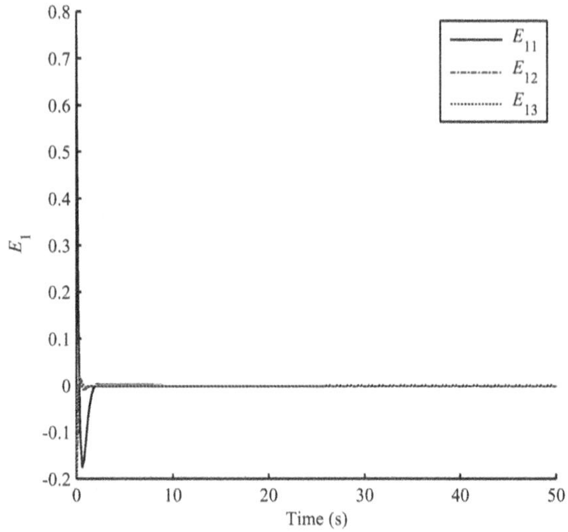

(deg) and $\beta_4 = 45$ (deg) are selected. Moreover, the misalignment error angles are assumed to be $\Delta\alpha = [0.2, 0.2, 0.1, 0.1]^{\mathrm{T}}$ (rad) and $\Delta\beta = [0.2, 0.2, 0.1, 0.1]^{\mathrm{T}}$ (rad). For all numerical examples presented in this section, control parameters are given in Table 6.1.

Fig. 6.3 Time response of the observation error $\boldsymbol{E}_2$ in CLF_IOC

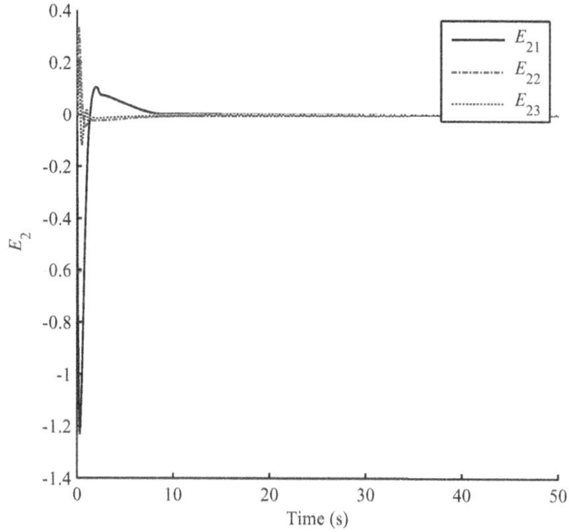

6.4.1 Comparison and Analysis Under Different Controllers

For the analysis of the effectiveness of ESO, Figs. 6.1, 6.2, 6.3 and 6.4 show time responses of observer errors obtained from the ESO in Eqs. (6.3) and (6.4) under two different control laws (CLF_IOC, RESCLF_QPC). As we can see, the presented ESO has estimated the actual states successfully, and the UUB stability in finite time of observer errors $\boldsymbol{E}_1 = [E_{11},\ E_{12},\ E_{13}]^1$ is achieved about 3 s ($T_1 = 3$). After another short period of time, roughly less than 7 s ($T_2 = 10 > T_1$), the observer errors $\boldsymbol{E}_2 = [E_{21},\ E_{22},\ E_{23}]^{\mathrm{T}}$ derived from the extended state can also converge to the original/residual region, as shown in Figs. 6.3 and 6.4. These convergences or characteristics were obtained in the presence of disturbances and uncertainties.

Figures 6.5, 6.6, 6.7 and 6.8 show the time responses of attitude MRPs and angular velocity, respectively. The system reaches the commanded target smoothly with a settling time less than 40 s under the proposed control methods. Moreover, the attitude in Figs. 6.9 and 6.10 further illustrated the effectiveness of the proposed methods. It should be pointed out that, attitude performances of the two schemes CLF_IOC and RESCLF_QPC are analogous to each other because some related parameters values are all the same, and optimal trajectories can be achieved in some degree under the two proposed optimal control strategies with similar cost functions. The biggest difference or key feature from the comparisons of the two controllers is the energy consumption saving issue, which will be analyzed in the next subsection.

The time responses of the control torques $\boldsymbol{u}$ and actual actuator control torques $\boldsymbol{\tau}$ provided by four reaction wheels are shown in Figs. 6.11, 6.12, 6.13 and 6.14. Note that the restriction on the actuator output torque magnitude 0.5 N.m is considered explicitly. Thanks to the quadratic programming based RESCLF_QPC controller, the specified actuator constraints have been met successfully. However, the

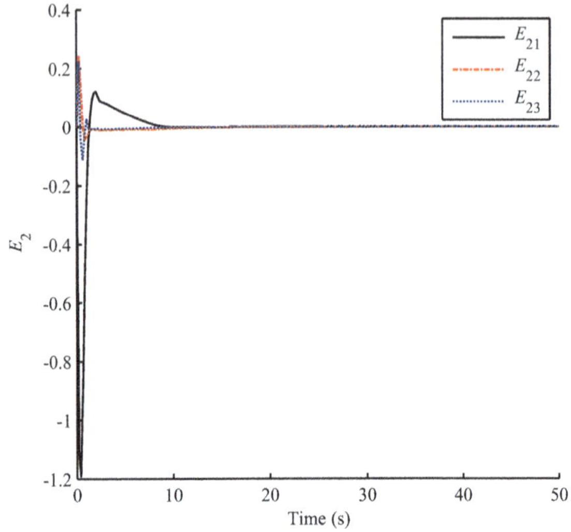

Fig. 6.4 Time response of the observation error $\boldsymbol{E}_2$ in RESCLF_QPC

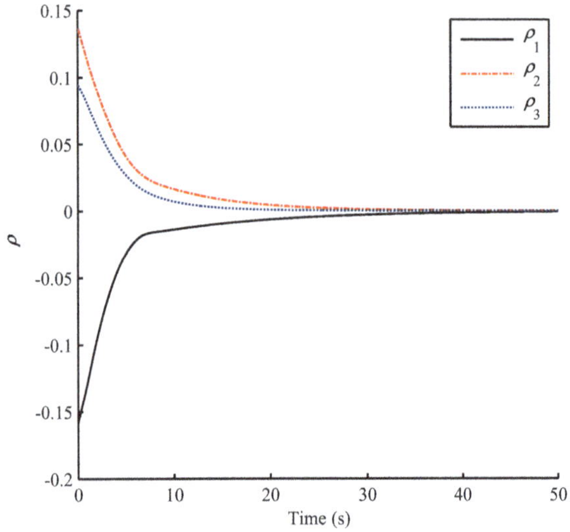

Fig. 6.5 Time response of MRPs in CLF_IOC

CLF_IOC scheme achieves relatively worse performance with respect to actuator constraints. As shown in Figs. 6.13 and 6.14, comparing the cases between CLF_IOC and RESCLF_QPC, the CLF_IOC scheme without considering actuator constraints results in saturation of reaction flywheel at the beginning of the attitude control process, especially for τ_1 and τ_3. Generally speaking, comparing the responses between the three control laws, it shows the necessity and good performance of the proposed controllers.

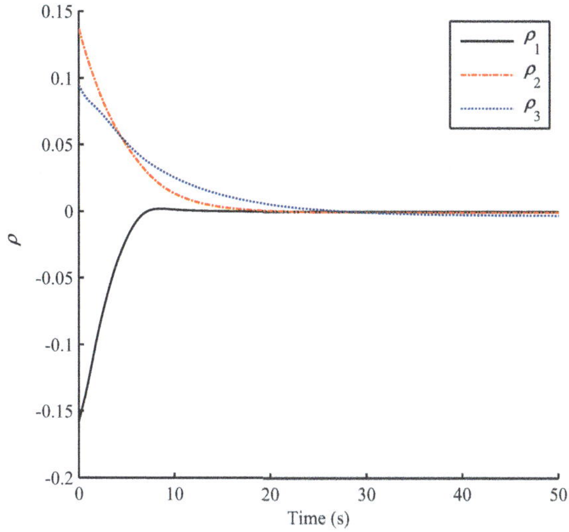

Fig. 6.6 Time response of MRPs in RESCLF_QPC

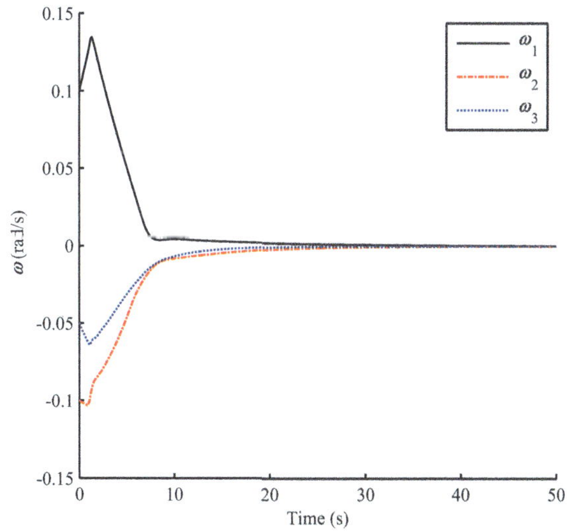

Fig. 6.7 Time response of angular velocity in CLF_IOC

6.4.2 Comparisons and Analyses of Energy Consumptions Under Different Controllers

Energy-efficient issue for spacecraft in orbit is of crucial importance to prolong the working life and achieve the target missions. Therefore, the energy consumption optimization problem should be also considered seriously for the spacecraft attitude control system design. To analyze the system total energy consumption, the energy index function derived from the optimal costing function is defined by $Enegy =$

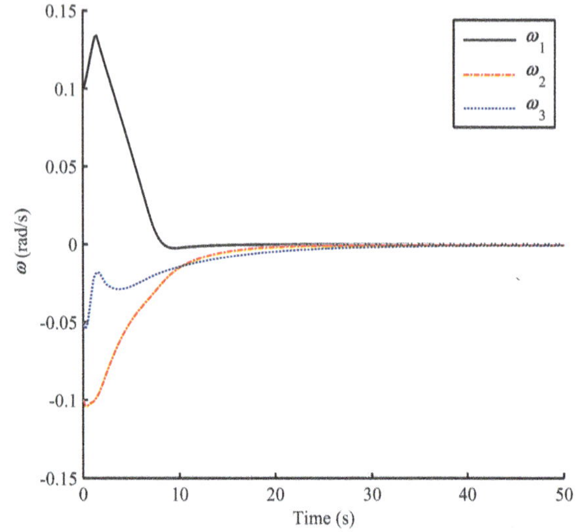

Fig. 6.8 Time response of angular velocity in RESCLF_QPC

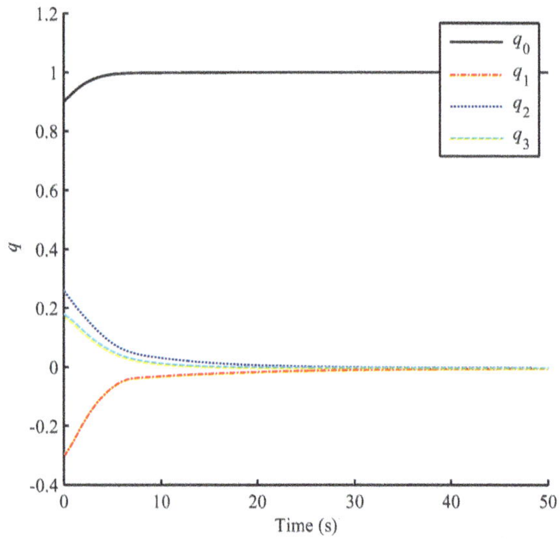

Fig. 6.9 Time response of quaternion in CLF_IOC

$\frac{1}{2}\int_0^T ||\tau||^2 \mathrm{d}t$, where T denotes the simulation time, and $T = 50$s is chosen in the simulation. The time responses of the energy consumption performance comparisons of the two strategies with external disturbances and model uncertainties are shown in Figs. 6.15 and 6.16. And it can be seen clearly that the energy consumption of the RESCLF_QPC is minimum and an obvious improvement over the CLF_IOC has been achieved, although the design of two schemes are both based on optimization method/theory. Such an improvement is benefitted and achieved by utilizing the optimization technique with actuator constraints.

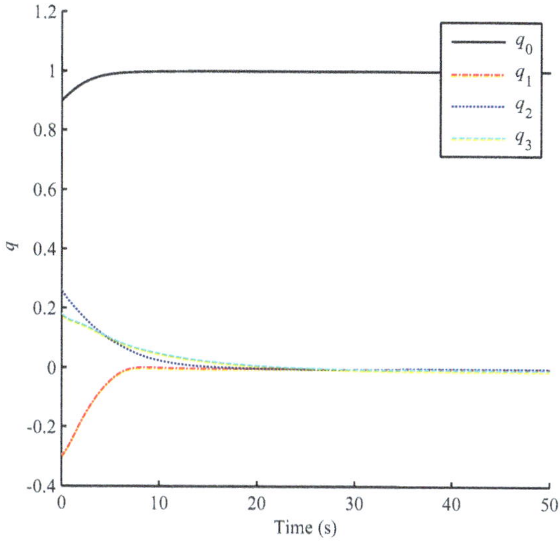

Fig. 6.10 Time response of quaternion in RESCLF_QPC

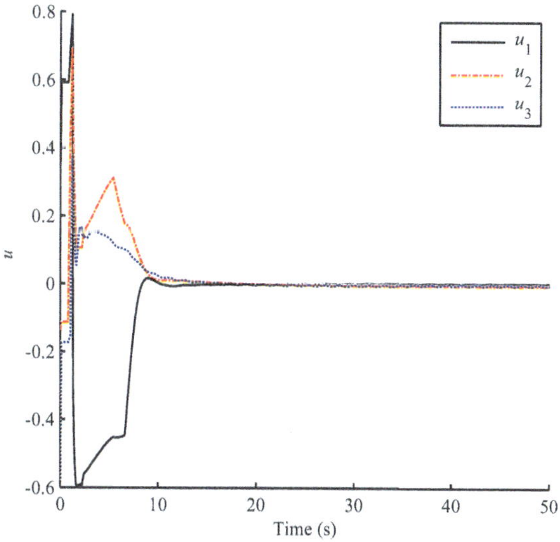

Fig. 6.11 Time response of control torque in CLF_IOC

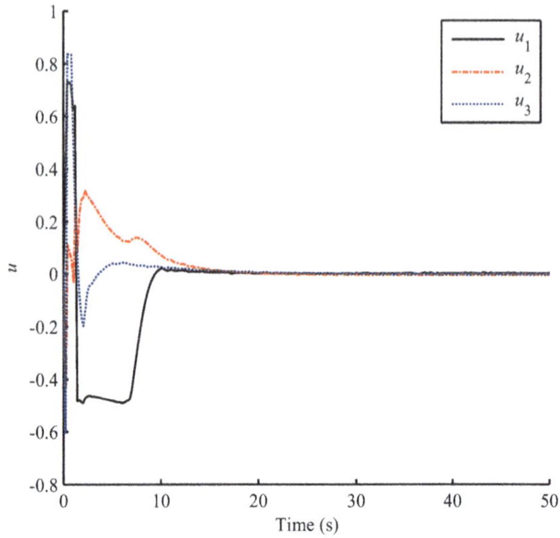

Fig. 6.12 Time response of control torque in RESCLF_QPC

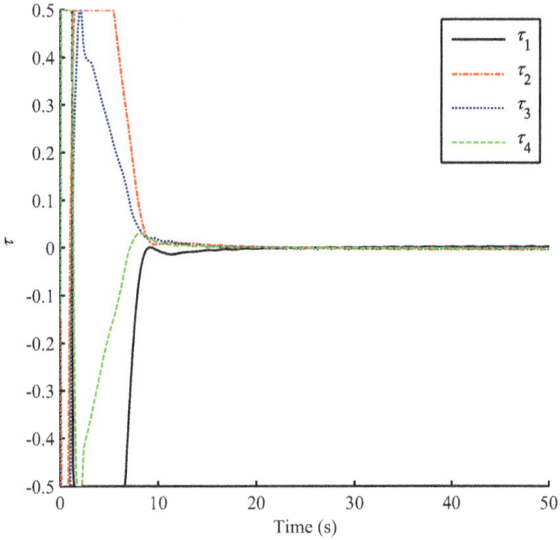

Fig. 6.13 Time response of actual actuator control torque in CLF_IOC

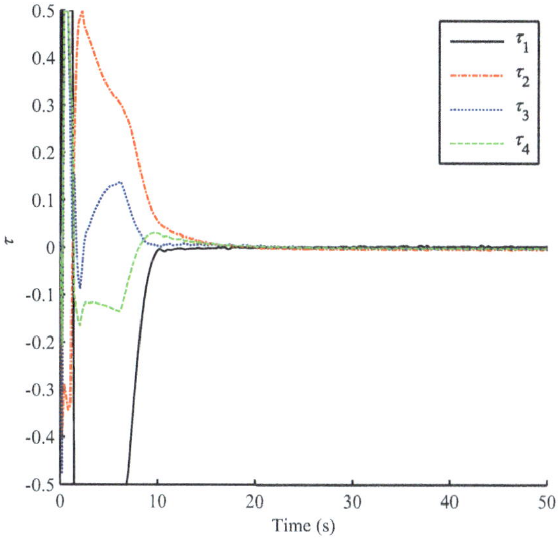

Fig. 6.14 Time response of actual actuator control torque in RESCLF_QPC

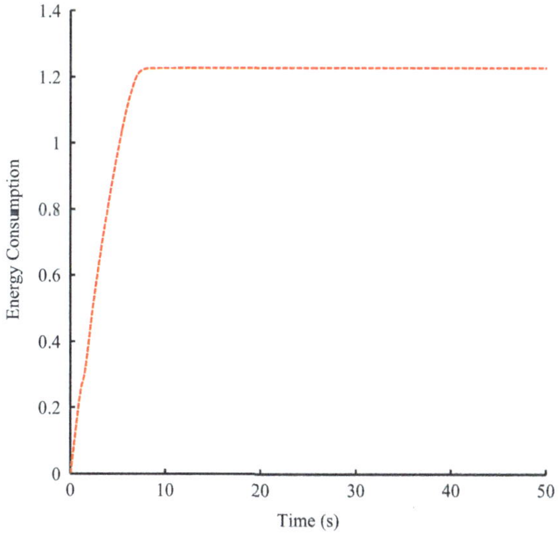

Fig. 6.15 Time response of energy consumption in CLF_IOC

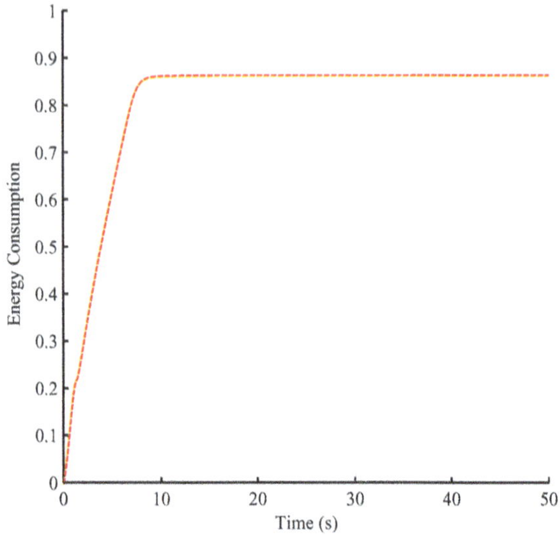

Fig. 6.16 Time response of energy consumption in RESCLF_QPC

6.5 Conclusions

In this chapter, CLF based robust optimal control laws incorporated with ESO are proposed for spacecraft attitude stabilization. In this context, the ESO is firstly proposed to compensate the specified total uncertainty, an inverse optimal CLF approach is involved to guarantee the asymptotic stability of the closed-loop system, and then optimal/minimum performance index can be achieved simultaneously. Furthermore, a RES-CLF based optimal control scheme is further investigated, and actuator saturation is restrained successfully utilizing the quadratic programming technique. Simulation results illustrate the necessity and better performance of the proposed schemes. However, it should be pointed out that actuators' faults such as loss of effectiveness or stuck are not considered explicitly in this chapter, which is one of the subjects for our future researches.

References

1. Han, J.Q. (2009) From PID to active disturbance rejection control. IEEE transactions on Industrial Electronics 56(3): 900–906
2. Xia, Y.Q, Zhu, Z., Fu, M.Y., Wang, S. (2011) Attitude tracking of rigid spacecraft with bounded disturbances. IEEE Transactions on Industrial Electronics 58(2): 647–659
3. Li S.L., Yang X., Yang D. (2009) Active disturbance rejection control for high pointing accuracy and rotation speed. Automatica 45(8): 1854–1860
4. Pukdeboon, C., Jitpattanakul, A. (2013) Finite-time anti-disturbance inverse optimal attitude tracking control of flexible spacecraft. Mathematical Problems in Engineering 2013: 1–13

5. Guo, B.Z., Zhao, Z.L. (2011) On the convergence of an extended state observer for nonlinear systems with uncertainty. Systems & Control Letters 60(6): 420–430
6. Krstic, M., Li, Z.H. (1998) Inverse optimal design of input-to-state stabilizing nonlinear controllers. IEEE Transactions on Automatic Control 43(3): 336–350
7. Luo, W.C., Chu, Y.C., Ling, K.V. (2005) H-infinity inverse optimal attitude-tracking control of rigid spacecraft. Journal of Guidance, Control, and Dynamics 28(3): 481–493
8. Luo, W.C., Chu, Y.C., Ling, K.V. (2005) Inverse optimal adaptive control for attitude tracking of spacecraft. IEEE Transactions on Automatic Control 50(11): 1639–1654
9. Ames, A.D., Galloway, K., Sreenath, K., Grizzle, J.W. (2014) Rapidly exponentially stabilizing control lyapunov functions and hybrid zero dynamics. IEEE Transactions on Automatic Control 59(4): 876–891
10. Freeman, R.A., Kokotovic, P.V. (1996) Robust nonlinear control design: State-space and lyapunov techniques
11. Ali, I., Radice, G., Kim, J. (2010) Backstepping control design with actuator torque bound for spacecraft attitude maneuver. Journal of Guidance, Control, and Dynamics 33(1): 254–259
12. Boskovic, J.D., Li, S., Mehra, R.K. (2001) Robust adaptive variable structure control of spacecraft under control input saturation. Journal of Guidance, Control, and Dynamics 24(1): 14–22
13. Bang, H., Tahk, M.J., Choi, H.D. (2003) Large angle attitude control of spacecraft with actuator saturation. Control Engineering Practice 11(9): 989–997
14. Hu, Q.L., Li, B., Zhang, Y.M. (2014) Nonlinear proportional-derivative control incorporating closed-loop control allocation for spacecraft. Journal of Guidance, Control, and Dynamics 37(3): 799–812
15. Galloway, K., Sreenath, K., Ames, A.D., Grizzle, J.W. (2015) Torque saturation in bipedal robotic walking through control lyapunov function-based quadratic programs. IEEE Access 3: 323–332
16. Xiao, B., Hu, Q.L., Zhang, Y.M., Huo, X. (2014) Fault-tolerant tracking control of spacecraft with attitude-only measurement under actuator failures. Journal of Guidance, Control, and Dynamics 37(3): 838–849
17. Khalil, H.K., Grizzle, J.W. (2002) Nonlinear systems, vol 3. Prentice hall Upper Saddle River, NJ
18. Park, Y. (2013) Inverse optimal and robust nonlinear attitude control of rigid spacecraft. Aerospace Science and Technology 28(1): 257–265
19. Pukdeboon, C., Zinober, A.S.I. (2012) Control lyapunov function optimal sliding mode controllers for attitude tracking of spacecraft. Journal of the Franklin Institute 349(2): 456–475
20. Ames, A.D., Powell, M. (2013) Towards the unification of locomotion and manipulation through control lyapunov functions and quadratic programs. In: Control of Cyber-Physical Systems, pp 219–240
21. Westervelt, E.R., Grizzle, J.W., Chevallereau, C., Choi, J.H., Morris, B. (2018) Feedback control of dynamic bipedal robot locomotion. CRC press
22. Rugh, W.J. (1996) Linear system theory, vol 2. prentice hall Upper Saddle River, NJ
23. Petersen, I.R., Barmish, B.R. (1984) Control effort considerations in the stabilization of uncertain dynamical systems. In: 1984 American Control Conference, IEEE, pp 490–495

Chapter 7
Spacecraft Attitude Fault-Tolerant Control Based on Iterative Learning Observer and Control Allocation

7.1 Introduction

For the over-actuated spacecraft ACS, it is generally equipped with redundant actuators in the specified configurations. Consider the constraints of mechanical precision and physical technology in the manufacturing and assembly of actuators (RWs, control moment gyros, thrusters and so on), the installing configurations of the redundant actuators often have some certain deviations or misalignments from the nominal configurations, due to the impact of strong shock and vibration during launch, and the influence of external disturbances. Additionally, actuator faults may occur after a long-time on-orbiting operation. Thus, a CA based FTC scheme will be proposed for the spacecraft ACS in this chapter.

Although some adaptive control, SMC and robust control based PFTC schemes have been presented in the existing literatures, the PFTC without faults estimation is only valid for presumed faults. In contrast to the PFTC, AFTC compensates for the faults by synthesizing or selecting the new controller on-line and requires the fault diagnosis mechanism. Patton et al. [1] proposed the robust FDD for thrusters' faults of the a satellite system Mars Express. Reference [2] presented an integrated fault diagnosis by state augmentation for the satellite ACS. An iterative neuron PID observer was proposed and used for the FDD of the spacecraft ACS in [3]. A on-line CA based FTC scheme was proposed for the autonomous spacecraft rendezvous, which consisting of a fault detector for robust and quick fault detection, two-stage hierarchical isolation strategy for fault isolation. Xiao et al. [4] has proposed a TSMC based observer which can observe the RWs' faults and external disturbances in finite time. Consider the limited computing capability of on-board computer, an ILO based FDD was designed in [5] to reconstruct both the time-varying and constant actuator faults for spacecraft ACS. In addition, these results only addressed a single type of actuator fault, and there are few control schemes able to handle configuration misalignment and the loss of actuator effectiveness simultaneously in the presence of external disturbances.

Q. Hu et al., *Control Allocation for Spacecraft Under Actuator Faults*,
https://doi.org/10.1007/978-981-16-0439-3_7

To make effective use of the redundancy of actuators for the over-actuated control systems, the CA technique is often utilized to distribute the virtual control signals to individual actuator in the best manner while taking into account the constraints. There are some interesting results available for CA scheme design in the last decades, such as the daisy chain allocation method [6], the linear or nonlinear programming method [7], and the dynamic control allocation method [8] and so on. But these above mentioned CA schemes don't considered the issue that there exist actuator faults, so some schemes are developed to allocate the control torque incorporated with actuator faults information. In [9], the robust least-square CA was proposed for the flight fault tolerant control system. Taking into account the unknown disturbances and inertia uncertainty, a novel FTC law incorporating on-line CA was developed in [10]. Further considering fault estimation error, the authors presented a robust CA method in [11]. A constrained CA method was proposed in [12] with the ability to comprise several control effectors into only three aerodynamic moments around the body axes.

This chapter introduces an observer-based control scheme with fault-tolerant capability to achieve the attitude stabilization of rigid spacecraft in the present of system uncertainties, such as actuator uncertainties and faults, external disturbances and input saturation. Firstly, an iterative learning observer is developed to estimate the torque deviation and guarantees that the estimation errors of the torque deviation and the angular velocity converge to some small residual sets. Then, the virtual control based on integral-type sliding mode manifold is developed, which is able to ensure that all signals of the closed-loop system are globally uniformly bounded. Finally, based on the virtual controller and the iterative learning observer, a robust CA algorithm is developed to distribute the virtual control law to each actuator in an optimal manner with fault-tolerant ability. The proposed FTC scheme is analytically verified and illustrated via simulation results. This could further effectively improve the control accuracy under the actuator faults and misalignments (even with a small value) for the spacecraft ACS. The appreciating properties of the presented method are generalized as follows.

(1) Proposing an iterative learning observer which contains the sign function to estimate the torque deviation caused by actuator faults. Particularly, the sign function used in the observer could decrease the computing complexity for only need to compare the size between the angular velocity of the space-craft and its estimate value.

(2) The faults estimation information is introduced in robust CA strategy, which could achieve better fault-tolerant ability than the existing methods.

(Remark: The main control schemes and contents of this chapter are from the published journal paper "Qinglei Hu, Guanglin Niu and Chenliang Wang, Spacecraft attitude fault-tolerant control based on iterative learning observer and control allocation, Aerospace Science and Technology, 2018, 75: 245–253". The authors appreciate the permission from the Elsevier Paris to reuse partly the results published in the relevant Journal.)

7.2 Problem Formulation

Similar to [13], the actual control torque generated by RWs with faults and configuration misalignment can be modeled by

$$\begin{aligned}\boldsymbol{u} &= (\boldsymbol{D}_0 + \Delta\boldsymbol{D})\,\boldsymbol{E}(t)\boldsymbol{\tau}_c = (\boldsymbol{D}_0 + \Delta\boldsymbol{D})\,\boldsymbol{\tau}_c + (\boldsymbol{D}_0 + \Delta\boldsymbol{D})\underbrace{(\boldsymbol{E}(t) - \boldsymbol{I}_{m\times m})\,\boldsymbol{\tau}_c}_{\boldsymbol{u}_f} \\ &- \boldsymbol{D}_0\boldsymbol{\tau}_c + (\boldsymbol{D}_0 + \Delta\boldsymbol{D})\,\boldsymbol{u}_f + \Delta\boldsymbol{D}\boldsymbol{\tau}_c\end{aligned} \tag{7.1}$$

where $\boldsymbol{D}_0 \in \mathbf{R}^{3\times m}$ is the nominal control torque distribution matrix with $m > 3$ the number of actuators, $\Delta\boldsymbol{D} \in \mathbf{R}^{3\times m}$ denotes the configuration matrix of misalignment, $\boldsymbol{E}(t) = \mathrm{diag}\,(e_1(t), e_2(t), \ldots, e_m(t)) \in \mathbf{R}^{m\times m}$ represents the loss of effectiveness for actuators with $0 \le e_i(t) \le 1 \quad (i = 1, 2, \ldots, m)$, $\boldsymbol{\tau}_c$ denotes the control signal commanded by the controller and delivered to actuators, and $\boldsymbol{u}_f$ denotes the deviation value caused by actuator faults. Note that $e_i(t) = 1$ indicates the ith actuator operates normally, $0 < e_i(t) < 1$ implies that the ith actuator partially loses its power, and $e_i(t) = 0$ indicates the ith actuator fails completely. Substituting Eq. (7.1) into Eq. (2.20), the spacecraft attitude dynamics can be rewritten as

$$\boldsymbol{J}\dot{\boldsymbol{\omega}} = -\boldsymbol{S}(\boldsymbol{\omega})\boldsymbol{J}\boldsymbol{\omega} + \boldsymbol{D}_0\boldsymbol{\tau}_c + (\boldsymbol{D}_0 + \Delta\boldsymbol{D})\,\boldsymbol{u}_f + \Delta\boldsymbol{D}\boldsymbol{\tau}_c + \boldsymbol{d} \tag{7.2}$$

The following assumption is made, which is mild for the external disturbance.

Assumption 7.1 The external disturbance $\boldsymbol{d}$ is bounded, i.e., there exists a constant $d_{\max}$ such that $\|\boldsymbol{d}\| \le d_{\max}$.

Consider the spacecraft ACS given by Eqs. (2.14) and (7.2), the control objective is to propose a control scheme such that all signals of the closed-loop system are uniformly ultimately bounded and the attitude vector $\boldsymbol{q}_v$ and relative angular velocity $\boldsymbol{\omega}$ converge asymptotically to zero, regardless of the actuator faults, external disturbances and actuator saturation.

7.3 Iterative Learning Observer Design

In view of Eq. (7.2), the torque deviation caused by actuator faults as well as configuration misalignment can be defined as $\boldsymbol{\tau}_f = (\boldsymbol{D}_0 + \Delta\boldsymbol{D})\,\boldsymbol{u}_f + \Delta\boldsymbol{D}\boldsymbol{\tau}_c$, and the attitude dynamics in Eq. (7.2) can be rewritten as

$$\boldsymbol{J}\dot{\boldsymbol{\omega}} = -\boldsymbol{S}(\boldsymbol{\omega})\boldsymbol{J}\boldsymbol{\omega} + \boldsymbol{D}_0\boldsymbol{\tau}_c + \boldsymbol{\tau}_f + \boldsymbol{d} \tag{7.3}$$

For any specified vector $\boldsymbol{x} \in \mathbf{R}^n$, let $\mathrm{sign}(\boldsymbol{x}) = \left[\mathrm{sign}\,(x_1), \mathrm{sign}\,(x_2), \ldots, \mathrm{sign}\,(x_n)\right]^{\mathrm{T}}$. Then, to estimate the torque deviation, the following iterative learning observer is proposed:

$$\begin{cases} \boldsymbol{J}\dot{\hat{\omega}}(t) = -\hat{\omega}(t) \times \boldsymbol{J}\hat{\omega} + \lambda(\omega - \hat{\omega}) + \boldsymbol{D}_0\boldsymbol{\tau}_c + \boldsymbol{v}(t) + \boldsymbol{L}\,\text{sign}(\omega - \hat{\omega}) \\ \boldsymbol{v}(t) = \boldsymbol{K}_1\boldsymbol{v}(t-T) + \boldsymbol{K}_2\,\text{sign}(\omega - \hat{\omega}) \end{cases} \tag{7.4}$$

where $\hat{\omega} \in \mathbf{R}^3$ and $\boldsymbol{v}(t) \in \mathbf{R}^3$ respectively denote the angular velocity estimation and the estimate of the torque deviation, T is the updating interval which is used as the sampling time interval in our work, λ is a positive scalar parameter, $\boldsymbol{L} \in \mathbf{R}^{3\times3}$ is a positive definite matrix whose evaluation depends on the upper bound of disturbance, and $\boldsymbol{K}_1 \in \mathbf{R}^{3\times3}$ and $\boldsymbol{K}_2 \in \mathbf{R}^{3\times3}$ are positive definite gain matrixes.

Assumption 7.2 The torque deviation $\boldsymbol{\tau}_f$ is bounded and satisfies

$$\left\|\boldsymbol{\tau}_f(t) - \boldsymbol{K}_1\boldsymbol{\tau}_f(t-T)\right\| \leq k_d$$

where $k_d > 0$ is a known scalar.

Assumption 7.3 ([4]) The angular velocity $\boldsymbol{\omega}(t)$ possesses the bounded first-order derivative. Besides, with a sufficiently small sampling time, $\boldsymbol{\omega}(t) \times \boldsymbol{J}\boldsymbol{\omega}(t)$ satisfies

$$\| - \boldsymbol{\omega}(t) \times \boldsymbol{J}\boldsymbol{\omega}(t) + \hat{\omega}(t) \times \boldsymbol{J}\hat{\omega}(t)\| \leq \kappa\|\tilde{\omega}(t)\|$$

for some known positive scalar satisfying $\kappa > 0$.

Then the following statement can be concluded as:

Theorem 7.1 *Consider the ACS described by Eqs. (2.14) and (7.3) and the iterative learning observer in Eq. (7.4) under Assumptions 7.1–7.3. Select the observer gains as*

$$\begin{cases} -\kappa + \lambda - \gamma_4 > 0 \\ 1 - [1 + 1/\gamma_4 + \delta\alpha_1]\,\|\boldsymbol{K}_1\|^2 \geq 0 \\ L_{\min} - d_{\max} \geq 0 \end{cases} \tag{7.5}$$

where $\gamma_i (i = 1, 2, 3, 4)$, $\alpha_1 = 1 + \gamma_1 + 1/\gamma_2$, and δ are positive constants, and $L_{\min}$ denotes the minimum eigenvalue of observer gain matrix $\boldsymbol{L}$ Then, the estimation errors of the torque deviation and the angular velocity converge to some small residual set.

Remark 7.1 It can be seen that the designed iterative learning observer ensures the convergence of estimation errors irrespective of the control input acting on the rigid spacecraft. Hence, whatever the controller is selected, the expected observation performance could be achieved.

Remark 7.2 The term $\boldsymbol{L}\,\text{sign}(\omega(t) - \hat{\omega}(t))$ in observer (7.4) is employed to improve the robustness with respect to external disturbances. The torque deviation reconstruction will be introduced into the CA strategy to improve the fault-tolerant capability of the whole system.

7.4 Control Strategy Design

In this section, the controller will be designed through two steps with the principle of control allocation. In the first step, the virtual control torque is developed to specify the total attitude control torque; while in the second step, a robust CA strategy on account of the imprecision in observation is designed to distribute the desired total control torque derived from the virtual controller to each actuator.

7.4.1 Virtual Feedback Controller Design

First of all, employing the SMC technique, we construct an integral-type sliding mode manifold as follows:

$$\boldsymbol{S} = \boldsymbol{J}\left[\boldsymbol{\omega}(t) - \boldsymbol{\omega}(t_0)\right] + \int_{t_0}^{\mathrm{T}} \left(\boldsymbol{\omega}(\chi) \times \boldsymbol{J}\boldsymbol{\omega}(\chi) - \boldsymbol{\tau}_{nom}\right) \mathrm{d}\chi \tag{7.6}$$

where $\boldsymbol{\tau}_{nom}$ will be defined latter. Note that the sliding manifold satisfies $\boldsymbol{S} = 0$ at the initial time $t = t_0$ and thus the reaching phase is eliminated. Then, a virtual controller can be given as

$$\boldsymbol{\tau} = \boldsymbol{\tau}_{nom} + \boldsymbol{\tau}_l \tag{7.7}$$

which is composed by a nominal control law $\boldsymbol{\tau}_{\text{nom}}$ and a switch controller $\boldsymbol{\tau}_l$ The nominal controller $\boldsymbol{\tau}_{\text{nom}}$ is given by

$$\boldsymbol{\tau}_{nom} = -k_v \tanh\left(\boldsymbol{\omega}(t)/k^2(t)\right) - k_p \boldsymbol{q}_v \tag{7.8}$$

where k_v and k_p are both positive scalar parameters satisfying $k_v \geq d_{\max}$,

$$\tanh\left(\boldsymbol{\omega}(t)/k^2(t)\right) = \left[\tanh\left(\omega_1(t)/k^2(t)\right), \tanh\left(\omega_2(t)/k^2(t)\right), \tanh\left(\omega_3(t)/k^2(t)\right)\right]^{\mathrm{T}},$$

with $\tanh(\cdot)$ denoting the standard hyperbolic tangent function, and $k^2(t)$ is a nonzero scalar sharpness function which satisfies $0 < k_{\min}^2 \leq k^2(t) \in L_\infty$ and $\dot{k} \in L_\infty$, and it can be selected as a constant k^2 in our work. On the other hand, $\boldsymbol{\tau}_l$ is a switch controller component used to compensate for the external disturbance and drives the system to the sliding manifold, given by

$$\boldsymbol{\tau}_l = \begin{cases} -\beta \frac{\boldsymbol{S}}{\|\boldsymbol{S}\|}, & \text{if } \boldsymbol{S} \neq \boldsymbol{0} \\ \boldsymbol{0}, & \text{if } \boldsymbol{S} = \boldsymbol{0} \end{cases} \tag{7.9}$$

with the parameter β satisfying $\beta \geq d_{\max}$.

Remark 7.3 In this step, the virtual controller is designed and will be transferred to CA strategy based on $\boldsymbol{\tau} = \boldsymbol{D}_0\boldsymbol{\tau}_c(t) + \boldsymbol{\tau}_f$ Thus, the attitude dynamics in Eq. (7.3) with the virtual controller can be rewritten as

$$\boldsymbol{J}\dot{\boldsymbol{\omega}} = -\boldsymbol{S}(\boldsymbol{\omega})\boldsymbol{J}\boldsymbol{\omega} + \boldsymbol{\tau} + \boldsymbol{d} \tag{7.10}$$

Then the following main result can be stated as:

Theorem 7.2 *The virtual controller design in Eqs. (7.6)–(7.8) can guarantee asymptotical stability of the control system in the presence of actuator faults, external disturbances and input constraint.*

Furthermore, considering the input constraint, the virtual controller could be bounded as

$$\|\boldsymbol{\tau}\| \leq k_v + k_p + \beta \leq \overline{u} \tag{7.11}$$

where $\overline{u}$ *is the upper bound of the command control input limitation. In view of (7.11), by selecting the proper controller parameters* k_v,k_p *and* β*. The input constraint issue can be handled. This completes the proof.*

7.4.2 Robust Control Allocation Design

Based on the above results, in this subsection the CA algorithm is generated to distribute the virtual control law to each actuator accounting for actuator faults and configuration misalignment in an optimal manner. Besides, the amplitude of control torque is limited for the actual physical constraints. Since there exists the estimation error in the proposed iterative learning observer, the imprecision of torque deviation observation $v(t)$ is taken into account. The relation between estimation of the torque deviation and its actual value can be expressed as

$$\boldsymbol{\tau}_f(t) = \boldsymbol{v}(t) + \tilde{\boldsymbol{e}}(t)$$

where $\tilde{e}(t)$ is the torque deviation estimation error. Further, the virtual controller is given by

$$\boldsymbol{\tau} = \boldsymbol{D}_0\boldsymbol{\tau}_c(t) + \boldsymbol{v}(t) + \tilde{\boldsymbol{e}}(t) \tag{7.12}$$

The optimization problem involves uncertainties in Eq. (7.12) can be described as

$$\boldsymbol{\tau}_c = \arg \min_{\|\boldsymbol{\tau}_c\| \leq \tau_{\max}} \max_{\|\tilde{\boldsymbol{e}}(t)\| \leq \theta} \{\boldsymbol{\tau}_c{}^{\mathrm{T}}\boldsymbol{M}\boldsymbol{\tau}_c + c\|\boldsymbol{D}_0\boldsymbol{\tau}_c + \boldsymbol{v}(t) + \tilde{\boldsymbol{e}}(t) - \boldsymbol{\tau}\|^2\} \tag{7.13}$$

where $\boldsymbol{M}$ is a positive-definite weighting matrix which can be denoted as $\boldsymbol{M} = \boldsymbol{D}_0^{\mathrm{T}} \operatorname{diag}\left(|v_1(t)|, |v_2(t)|, |v_3(t)|\right) \boldsymbol{D}_0$, c is a positive scalar parameter, $\tau_{\max}$ denotes the upper bound of the control torque, and the upper bound of the estimation error is defined as $\theta = \sqrt{\frac{\sigma}{\delta}}$.

With the definition $\boldsymbol{a} = \boldsymbol{\tau} - \boldsymbol{v}(t)$, the second term in (7.13) is given as

$$\|\boldsymbol{D}_0\boldsymbol{\tau}_c + \boldsymbol{v}(t) + \tilde{\boldsymbol{e}}(t) - \boldsymbol{\tau}\| \le \|\boldsymbol{D}_0\boldsymbol{\tau}_c - \boldsymbol{a}\| + \|\tilde{\boldsymbol{e}}(t)\| \le \|\boldsymbol{D}_0\boldsymbol{\tau}_c - \boldsymbol{a}\| + \theta \quad (7.14)$$

The upper bound in (7.14) is achievable if $\tilde{\boldsymbol{e}}(t) = \frac{\boldsymbol{D}_0\boldsymbol{\tau}_c - \boldsymbol{a}}{\|\boldsymbol{D}_0\boldsymbol{\tau}_c - \boldsymbol{a}\|}\theta$. Under this, it can be checked that

$$\|\boldsymbol{D}_0\boldsymbol{\tau}_c + \boldsymbol{v}(t) + \tilde{\boldsymbol{e}}(t) - \boldsymbol{\tau}\| = \|\boldsymbol{D}_0\boldsymbol{\tau}_c - \boldsymbol{a}\| + \theta \quad (7.15)$$

Then, the robust CA of Eq. (7.13) can be rewritten as

$$\boldsymbol{\tau}_c = \arg \min_{\|\boldsymbol{\tau}_c\| \le \tau_{\max}} \{\boldsymbol{\tau}_c{}^{\mathrm{T}} \boldsymbol{M} \boldsymbol{\tau}_c + c(\|\boldsymbol{D}_0\boldsymbol{\tau}_c - \boldsymbol{a}\| + \theta)^2\} \quad (7.16)$$

The solution of the above second-order cone programming problem can be obtained by solving the second-order cone programming problem formulation as follows:

$$\min_{\boldsymbol{\tau}_c, y_i} y_1$$

$$\text{subject to} \begin{cases} \|\operatorname{col}(y_2, y_3)\| \le y_1 \\ \left\|\boldsymbol{M}^{1/2}\boldsymbol{\tau}_c\right\| \le y_2 \\ \|\boldsymbol{D}_0\boldsymbol{\tau}_c - \boldsymbol{a}\| \le \frac{y_3}{\sqrt{c}} - \theta \\ \|\boldsymbol{\tau}_c\| \le \tau_{\max} \end{cases} \quad (7.17)$$

where y_i $(i = 1, 2, 3)$ are some variables, and $\operatorname{col}(y_2, y_3)$ is a column vector composed of y_2 and y_3. Note that some optimization softwares, such as YALMIP and CVX, can be employed to solve the above second-order cone programming problem.

The algorithm of solving the proposed robust CA is summarized as follows:

- **Step 1.** Solving the second-order cone programming problem (7.17) *equation reference goes here* by software (such as YALMIP) to attain the optimal y_2 and.
- **Step 2.** Solving the second-order cone programming problem (7.17) Substituting the optimal y_2 and y_3 in the first inequality of Eq. (7.17) to obtain the solution $\boldsymbol{\tau}_c$ The structure of our proposed scheme is shown in Fig. 7.1.

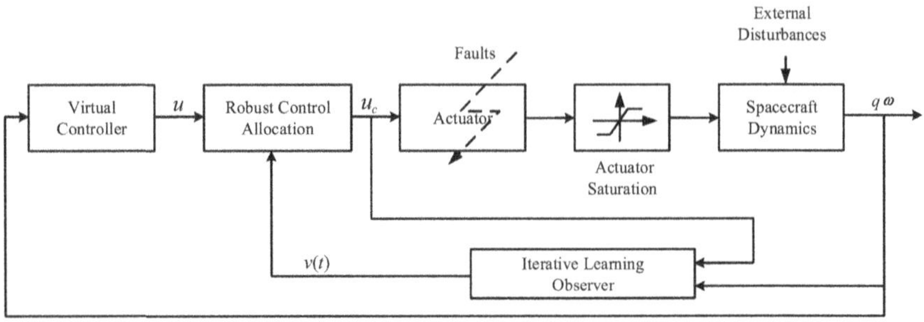

Fig. 7.1 Structure of the proposed fault-tolerant attitude control scheme

7.5 Simulation Results

In this section, some numerical simulations are performed to illustrate the effectiveness of the proposed scheme. In the simulation, the inertia matrix is given by

$$\boldsymbol{J} = \begin{bmatrix} 20 & 0 & 0.9 \\ 0 & 17 & 0 \\ 0.9 & 0 & 15 \end{bmatrix} \text{kg} \cdot \text{m}^2$$

The initial values of the attitude and angular velocity is chosen as $\boldsymbol{\omega}(0) = [0.1, -0.1, -0.05]^{\mathrm{T}}$ (rad/s) and $\boldsymbol{q}(0) = [0.9, -0.3, 0.26, 0.18]^{\mathrm{T}}$. Similar to that of the other chapters, the nominal control torque distribution matrix is given by

$$\boldsymbol{D}_0 = \begin{bmatrix} 1 & 0 & 0 & 1/\sqrt{3} \\ 0 & 1 & 0 & 1/\sqrt{3} \\ 0 & 0 & 1 & 1/\sqrt{3} \end{bmatrix}$$

RWs are employed as actuators, and a common configuration of four reaction wheels is that the rotation axes of three RWs are orthogonal to the spacecraft ontology shaft and the fourth reaction wheels is installed in the equiangular direction with the ontology three axis. In particular, considering the actuator saturation, the available control torque is supposed to satisfy $|u_i| \leq 0.5$ (N.m), where $i = 1, 2, 3$. The external disturbance is set as $\boldsymbol{d} = 0.002[\sin(t), \cos(t), \sin(2t)]^{\mathrm{T}}$. All the simulations are conducted by using the MATLAB/Simulink software package, and the updating interval of iterative learning observer is selected as 0.01s. Additionally, YALMIP and SEDUMI are employed for solution of second-order cone programming problem. The design parameters are given as follows.

For the designed observer, the initial value of angular velocity and fault estimation is set as $\hat{\boldsymbol{\omega}}(0) = \boldsymbol{\omega}(0)$ and $\boldsymbol{v}(0) = \boldsymbol{q}(0)$. The other gains of the observer are given by $\boldsymbol{L} = [0.005, 0, 0; 0, 0.005, 0; 0, 0, 0.005]$, $\boldsymbol{K}_1 = [1, 0, 0; 0, 1, 0; 0, 0, 1]$, $\boldsymbol{K}_2 = [0.0005, 0, 0; 0, 0.001, 0; 0, 0, 0.001]$, $\boldsymbol{K}_3 = [0.0003, 0, 0; 0, 0.0006, 0; 0, 0, 0.0006]$ and $\lambda = 30$. Moreover, the intermediate parameters are selected as $\sigma =$

0.00032, $\kappa = 5$, $\gamma_1 = 1$, $\gamma_2 = 1$, $\gamma_3 = 1$, $\gamma_4 = 20$, $\alpha_1 = 3$, $\alpha_2 = 2$, $\alpha_3 = 3$ and $\delta = 0.01$. The controller parameters and CA parameters are chosen as $k_p = 15$, $k_v = 20$, $\beta = 0.01$ and $c = 50$, $\theta = 0.001$, respectively. In order to demonstrate the effectiveness of the proposed scheme, two cases under different fault scenarios are considered.

7.5.1 *Case 1*

In this case, the fault scenario is set as

$$e_1(t) = \begin{cases} 1 & 0 \le t < 30\,\text{s} \\ 0.8 & t \ge 30\,\text{s} \end{cases}$$

$$e_2(t) = \begin{cases} 1 & 0 \le t < 70\,\text{s} \\ 0.7 & t \ge 70\,\text{s} \end{cases}$$

$$e_3(t) = \begin{cases} 1 & 0 \le t < 40\,\text{s} \\ 0.5 & t \ge 40\,\text{s} \end{cases}$$

$$e_4(t) = \begin{cases} 1 & 0 \le t < 40\,\text{s} \\ 0.9 + 0.1\sin(0.2t) & t \ge 40\,\text{s} \end{cases}$$

From Fig. 7.2, it can be achieved that the observer could reconstruct the torque deviation values, regardless of the faults. In Figs. 7.3 and 7.4, the high precision of the torque deviation estimation error and the angular velocity estimation error is shown, even when the faults occur. The time responses of the attitude are shown in Figs. 7.5, 7.6, 7.7 and 7.8. It is noted that an acceptable desirable orientation response is achieved. From Fig. 7.5, q_0 stabilize to 1 and $\boldsymbol{q}_v$ stabilize to the origin within 60s. And similar conclusions can be made from Figs. 7.6, 7.7 and 7.8. The time response of the control torque and actual thrusts are shown in Figs. 7.9 and 7.10. In Fig. 7.10, one can achieve that the actual control torques and the actuator torques is suitable for a spacecraft and the saturation limitation 0.5 N.m is never violated. Moreover, it can be obtained from Fig. 7.11 that a significant energy savings is achieved by the proposed scheme.

7.5.2 *Case 2*

In this subsection, the fault scenario is set as

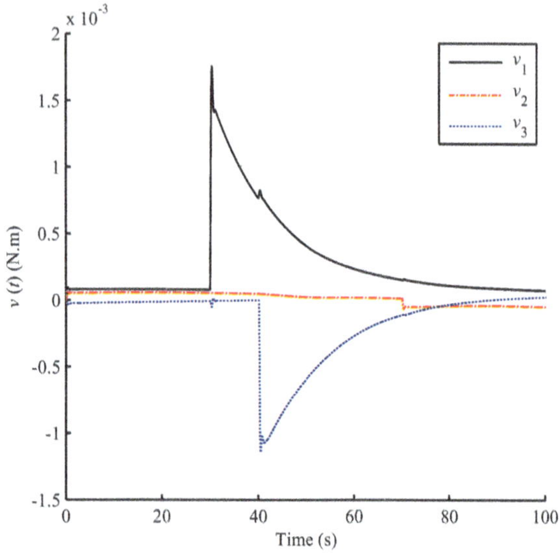

Fig. 7.2 The response of torque deviation estimation

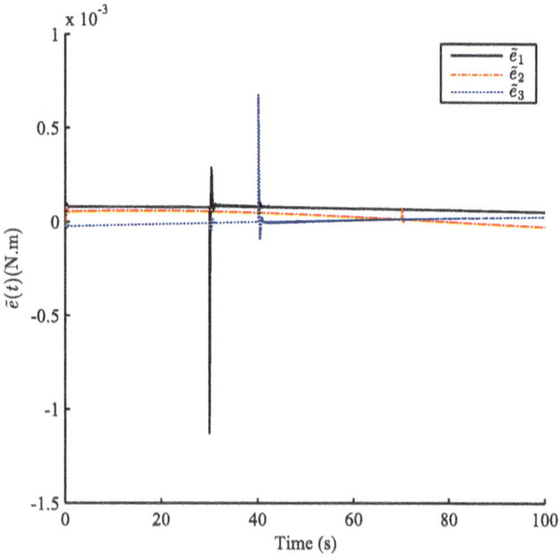

Fig. 7.3 The estimation error of torque deviation

$$e_1(t) = \begin{cases} 1 & 0 \le t < 30\,\text{s} \\ 0.5 & t \ge 30\,\text{s} \end{cases}$$

$$e_2(t) = \begin{cases} 1 & 0 \le t < 70\,\text{s} \\ 0.4 & t \ge 70\,\text{s} \end{cases}$$

$$e_3(t) = \begin{cases} 1 & 0 \le t < 40\,\text{s} \\ 0 & t \ge 40\,\text{s} \end{cases}$$

$$e_4(t) = \begin{cases} 1 & 0 \le t < 40\,\text{s} \\ 0.85 + 0.07\sin(0.2t) & t \ge 40\,\text{s} \end{cases}$$

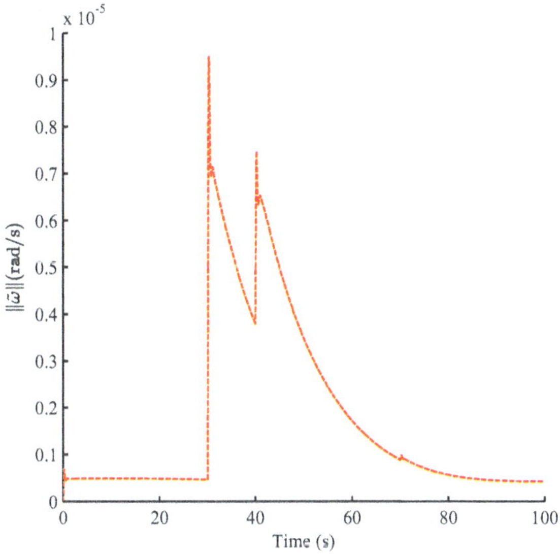

Fig. 7.4 The norm of angular velocity estimation error

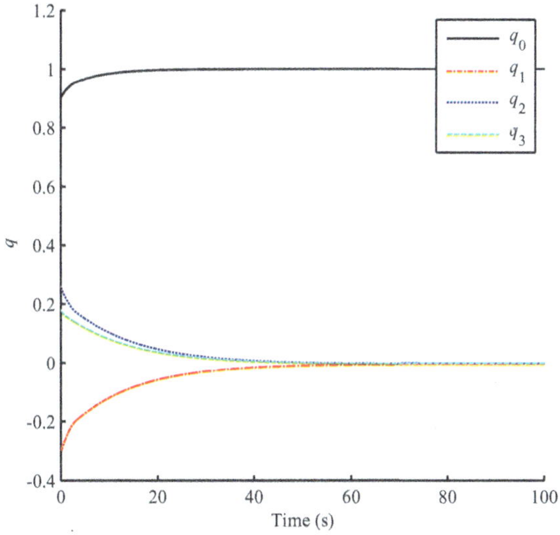

Fig. 7.5 The response of quaternion

As shown in Fig. 7.12, the observer could reconstruct the torque deviation values, even under worse conditions. The torque deviation estimation error and the angular velocity estimation error are shown in Figs. 7.13 and 7.14. Figures 7.15, 7.16, 7.17 and 7.18 show that the proposed integral-type SMC law combined with robust CA strategy guarantees that the attitudes can converge to the equilibriums. As can be seen from Figs. 7.19 and 7.20, the robust CA provides excellent performance after the faults occur, with fast response as well as high accuracy. Particularly, because the proposed robust CA takes into account of the actuator faults as well as the estimation

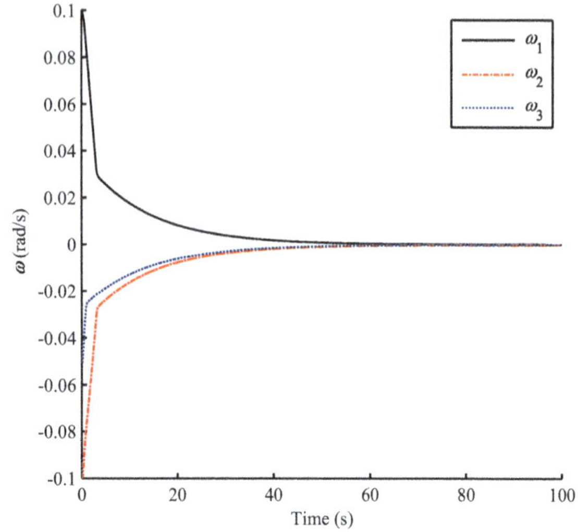

Fig. 7.6 The response of angular velocity

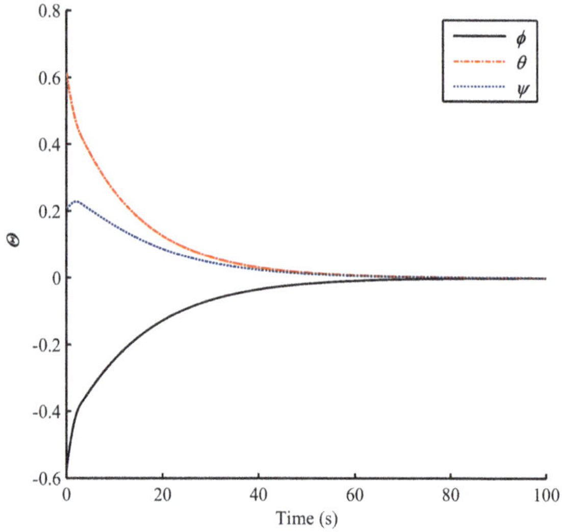

Fig. 7.7 The response of Euler angle

error, the associated control strategy can tolerate the fault when the whole system has reached stabilization. Moreover, it can be seen from Fig. 7.21 that the energy consumption of the proposed control strategy is acceptable.

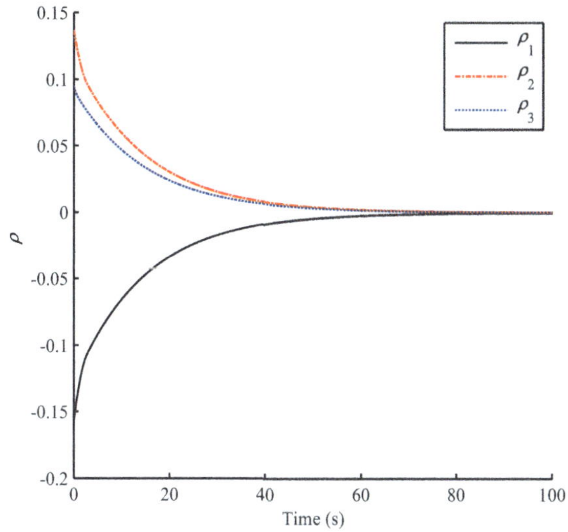

Fig. 7.8 The response of MRPs

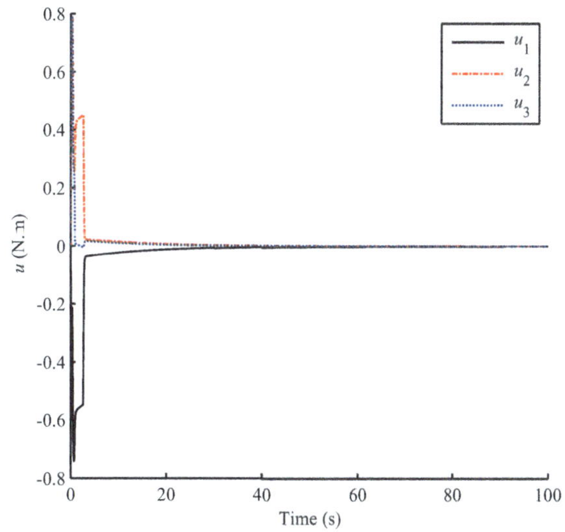

Fig. 7.9 The response of control torque

7.6 Conclusions

This chapter has addressed the attitude stabilization problem of a rigid spacecraft with actuator faults, actuator misalignment, input saturation and external disturbances as well. An iterative learning observer has been presented to estimate the torque deviation caused by faults and configuration misalignment, which ensures rapid convergence of the estimation errors as well as high precision. From both the theoretical analysis and simulation results, this observer is able to estimate and compensate

Fig. 7.10 The response of actual control torque provided by the actuators

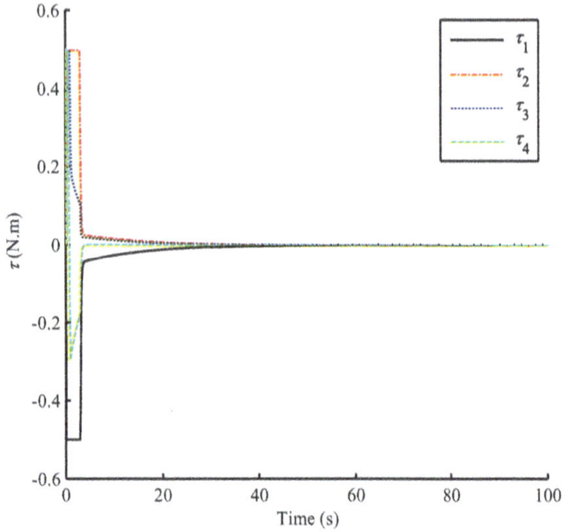

Fig. 7.11 The response of energy consumption

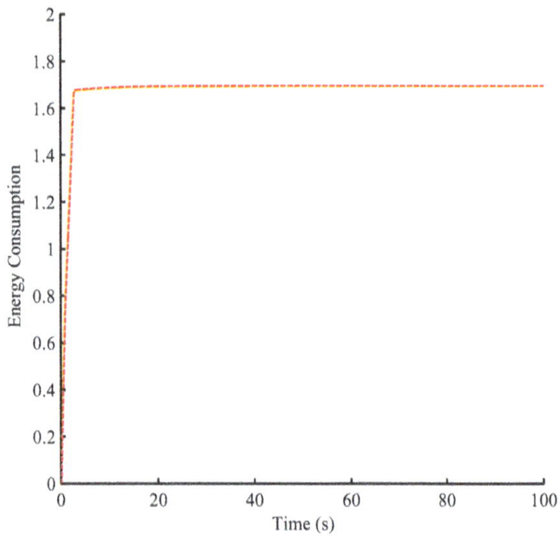

for the specified synthetic uncertainty stemming from actuator failures and model deviations accurately, even though driven by different control approaches. Using the torque deviation estimation and the designed virtual control law with integral-type sliding mode manifold, a robust CA strategy has been developed to distribute the control signals to the actuator with rigorous convergence analysis, to compensate for the probable existing observation error derived from iterative observer in real-time. The key characteristic of the proposed strategies is that the whole closed-loop fault tolerant control system can be guaranteed theoretically to be stable. Numerical

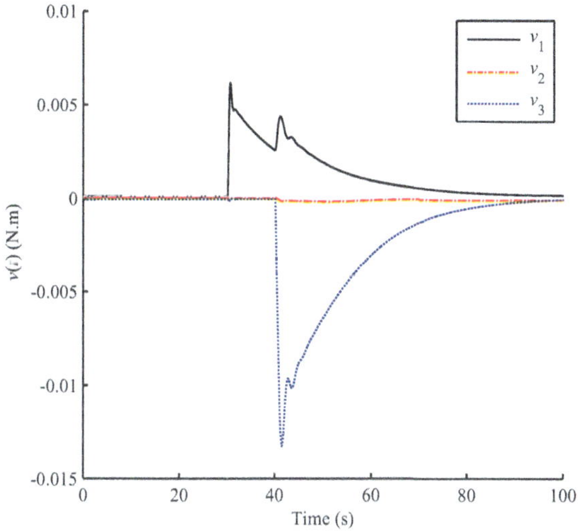

Fig. 7.12 The response of torque deviation estimation

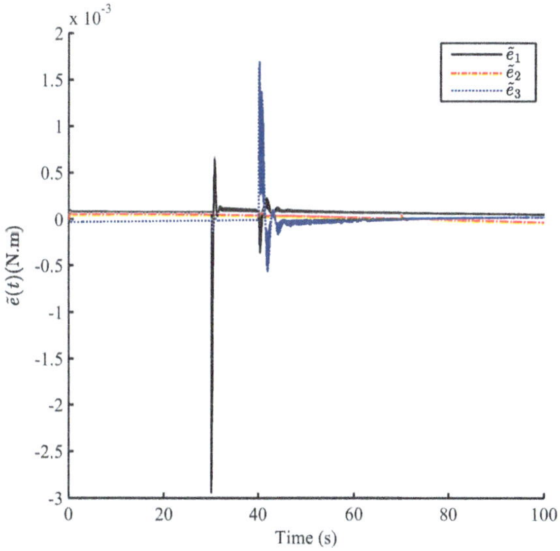

Fig. 7.13 The estimation error of torque deviation

simulations are performed to validate the fault-tolerant capability of the proposed observer and the control strategy in the presence of actuator faults as well as disturbances. However, the proposed methods only provide the so called synthetic failure and/or uncertainty of the system, but cannot isolate and identify the individual failure of the actuators. In addition, the closed-loop stability of CA under the fault diagnosis mechanism/method for the spacecraft attitude system is also an open problem, which is a significant issue for the requirements of practical engineering. In the next chapter, a closed-loop CA incorporating with a SPD control law will be presented

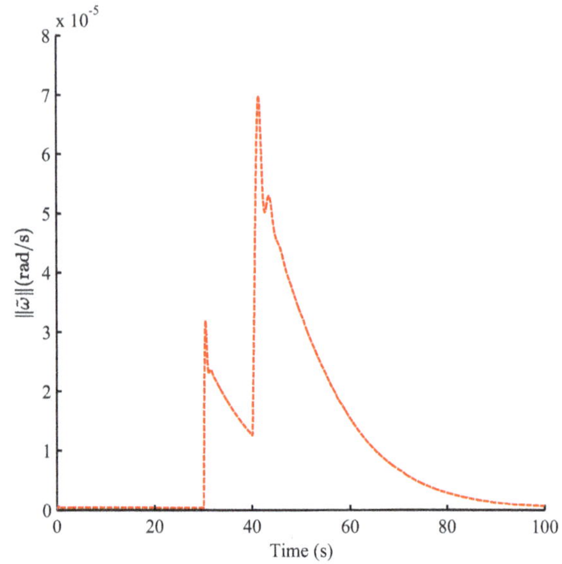

Fig. 7.14 The norm of angular velocity estimation error

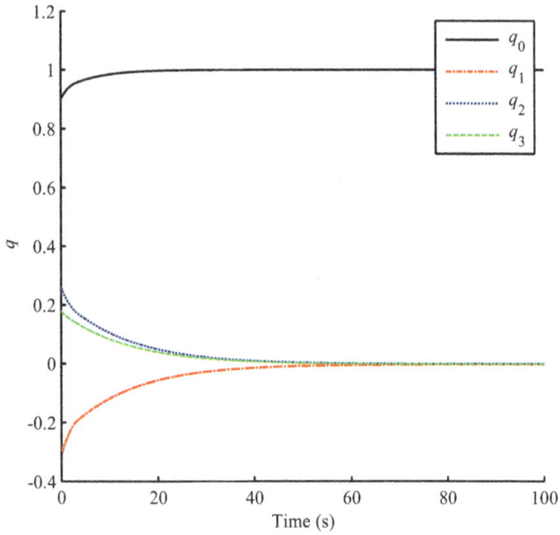

Fig. 7.15 The response of quaternion

for the spacecraft ACS, and the closed-loop CA system is proved to be asymptotic stability by using the discrete-time system theory.

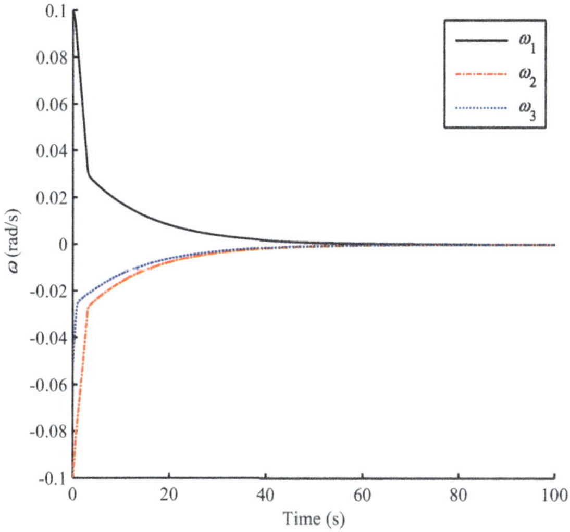

Fig. 7.16 The response of angular velocity

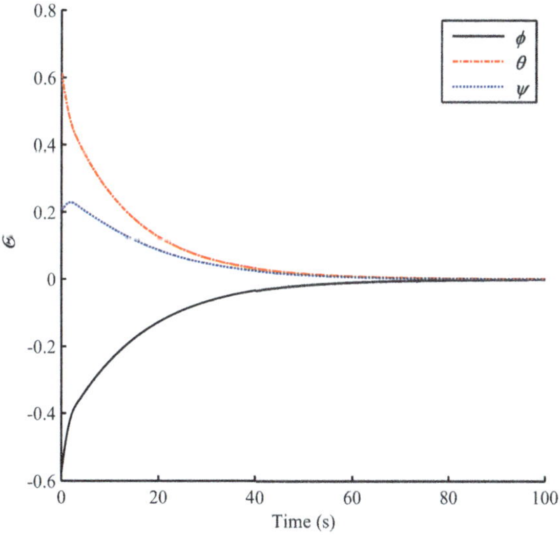

Fig. 7.17 The response of Euler angle

Appendices

Appendix 1

Proof (of Theorem 7.1) Define the estimation error of the torque deviation as

$$\tilde{e}(t) = \boldsymbol{\tau}_f(t) - \boldsymbol{v}(t) \tag{7.18}$$

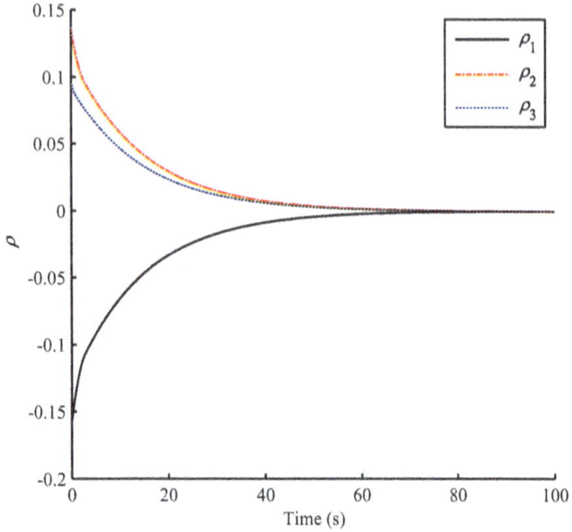

Fig. 7.18 The response of MRPs

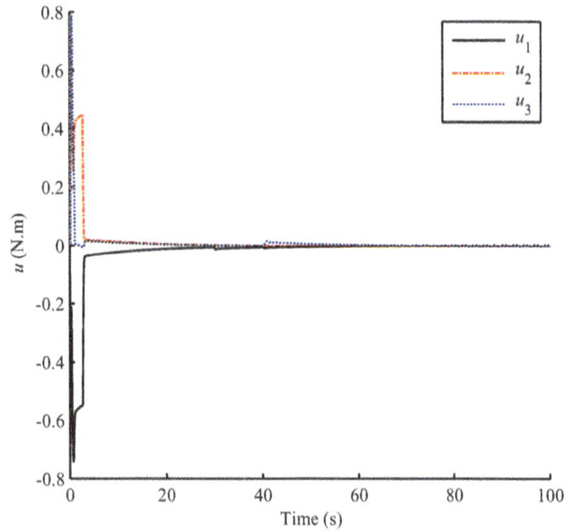

Fig. 7.19 The response of control torque

with $\boldsymbol{f}(t) = \boldsymbol{\tau}_f(t) - \boldsymbol{K}_1\boldsymbol{\tau}_f(t-T)$. Under this, the torque deviation estimation error in Eq. (7.18) can be rewritten as

$$\begin{aligned}\tilde{\boldsymbol{e}}(t) &= \boldsymbol{K}_1\boldsymbol{\tau}_f(t-T) + \boldsymbol{f}(t) - \boldsymbol{v}(t) \\ &= \boldsymbol{K}_1\tilde{\boldsymbol{e}}(t-T) - \boldsymbol{K}_2\,\mathrm{sign}[\tilde{\omega}(t)] + \boldsymbol{f}(t)\end{aligned} \tag{7.19}$$

It can be proved that

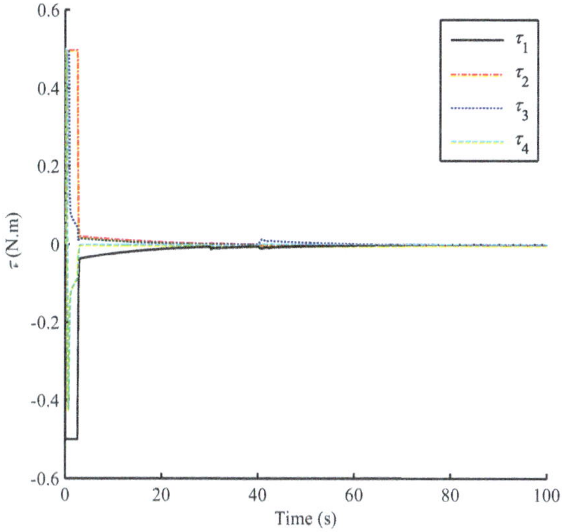

Fig. 7.20 The response of actual control torque provided by the actuators

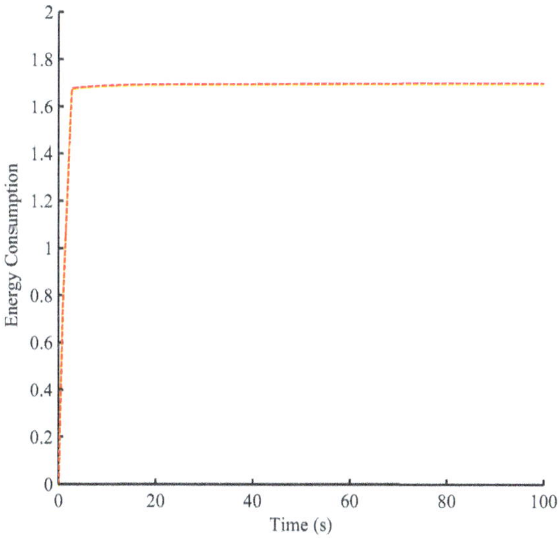

Fig. 7.21 The response of energy consumption

$$\begin{aligned}\tilde{e}^{\mathrm{T}}(t)\tilde{e}(t) =& \tilde{e}^{\mathrm{T}}(t-T)\boldsymbol{K}_1^{\mathrm{T}}\boldsymbol{K}_1\tilde{e}(t-T) - 2\tilde{e}^{\mathrm{T}}(t-T)\boldsymbol{K}_1^{\mathrm{T}}\boldsymbol{K}_2\sin[\tilde{\omega}(t)] \\ &+ 2\tilde{e}^{\mathrm{T}}(t-T)\boldsymbol{K}_1^{\mathrm{T}}\boldsymbol{f}(t) + \mathrm{sign}^{\mathrm{T}}[\tilde{\omega}(t)]\boldsymbol{K}_2^{\mathrm{T}}\boldsymbol{K}_2\,\mathrm{sign}[\tilde{\omega}(t)] \\ &- 2\,\mathrm{sign}^{\mathrm{T}}[\tilde{\omega}(t)]\boldsymbol{K}_2^{\mathrm{T}}\boldsymbol{f}(t) + \boldsymbol{f}^{\mathrm{T}}(t)\boldsymbol{f}(t)\end{aligned} \tag{7.20}$$

With the well-known Young's inequality, the following inequalities in Eq. (7.20) can be guaranteed

$$\begin{cases} -2\tilde{\boldsymbol{e}}^{\mathrm{T}}(t-T)\boldsymbol{K}_1{}^{\mathrm{T}}\boldsymbol{K}_2\mathrm{sign}[\tilde{\omega}(t)] \le \gamma_1\tilde{\boldsymbol{e}}^{\mathrm{T}}(t-T)\boldsymbol{K}_1{}^{\mathrm{T}}\boldsymbol{K}_1\tilde{\boldsymbol{e}}(t-T) \\ \qquad + \dfrac{1}{\gamma_1}\mathrm{sign}^{\mathrm{T}}[\tilde{\omega}(t)]\boldsymbol{K}_2{}^{\mathrm{T}}\boldsymbol{K}_2\mathrm{sign}[\tilde{\omega}(t)] \\ 2\tilde{\boldsymbol{e}}^{\mathrm{T}}(t-T)\boldsymbol{K}_1{}^{\mathrm{T}}\boldsymbol{f}(t) \le \gamma_2\tilde{\boldsymbol{e}}^{\mathrm{T}}(t-T)\boldsymbol{K}_1{}^{\mathrm{T}}\boldsymbol{K}_1\tilde{\boldsymbol{e}}(t-T) + \dfrac{1}{\gamma_2}\boldsymbol{f}^{\mathrm{T}}(t)\boldsymbol{f}(t) \\ -2\mathrm{sign}^{\mathrm{T}}[\tilde{\omega}(t)]\boldsymbol{K}_2{}^{\mathrm{T}}\boldsymbol{f}(t) \le \gamma_3\mathrm{sign}^{\mathrm{T}}[\tilde{\omega}(t)]\boldsymbol{K}_2{}^{\mathrm{T}}\boldsymbol{K}_2\mathrm{sign}[\tilde{\omega}(t)] + \gamma_3\boldsymbol{f}^{\mathrm{T}}(t)\boldsymbol{f}(t) \end{cases}$$

Substituting the above inequalities into Eq. (7.20) yields

$$\begin{aligned} \tilde{\boldsymbol{e}}^{\mathrm{T}}(t)\tilde{\boldsymbol{e}}(t) \le & \alpha_1\tilde{\boldsymbol{e}}^{\mathrm{T}}(t-T)\boldsymbol{K}_1{}^{\mathrm{T}}\boldsymbol{K}_1\tilde{\boldsymbol{e}}(t-T) + \alpha_3 f^{\mathrm{T}}(t)f(t) \\ & + \alpha_2\mathrm{sign}^{\mathrm{T}}[\tilde{\omega}(t)]\boldsymbol{K}_2{}^{\mathrm{T}}\boldsymbol{K}_2\mathrm{sign}[\tilde{\omega}(t)] \end{aligned} \tag{7.21}$$

where $\alpha_1 = 1 + \gamma_1 + 1/\gamma_2$, $\alpha_2 = 1 + 1/\gamma_1$ and $\alpha_3 = 1 + \gamma_2 + \gamma_3$ are positive constants.

On the other hand, the estimation error of the angular velocity $\tilde{\omega}(t) = \omega(t) - \hat{\omega}(t)$ can be given as

$$\boldsymbol{J}\dot{\tilde{\omega}}(t) = -\hat{\omega}(t) \times \boldsymbol{J}\hat{\omega}(t) + \omega(t) \times \boldsymbol{J}\omega(t) - \lambda\tilde{\boldsymbol{e}}(t) - \boldsymbol{L}\mathrm{sign}\left[\tilde{\omega}(t)\right] + \boldsymbol{d} \tag{7.22}$$

Select the Lyapunov candidate as follows

$$V_1 = \frac{1}{2}\tilde{\omega}\boldsymbol{J}\tilde{\omega} + \int_{t-T}^{\mathrm{T}} \tilde{\boldsymbol{e}}^{\mathrm{T}}(r)\tilde{\boldsymbol{e}}(r)dr \tag{7.23}$$

Taking the derivative of V_1 respect to time and substituting Eq. (7.22), one has

$$\begin{aligned} \dot{V}_1 =& \tilde{\omega}[-\omega \times \boldsymbol{J}\omega + \hat{\omega} \times \boldsymbol{J}\hat{\omega} - \lambda\tilde{\omega} + \tilde{\boldsymbol{e}}(t) - \boldsymbol{L}\,\mathrm{sign}(\tilde{\omega}) + \boldsymbol{d}] \\ & + \tilde{\boldsymbol{e}}^{\mathrm{T}}(t)\tilde{\boldsymbol{e}}(t) - \tilde{\boldsymbol{e}}^{\mathrm{T}}(t-T)\tilde{\boldsymbol{e}}(t-T) \\ \le & (\kappa - \lambda + \gamma_4)\left\|\tilde{\omega}\right\|^2 + \left(1 + \frac{1}{\gamma_4}\right)\tilde{\boldsymbol{e}}^{\mathrm{T}}(t)\tilde{\boldsymbol{e}}(t) \\ & - \tilde{\boldsymbol{e}}^{I}(t-T)\tilde{\boldsymbol{e}}(t-T) - (L_{\min} - d_{\max})\left\|\tilde{\omega}\right\| \end{aligned} \tag{7.24}$$

If $L_{\min}$ is selected sufficiently large such that $L_{\min} \ge d_{\max}$, then the following inequality can be achieved

$$\begin{aligned} \dot{V}_1 \le & -(-\kappa + \lambda - \gamma_4)\|\tilde{\omega}\|^2 - \delta\|\tilde{\boldsymbol{e}}(t)\|^2 \\ & - [1 - (1 + 1/\gamma_4 + \delta)\,\alpha_1\|\boldsymbol{K}_1\|^2]\|\tilde{\boldsymbol{e}}(t-T)\|^2 \\ & + (1 + 1/\gamma_4 + \delta)\,[\alpha_3\boldsymbol{f}^{\mathrm{T}}(t)\boldsymbol{f}(t) + \alpha_2\boldsymbol{K}_2{}^{\mathrm{T}}\boldsymbol{K}_2] \end{aligned} \tag{7.25}$$

where $\sigma = (1 + 1/\gamma_4 + \delta)\left\{\alpha_3 k_d^2 + \alpha_2\left\|\boldsymbol{K}_2\right\|^2\right\}$.

Letting $1 - (1 + 1/\gamma_4 + \delta)\,\alpha_1\left\|\boldsymbol{K}_1\right\|^2 \ge 0$, Eq. (7.25) can be further simplified as

$$\dot{V}_1 \leq -(-\kappa + \lambda - \gamma_4) \|\tilde{\boldsymbol{\omega}}\|^2 - \delta \|\tilde{\boldsymbol{e}}(t)\|^2 + \sigma \tag{7.26}$$

It is obvious that if $\|\tilde{\boldsymbol{\omega}}\| > \sqrt{\frac{\sigma}{-\kappa+\alpha-\gamma_4}}$ or $\|\tilde{\boldsymbol{e}}(t)\| > \sqrt{\frac{\sigma}{\delta}}$, then $\dot{V}_1 < 0$ and thus V_1 decreases. Employing Theorem 4.18 in [24], the estimated errors are ultimately uniformly bounded by

$$\|\tilde{\boldsymbol{\omega}}\| \leq \sqrt{\frac{\sigma}{-\kappa + \alpha - \gamma_4}}, \ \|\tilde{\boldsymbol{e}}(t)\| \leq \sqrt{\frac{\sigma}{\delta}} \tag{7.27}$$

The above residual sets can be made arbitrarily small along with the increasing values of $-\kappa + \alpha - \gamma_4$ and δ.

Appendix 2

Proof (of Theorem 7.2) Taking the derivative of the sliding manifold in Eq. (7.6) yields

$$\dot{\boldsymbol{S}} = \boldsymbol{J}\dot{\boldsymbol{\omega}} + \boldsymbol{\omega} \times \boldsymbol{J}\boldsymbol{\omega} - \boldsymbol{\tau}_{nom} \tag{7.28}$$

and substituting Eq. (7.10) into Eq. (7.28), it can be obtained that

$$\dot{\boldsymbol{S}} = \boldsymbol{\tau} + \boldsymbol{d} - \boldsymbol{\tau}_{nom} \tag{7.29}$$

Define

$$V_2 = \frac{1}{2}\boldsymbol{S}^{\mathrm{T}}\boldsymbol{S} \tag{7.30}$$

Then, we have

$$\begin{aligned} \dot{V}_2 &= \boldsymbol{S}^{\mathrm{T}}\dot{\boldsymbol{S}} \\ &= \boldsymbol{S}^{\mathrm{T}}(\boldsymbol{\tau} + \boldsymbol{d} - \boldsymbol{\tau}_{nom}) \\ &\leq (-\beta + d_{\max})\|\boldsymbol{S}\| \\ &\leq 0 \quad if\, \boldsymbol{S} \neq \boldsymbol{0} \end{aligned} \tag{7.31}$$

from which it can be shown that $\boldsymbol{S} = \boldsymbol{0}$ after a finite time instant, i.e., the sliding motion can be obtained.

Then, we analyze the sliding motion on the sliding manifold. Define

$$V_3 = \frac{1}{2}\boldsymbol{\omega}^{\mathrm{T}}\boldsymbol{J}\boldsymbol{\omega} + 2k_p(1 - q_0) \tag{7.32}$$

Taking the time derivative of Eq. (7.32), substituting Eq. (7.10) and considering the fact that $\boldsymbol{\omega}^{\mathrm{T}} \tanh\left(\boldsymbol{\omega}(t)/k^2(t)\right) \geq 0$, it follows that

$$\begin{aligned}\dot{V}_3 =&\boldsymbol{\omega}^{\mathrm{T}} \boldsymbol{J}\dot{\boldsymbol{\omega}} - 2k_p\dot{q}_0 \\ =&\boldsymbol{\omega}^{\mathrm{T}}\left(-\boldsymbol{S}(\boldsymbol{\omega})\boldsymbol{J}\boldsymbol{\omega} + \boldsymbol{d} + \boldsymbol{\tau}_{nom}\right) + k_p\boldsymbol{q}_v^{\mathrm{T}}\boldsymbol{\omega} \\ \le&\left(-k_v + d_{\max}\right)\|\boldsymbol{\omega}\|\end{aligned} \tag{7.33}$$

From Eq. (7.33), it can be obtained that $\dot{V}_3 \le 0$. Subsequently, q_0, $\boldsymbol{q}_v$ and $\boldsymbol{\omega}$ are globally bounded, and it is obvious that $\ddot{V}_3$ is bounded. In view of Barbalat's Lemma, it is clear that $\lim_{t\to\infty}\boldsymbol{\omega}(t) = 0$. From Eqs. (7.10) and (7.28), taking account of $\boldsymbol{S} = \boldsymbol{0}$, it could be obtained that

$$\begin{aligned}\boldsymbol{J}\ddot{\boldsymbol{\omega}} &= -\boldsymbol{S}(\boldsymbol{\omega})\boldsymbol{J}\boldsymbol{\omega} - S(\boldsymbol{\omega})\boldsymbol{J}\dot{\boldsymbol{\omega}} - k_v\frac{d}{dt}\tanh\left(\boldsymbol{\omega}/k^2(t)\right) - k_p\dot{\boldsymbol{q}}_v \\ &= -\boldsymbol{S}(\dot{\boldsymbol{\omega}})\boldsymbol{J}\boldsymbol{\omega} - S(\boldsymbol{\omega})\boldsymbol{J}\boldsymbol{\omega} - k_v\,\mathrm{sech}^2\left(\boldsymbol{\omega}/k^2(t)\right)\frac{\dot{\boldsymbol{\omega}}k(t) - 2\boldsymbol{\omega}\dot{k}(t)}{k^3(t)}\end{aligned}$$

From the above equation, it can be shown that $\ddot{\boldsymbol{\omega}}$ is bounded. Then, employing Barbalat's Lemma again, together with $\lim_{t\to\infty}\boldsymbol{\omega}(t) = 0$ and uniform continuity of $\dot{\boldsymbol{\omega}}$, it will lead to $\lim_{t\to\infty}\dot{\boldsymbol{\omega}}(t) = 0$. From Eq. (7.33), we have $\lim_{t\to\infty}\boldsymbol{\tau}_{nom}(t) = \boldsymbol{0}$ Therefore, from Eq. (7.8), it implies that $\lim_{t\to\infty}\boldsymbol{q}_v(t) = 0$.

References

1. Patton, R.J., Uppal, F.J., Simani, S., Polle, B. (2010) Robust fdi applied to thruster faults of a satellite system. Control Engineering Practice 18(9): 1093–1109
2. Gao, C., Duan, G. (2014) Fault diagnosis and fault tolerant control for nonlinear satellite attitude control systems. Aerospace Science and Technology 33(1): 9–15
3. Wu, Q., Saif, M. (2006) Robust fault diagnosis for a satellite large angle attitude system using an iterative neuron pid (inpid) observer. In: Proceedings of American Control Conference, Minneapolis, MN, USA, pp 5710–5715
4. Xiao, B., Hu, Q.L., Wang, D.W. (2013) Spacecraft attitude fault tolerant control with terminal sliding-mode observer. Journal of Aerospace Engineering 28(1): 1–15
5. Chen, W., Saif, M. (2007) Observer-based fault diagnosis of satellite systems subject to time-varying thruster faults. Journal of dynamic systems, measurement, and control 129(3): 352–356
6. Buffington, J.M., Enns, D.F. (1996) Lyapunov stability analysis of daisy chain control allocation. Journal of Guidance, Control, and Dynamics 19(6): 1226–1230
7. Buffington, J.M., Enns, D.F., Teel, A.R. (1998) Control allocation and zero dynamics. Journal of Guidance, Control, and Dynamics 21(3): 458–464
8. Härkegård, O., Glad, S.T. (2005) Resolving actuator redundancy–optimal control vs. control allocation. Automatica 41(1): 137–144
9. Cui, L., Yang, Y. (2011) Disturbance rejection and robust least-squares control allocation in flight control system. Journal of Guidance, Control, and Dynamics 34(6): 1632–1643
10. Shen, Q., Wang, D., Zhu, S., Poh, E.K. (2015) Inertia-free fault-tolerant spacecraft attitude tracking using control allocation. Automatica 62: 114–121
11. Shen, Q., Wang, D., Zhu, S., Poh, E.K. (2017) Robust control allocation for spacecraft attitude tracking under actuator faults. IEEE Transactions on Control Systems Technology 25(3): 1068–1075

12. De Almeida, F.A. (2016) Robust off-line control allocation. Aerospace Science and Technology 52: 1–9
13. Murugesan, S., Goel, P.S. (1987) Fault-tolerant spacecraft attitude control system. Sadhana-Academy Proceeding in Engineering Sciences 11(12): 232–261

Chapter 8
Nonlinear Proportional-Derivative Control Incorporating Closed-Loop Control Allocation for Spacecraft

8.1 Introduction

As is well-known, accurate and reliable attitude stabilization is one of the most important issues and widely studied applications in the advanced spacecraft ACS design. For orbiting spacecraft, dynamics are strongly nonlinear in nature and are also affected by various external disturbances from the environment that influence the mission objectives significantly. In addition, actuator uncertainties induced by misalignment during installation and torque magnitude measurement errors further increase the design complexity and difficulty. In the practical aerospace engineering, the actuator saturation issue which is derived from the physical limitation on the actuator, should be also taken into consideration during the control law design. It may result in severe discrepancies or derivations between the desired control signals and the actual control output produced by the actuators, and the discrepancies or saturation phenomenon may reduce the control system performances and even cause the instability of the whole control system if it could be well handled and solved. All of the above mentioned problems cause considerable difficulties and challenges in the spacecraft ACS design, for meeting high-precision pointing requirements and desired control performance during the aerospace missions. In the last decades, there exist many results about the spacecraft ACS, such as nonlinear feedback control, robust control, adaptive control and so on [1–6]. Generally speaking, most or part of the previous works design by some advanced control techniques can handle external disturbances and uncertainties to a certain degree, but can hardly be applied to the practical engineering application. In view of its simplicity and good performance for practical applications, the PD or PD-like controls have been extensively studied and applied as well [7–9]. But, the actuator saturation problem can't be considered in the above results. To solve this practical problem, some interesting methods and results were also presented by taken the actuators' output magnitude constraints into account, such as an inverse tangent-based tracking function and backstepping method based nonlinear control law [10], explicit saturation function based robust attitude

Q. Hu et al., *Control Allocation for Spacecraft Under Actuator Faults*,
https://doi.org/10.1007/978-981-16-0439-3_8

controller [11], standard hypertangent function based PD control scheme [12] and so on.

CA is one natural and necessary solution for achieving a desired control objective by distributing appropriately the lower dimension of the total control signals synthesized by the baseline controller to each individual actuator among the higher-dimension redundant actuators. According to the previous chapters in this book, one can discover that most of the existing CA schemes are studied and worked as an open-loop system in some sense, and the stability of the closed-loop system under combination of CA with the baseline controller is not taken into consideration [13, 14], due to the inherent difficulty for such a design task. There are few research works available in this relative area. As some early attempts along this line, a sufficient asymptotic stability condition for the zero dynamics of systems with daisy chain CA was given initially in [15] and for more common CA methods presented in [16] by the same authors. In [14], a closed-loop CA based on the cascaded generalized inverse method is established and a necessary sufficient condition for the stability of the satellite system is presented. The authors will attempt solve this problem by utilizing the discrete-time system theory in this chapter.

In this chapter, the SPD control law will be design for achieving the desired spacecraft attitude stabilization mission under input saturation, and a closed-loop CA will be proposed for dealing with the problem of distributing the total virtual control command synthesized by the SPD controller into the individual actuator properly by using constrained optimal quadratic programming algorithms. During the design of the closed-loop CA, the actuators' constraints such as position and rate constraints are taken into account. It should be noted that the significant achievement of this chapter is that the closed-loop system driven by the designed SPD baseline controller and the CA scheme can be proved and ensured to be asymptotically stable. The key features or contributions of the designed control scheme are that it ensures (1) asymptotical stability of the closed-loop system; (2) convergence of both attitude and bounded control torques; and (3) simple design procedure and structure of the proposed scheme. The feature of simple design and structure is of great interest for aerospace industry with real-time implementation when onboard computing power is limited. The benefits of the proposed control method are analytically authenticated and also validated via simulation studies.

(*Remark: The main control schemes and contents of this chapter are from the published journal paper "Qinglei Hu, Bo Li and Youmin Zhang, Nonlinear proportional-derivative control incorporating closed-loop control allocation for spacecraft, Journal of Guidance, Control, and Dynamics, 2014, 37(3): 799–812". The authors appreciate the permission from the American Institute of Aeronautics and Astronautics to reuse partly the results published in the relevant Journal.*)

8.2 Problem Formulation

Consider the spacecraft ACS given by Eqs. (2.17) and Eqs.(2.20) without external disturbances under the proceeding reasonable assumptions. Design a control input $\boldsymbol{u}(t) \in \boldsymbol{\Omega}_{\mathrm{u}}$ satisfying some specified constraints such that, for all physically realizable initial conditions, the states of the closed-loop system are globally asymptotically stable, which can be expressed as follows:

$$\lim_{t\to\infty} \rho(t) = 0, \ \lim_{t\to\infty} \omega(t) = 0$$

Assumption 8.1 *Due to physical limitations on the actuators, for simplicity, it is assumed that actuator output torques have the same constraint value* τ_{max}, i.e.

$$\boldsymbol{\tau}(t) \in \boldsymbol{\Omega}_{\emptyset} := \left\{\boldsymbol{\tau} \in \mathbf{R}^{m} \,||\tau_{\mathrm{i}}| \leq \tau_{\max},\ i = 1,\ 2,\ \ldots,\ m\right\} \tag{8.1}$$

where m is the number of actuators considered. That is to say, the virtual control demand designed from the control law ought to be subject to the following constraint:

$$\boldsymbol{u}(t) \in \boldsymbol{\Omega}_{\mathrm{u}} := \left\{\boldsymbol{u} \in \mathbf{R}^{n} \,||u_{\mathrm{i}}| \leq u_{\max}, i = 1,\ 2,\ \ldots,\ n\right\} \tag{8.2}$$

where $u_{\max}$ is a common maximum magnitude of the control signals. where n $(n < m)$ is the number of control inputs/signals to be sent to actuators and typically $n = 3$ as defined in this chapter, whereas the over-actuated actuators configuration normally used for spacecraft and considered in this chapter is $m = 4$, as will be illustrated later. The maximum magnitude of the control signals $u_{\max}$ can be determined by the physical saturation limitations on available over-actuated actuators. Note that, in the case that $n = m$, there will be no redundant actuators and $u_{\max}$ will be the same as $\tau_{\max}$, and then the CA module in the control system will become not necessary, and the spacecraft attitude control problem will then become a conventional control system design with only baseline controller needs to be designed.

8.3 Attitude Control Law Design

In view of exploiting available actuator redundancy for effective ACS design, attitude control law design in this chapter will use the principle of baseline controller and CA framework with the following two design steps [16]: (1) design a control law specifying which total control effort to be produced (net torques, etc.), known as virtual control law or baseline control law; and (2) design a CA scheme (or control allocator) that maps the total control demand determined by the baseline control law onto individual actuator settings (such as RW torques, etc.).

8.3.1 SPD-based Virtual/Baseline Control Law Design

An equivalent representation of Eq. (2.20) can be written as:

$$\boldsymbol{J}\dot{\boldsymbol{\omega}}(t) = -\boldsymbol{S}(\boldsymbol{\omega}(t))\boldsymbol{J}\boldsymbol{\omega}(t) + \boldsymbol{u}(t) \tag{8.3a}$$

$$\boldsymbol{u}(t) = \boldsymbol{D}\boldsymbol{\tau}(t) \tag{8.3b}$$

where $\boldsymbol{D} \in \mathbf{R}^{3\times m}$ is used to describe distribution of the physical actuators for representing the influence of each actuator on the angular acceleration of the spacecraft, and the full row rank of $\boldsymbol{D}$ is required to guarantee for the controllable of the over-actuated attitude system in normal case.

In what follows, such an attitude control law for spacecraft as given in Eqs. (2.17) and (8.1) will be developed to achieve the control objective given in latter section. For the following controller design, it is assumed that these states $\boldsymbol{\rho}(t)$ and $\boldsymbol{\omega}(t)$ can be obtained accurately from some measuring elements fixed on the orbiting spacecraft.

Motivated by the work given in [12], the first main result of this paper is summarized and presented as following theorem.

Theorem 8.1 *Given the spacecraft attitude stabilization control system modeled by Eqs. (2.17) and* (8.1), *there exist controller gains and to ensure the system is globally asymptotically stable under the following SPD control law:*

$$\boldsymbol{u}(t) = -(1-\beta)u_{\max}\mathrm{Tanh}(\boldsymbol{\omega}(t)/\alpha^2(t)) - \beta u_{\max}\frac{\boldsymbol{\rho}}{\sqrt{1+\boldsymbol{\rho}^{\mathrm{T}}\boldsymbol{\rho}}} \tag{8.4}$$

where β is a constant control gain with $\beta \in [0,\ 1]$, $\alpha^2(t)$ is a strictly bounded above-zero scalar sharpness function[17] that governs the magnitude of the control rates, and $0 < \alpha_{\min}^2 \le \alpha^2(t) \in L_\infty$, $\dot{\alpha} \in L_\infty$. Note that the control input saturation limit is represented by a proper value of u_{max} and the virtual control torques are subject to the input constraints in Eq. (8.2). *In addition, the vector* $\mathrm{Tanh}(\boldsymbol{\omega}(t)) \in \mathbf{R}^3$ *is defined as*

$$\mathrm{Tanh}(\boldsymbol{\omega}(t)) \triangleq \left[\tanh(\omega_1(t))\ \tanh(\omega_2(t))\ \tanh(\omega_R^3(t))\right]^{\mathrm{T}},$$

where $\mathrm{Tanh}(\cdot)$ *denotes the standard hyperbolic tangent function.*

Proof Considering the following Lyapunov function candidate:

$$V = \frac{1}{2}\boldsymbol{\omega}^{\mathrm{T}}\boldsymbol{J}\boldsymbol{\omega} + 4\beta u_{\max}\left(1 - \frac{1}{\sqrt{1+\boldsymbol{\rho}^{\mathrm{T}}\boldsymbol{\rho}}}\right) \tag{8.5}$$

where t is dropped from $\boldsymbol{\omega}(t)$ and $\alpha^2(t)$ for brevity in the derivation below and

$$1 - \frac{1}{\sqrt{1+\boldsymbol{\rho}^{\mathrm{T}}\boldsymbol{\rho}}} \ge 0$$

is obviously guaranteed.

Differentiation of V with respect to time with Eq. (2.20) and the control law Eq. (8.4) yields

$$\begin{aligned}\dot{V} &= \boldsymbol{\omega}^{\mathrm{T}} \boldsymbol{J} \dot{\boldsymbol{\omega}} + 4\beta u_{\max} \left(\frac{\boldsymbol{\rho}^{\mathrm{T}} \dot{\boldsymbol{\rho}}}{(1+\boldsymbol{\rho}^{\mathrm{T}}\boldsymbol{\rho})\sqrt{1+\boldsymbol{\rho}^{\mathrm{T}}\boldsymbol{\rho}}} \right) \\ &= \boldsymbol{\omega}^{\mathrm{T}} \boldsymbol{u} + \beta u_{\max} \boldsymbol{\omega}^{\mathrm{T}} \frac{\boldsymbol{\rho}}{\sqrt{1+\boldsymbol{\rho}^{\mathrm{T}}\boldsymbol{\rho}}} \\ &= -(1-\beta) u_{\max} \boldsymbol{\omega}^{\mathrm{T}} \mathrm{Tanh}(\boldsymbol{\omega}/\alpha^2)\end{aligned}$$

Because $\boldsymbol{\omega}^{\mathrm{T}} \mathrm{Tanh}(\boldsymbol{\omega}/\alpha^2) \geq 0$ is always guaranteed for all ω from the property of the standard hyperbolic tangent function, one can obtain

$$\dot{V} = -(1-\beta) u_{\max} \boldsymbol{\omega}^{\mathrm{T}} \mathrm{Tanh}(\boldsymbol{\omega}/\alpha^2) \leq 0 \tag{8.6}$$

which implies that V and $\boldsymbol{\omega}$ are bounded for all time.

Integrating both sides of Eq. (8.6) with respect to time, one can obtain

$$V_{\infty} - V_0 = -(1-\beta) u_{\max} \int_0^{\infty} \boldsymbol{\omega}^{\mathrm{T}} \mathrm{Tanh}(\boldsymbol{\omega}/\alpha^2) \mathrm{d}t \leq 0$$

where the symbols

$$V_{\infty} = \lim_{t \to \infty} V(t) \text{ and } V_0 = V(0)$$

are utilized for brevity. Because the term $\boldsymbol{\omega}^{\mathrm{T}} \mathrm{Tanh}(\boldsymbol{\omega}/\alpha^2) \geq 0$ is uniformly continuous and

$$\int_0^{\infty} \boldsymbol{\omega}^{\mathrm{T}} \mathrm{Tanh}(\boldsymbol{\omega}/\alpha^2) \mathrm{d}t$$

is finite, Barbalat's lemma dictates that ω converges to the origin, that is,

$$\lim_{t \to \infty} \omega(t) = 0.$$

Since $\dot{V} \equiv 0$ means that $\boldsymbol{\omega} \equiv 0$, from the closed-loop system formed by Eqs. (2.17), one has $\boldsymbol{\rho} \equiv 0$. Utilizing LaSalle's invariance principle [18], one can obtain that the closed-loop attitude system of the rigid spacecraft is globally asymptotically stable. □

Remark 8.1 Using the property of the standard hyperbolic tangent function and the fact that

$$\left| \frac{x_{\mathrm{i}}}{\sqrt{1+\boldsymbol{x}^{\mathrm{T}}\boldsymbol{x}}} \right| \leq 1, \ i = 1, \ 2, \ 3$$

where $\boldsymbol{x}$ represents $\boldsymbol{\rho}$ in the SPD control law in Eq. (8.4), along with the chosen bounds on the parameter β for ensuring that the SPD control law satisfies the saturation constraints in Eq. (8.2). Therefore, a name of SPD nonlinear control law in view of both the attitude P and velocity D variables and following explicit saturation constraints S are all considered in such a nonlinear control law formulation:

$$\begin{aligned} |u_{\mathrm{i}}| &\leq \left|(1-\beta)u_{\max}\mathrm{Tanh}(\omega_{\mathrm{i}}/\alpha^2)\right| + \left|\beta u_{\max}\frac{\rho_{\mathrm{i}}}{\sqrt{1+\boldsymbol{\rho}^{\mathrm{T}}\boldsymbol{\rho}}}\right| \\ &\leq |(1-\beta)u_{\max}| + |\beta u_{\max}| \leq u_{\max} \end{aligned} \tag{8.7}$$

Then, the control input saturation avoidance can be guaranteed by selecting a proper value of $u_{\max}$ which is determined by the physical limitations on each available actuator.

8.3.2 A Closed-Loop Control Allocation Design with Optimization Method

From the above analysis, even if the attitude stabilization target can be achieved by virtual control law Eq. (8.4), other practical problems, such as energy saving (minimizing energy consumption), practical control capability constraints due to actuator amplitude and rate saturation limits, etc., should be taken into account for simultaneous control system design. CA is necessary and useful for the control of over-actuated systems, including spacecraft systems. It deals with distributing the desired total control command determined by the virtual control law equivalently to the individual over-actuated actuators. Generally, the CA problem is formulated as an optimization problem systematically handling: redundant sets of actuators, actuator constraints and power consumption minimization, wear/tear and other undesirable effects. Since it is a practical and powerful approach for solving control problems for systems with over-actuated actuators, there are extensive literatures on CA which discuss different algorithms, approaches and applications. Page and Steinberg in [13] compare the closed- and open-loop performance of 16 different CA algorithms. Bodson provides a comprehensive evaluation of different constrained numerical optimization methods for control allocation in [19]. Using CA, the actuator selection task is separated from the regulation (baseline control design) task in the overall control system design. A common sense in the early stage of study for CA combining baseline/virtual control assumes that there is no effect of the stability of overall closed-loop system based on such a separated design for baseline controller and control allocator since the closed-loop system's stability is mainly guaranteed by the baseline controller[20, 21]. Such a common sense, in fact, was based on the assumption that the control allocation scheme does not affect system stability, which also implied the inherent stability of the control allocator. Such an assumption is in fact not valid and must be approved in the similar way as stability issue with baseline controller. This is one motivation of the work to be presented in this paper for designing a control system with an integrated baseline control and control allocation scheme with guaranteed overall closed-loop system stability. For such a purpose and as a further extension to several existing works which consider the closed-loop system stability analysis in [13–15, 22], in which the stability/stabilization issue of the overall closed-loop system under the combination of CA with the baseline controller is still an unsolved or at least not well solved problem, a new dynamic CA scheme using a constrained optimal quadratic programming is developed in this study, in an effort to provide an overall closed-loop system stability guarantee with the combined baseline control law designed in Eq. (8.4) based on the SPD nonlinear control. The idea of the system stability guarantee for entire/overall closed-loop system is to be achieved (1) with the stability guarantee of the closed-loop system without consideration of over-actuated actuators and, therefore, the preceding SPD controller is proposed in this work, and (2) the stability guarantee of the CA part is achieved by the proposed closed-loop CA scheme to be presented in this section. The developed CA scheme and proof of the overall closed-loop system stability guarantee will be presented in the sequel.

Considering the actuator amplitude and rate constraints, and defining the following compact set for each RW as:

$$\boldsymbol{\Omega}_1 = \{\tau_i \in \mathbf{R}^m | \tau_{\min} \le \tau_i \le \tau_{\max}, i = 1, 2, 3, \ldots, m\} \tag{8.8a}$$

$$\boldsymbol{\Omega}_2 = \{\dot{\tau}_i \in \mathbf{R}^m | \eta_{\min} \le \dot{\tau}_i \le \eta_{\max}, i = 1, 2, 3, \ldots, m\} \tag{8.8b}$$

where $\tau_{\min}$ and $\tau_{\max}$, are the lower and upper actuator amplitude limitations, $\eta_{\min}$ and $\eta_{\max}$ denote the minimal and maximal actuator rate limitations, respectively.

Since the control allocator is part of a digital control system, it is reasonable to approximate the time derivative as:

$$\dot{\boldsymbol{\tau}}(t) \approx \frac{\boldsymbol{\tau}(t) - \boldsymbol{\tau}(t-T)}{T} \tag{8.9}$$

where T is the sampling time, and $\boldsymbol{\tau}(t-T)$ is the control torques in the previous sampling instant. Then one can get the overall actuator constraints as:

$$\underline{\boldsymbol{\tau}}(t) \le \boldsymbol{\tau}(t) \le \bar{\boldsymbol{\tau}}(t) \tag{8.10}$$

with $\underline{\tau_i}(t) = \max\{\tau_i(t)_{\min}, \tau_i(t-T) + T\min\{\eta_{\min}\}\}$ and $\bar{\tau}_i(t) = \min\{\tau_i(t)_{\max}, \tau_i(t-T) + T\max\{\eta_{\max}\}\}$.

8.3.2.1 An Open-Loop Constrained Optimal Control Allocation Design

Firstly, an open-loop control allocation scheme, such as Pseudo-Inverse (PI) method [20, 23, 24], is studied with the system block diagram shown in Fig. 8.1, where $\boldsymbol{D}^{\dagger} = \boldsymbol{D}^{\mathrm{T}}(\boldsymbol{D}\boldsymbol{D}^{\mathrm{T}})^{-1}$. In view of the above-mentioned problem and constraints, then

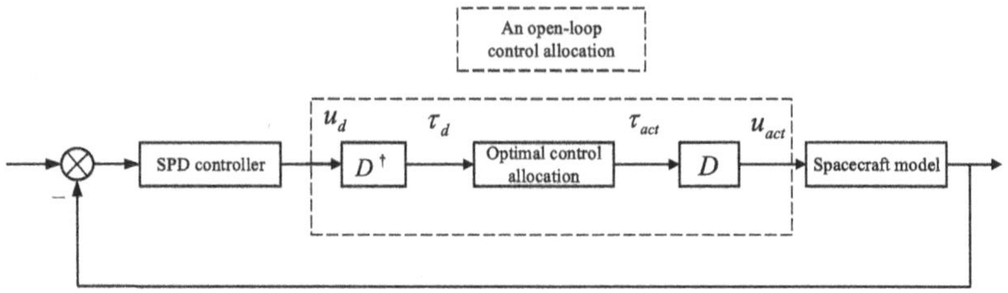

Fig. 8.1 Block diagram for spacecraft attitude control system with an open-loop control allocation scheme

the proposed control allocation algorithm can be formulated as a constrained optimal quadratic programming problem as follows:

$$\begin{aligned} J = & \arg \min_{\boldsymbol{\tau}(t)\in\Omega_0} \{||\boldsymbol{W}_0\boldsymbol{\tau}_{\text{act}}(t)||^2 + ||\boldsymbol{W}_1[\boldsymbol{\tau}_{\text{act}}(t) - \boldsymbol{\tau}_{\text{d}}(t)]||^2 \\ & + ||\boldsymbol{W}_2[\boldsymbol{\tau}_{\text{act}}(t) - \boldsymbol{\tau}_{\text{act}}(t-T)]||^2\} \end{aligned} \tag{8.11a}$$

$$\boldsymbol{\Omega}_0 = \arg \min_{\underline{\tau}(t)\le\tau_{\text{act}}(t)\le\bar{\tau}(t)} ||\boldsymbol{u}_{\text{d}}(t) - \boldsymbol{D}\boldsymbol{\tau}_{\text{act}}(t)|| \tag{8.11b}$$

where $\boldsymbol{W}_0$, $\boldsymbol{W}_1$ and $\boldsymbol{W}_2$ denote relevant diagonal positive-defined weighting matrices, $\boldsymbol{u}_{\text{d}}(t)$ and $\boldsymbol{\tau}_{\text{d}}(t)$ denote the desired system control torques and actuator torques, $\boldsymbol{u}_{\text{act}}(t)$ and $\boldsymbol{\tau}_{\text{act}}(t)$ denote the actual system control torques and actuator torques.

In order to facilitate for solving the constrained optimal problem, the choice of the elements $\boldsymbol{r}_0$ and $\boldsymbol{r}_1$ in diagonal positive-defined gain matrices $\boldsymbol{R}_0$ and $\boldsymbol{R}_1$ are selected, respectively:

$$\begin{aligned} \boldsymbol{r}_0 &= \begin{cases} \frac{\bar{\tau}_{act}(t)}{\tau_{\text{act}}(t)} & \tau_{\text{act}}(t) \ge \bar{\tau}_{\text{act}}(t) \\ 1 & \underline{\tau}_{\text{act}}(t) \le \tau_{\text{act}}(t) \le \bar{\tau}_{\text{act}}(t) \\ \frac{\underline{\tau}_{\text{act}}(t)}{\tau_{\text{act}}(t)} & \tau_{\text{act}}(t) \le \underline{\tau}_{\text{act}}(t) \end{cases} \\ \boldsymbol{r}_1 &= \begin{cases} \frac{\bar{\tau}_{\text{d}}(t)}{\tau_{\text{d}}(t)} & \tau_{\text{d}}(t) \ge \bar{\tau}_{\text{d}}(t) \\ 1 & \underline{\tau}_{\text{d}}(t) \le \tau_{\text{d}}(t) \le \bar{\tau}_{\text{d}}(t) \\ \frac{\underline{\tau}_{\text{d}}(t)}{\tau_{\text{d}}(t)} & \tau_{\text{d}}(t) \le \underline{\tau}_{\text{d}}(t) \end{cases} \end{aligned} \tag{8.12}$$

where the gain matrices can be defined as $\boldsymbol{R}_0 = \text{diag}(\boldsymbol{r}_0)$ and $\boldsymbol{R}_1 = \text{diag}(\boldsymbol{r}_1)$, which is to make the synthesized control torques $\boldsymbol{\tau}_{\text{act}}(t)$ and the desired torques $\boldsymbol{\tau}_{\text{d}}(t)$ in one limited region as defined in Eq.(8.10), and the more practical advantage of this definition is for solving the constrained optimal problem simply and conveniently.

Accordingly, the optimal quadratic programming problem described by Eq. (8.11) can then be modified as:

$$
\begin{aligned}
J = & \arg \min_{\tau(t)\in\Omega_0} \{||\boldsymbol{R}_0 \boldsymbol{W}_0 \boldsymbol{\tau}_{\text{act}}(t)||^2 + ||\boldsymbol{R}_1 \boldsymbol{W}_1[\boldsymbol{\tau}_{\text{act}}(t) - \boldsymbol{\tau}_{\text{d}}(t)]||^2 \\
& + ||\boldsymbol{W}_2[\boldsymbol{\tau}_{\text{act}}(t) - \boldsymbol{\tau}_{\text{act}}(t-T)]||^2\} \\
& \text{s.t. } \boldsymbol{u}_{\text{d}}(t) = \boldsymbol{D}\boldsymbol{\tau}_{\text{act}}(t)
\end{aligned} \tag{8.13}
$$

To solve this CA problem, the following lemma and corollaries will be helpful in the subsequent analysis.

Lemma 8.1 *[25]: For a matrix $\boldsymbol{A} \in \mathbf{R}^{m\times n}$, where $m < n$, with rank m, and $m = 3$, $n = 4$ in particular for this work, the minimum norm problem*

$$
\begin{aligned}
& \min_{x} ||\boldsymbol{x}||_2 \\
& \text{s.}t.\ \ \boldsymbol{x} = \boldsymbol{A}\boldsymbol{y}
\end{aligned}
$$

has the solution

$$
\boldsymbol{y} = \boldsymbol{A}^{\dagger}\boldsymbol{x}
$$

where the symbol † denotes the pseudo-inverse operator defined as $\boldsymbol{A}^{\dagger}=\boldsymbol{A}^{\text{T}}(\boldsymbol{A}\boldsymbol{A}^{\text{T}})^{-1}$.

Corollary 8.1 ([23]) *The weighting minimum norm problem*

$$
\begin{aligned}
& \min_{x}\ ||\boldsymbol{W}(\boldsymbol{x} - \boldsymbol{x}_0)||_2 \\
& \text{s.}t.\boldsymbol{A}\boldsymbol{x} = \boldsymbol{y}
\end{aligned}
$$

where $\boldsymbol{W}$ is a nonsingular weighting matrix, and has the following solution

$$
\boldsymbol{x} = \boldsymbol{F}\boldsymbol{x}_0 + \boldsymbol{G}\boldsymbol{y},\ \ \boldsymbol{F} = \boldsymbol{I} - \boldsymbol{G}\boldsymbol{A},\ \ \boldsymbol{G} = \boldsymbol{W}^{-1}(\boldsymbol{A}\boldsymbol{W}^{-1})^{\dagger}
$$

Corollary 8.2 ([23]) *The cost function*

$$
\min_{x}\ ||\boldsymbol{W}_1(\boldsymbol{x} - \boldsymbol{x}_1)||^2 + ||\boldsymbol{W}_2(\boldsymbol{x} - \boldsymbol{x}_2)||^2
$$

has the same minimizing argument as

$$
\min_{x}\ ||\boldsymbol{W}(\boldsymbol{x} - \boldsymbol{x}_0)||_2
$$

where $\boldsymbol{W} = (\boldsymbol{W}_1^2 + \boldsymbol{W}_2^2)^{1/2}$, $\boldsymbol{x}_0 = \boldsymbol{W}^{-2}(\boldsymbol{W}_1^2\boldsymbol{x}_1 + \boldsymbol{W}_2^2\boldsymbol{x}_2)$.

Then, let the above lemma and corollaries hold, and the control allocation problem of Eq. (8.13) has the solution as following:

$$
\tau_{\text{act}}(t) = \boldsymbol{E}\tau_{\text{d}}(t) + \boldsymbol{F}\tau_{\text{act}}(t-T) + \boldsymbol{G}\boldsymbol{u}_{\text{d}}(t) \tag{8.14}
$$

where $W = ((R_0 W_0)^2 + (R_1 W_1)^2 + W_2^2)^{1/2}$, $E = (I - GD)W^{-2}(R_1 W_1)^2$, $F = (I - GD)W^{-2}W_2^2$, $G = W^{-1}(DW^{-1})^{\dagger}$.

In view of Eq. (8.14), the real actuator torques are the weighted sum of the three types of torques defined as above. The benefit of this CA solution is that decreased effect of the noises or disturbances and high performance can be achieved [26].

Remark 8.2 Equation (8.11) is a cost function to be minimized subject to the linear and nonlinear constraints. From this cost function, the total power consumption is represented by the first term in the criterion, in which if the larger weighting matrix W_0 is chosen, the smaller energy consumption will be achieved. For the second term, it is used to minimize the error of the CA such that the control torques would be convergent to the desired command quickly when a large weighting matrix W_1 is selected. In addition, to minimize the changing rate of the control torques, the third term is chosen such that the response of the actuator will be smooth and steady if one can set a proper weighting matrix W_2. In fact, the three terms defined in Eq. (8.11a) do not contradict each other due to the fact that these weighting matrixes are used for representing their importance or priority or weight in this specified overall optimal object function.

8.3.2.2 A Closed-Loop Constrained Optimal Control Allocation Design

In order to apply the developed optimal CA scheme to a closed-loop system, a novel CA scheme denoted as closed-loop CA is proposed for spacecraft attitude stabilization control system in this subsection for the purpose of stability guarantee. In fact, similar to control theory for feedback, closed-loop CA is trying to use this powerful and useful concept for guaranteeing the expected control signals distribution among redundant actuators and expected convergence for the stability guarantee, that is to say, the process of $u_{act} \to u_d$ is stable asymptotically, which is the main result/conclusion in this subsection. This is not possible for open-loop CA since there is no such an internal feedback, although optimization through three weighting functions/matrices has been used for CA design. The overall system block diagram with details on such a CA with feedback scheme is presented in Fig. 8.2, in which z^{-1} is a delay or discretization operator and e is the CA error of the actuators from the actuator torques before and after the "optimal CA" module.

In general, the attitude control problem naturally falls into a control system design with CA module to be included for a practical and effectively over-actuated system. Although the stability/stabilization of the baseline controller could be achieved easily, such a design framework is problem driven and is not a simple nor straightforward combination of existing controllers design for each individual baseline control law and CA; at the same time, a stability-guaranteed overall system is achieved, which has been considered rarely/ambiguous or not been well achieved in most existing works in the field. From this viewpoint, the key result of this overall closed-loop CA system is presented as following:

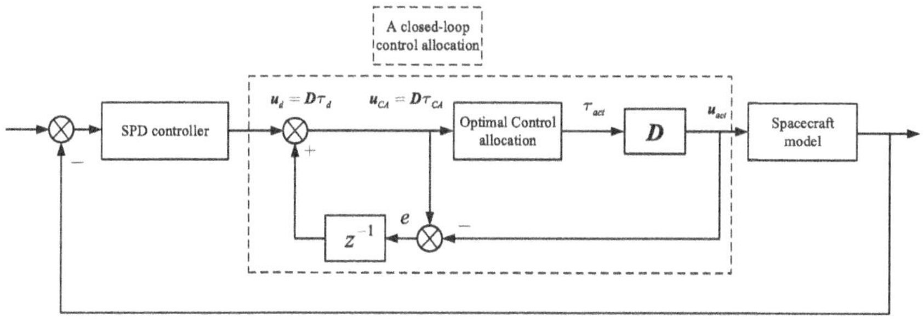

Fig. 8.2 Block diagram for spacecraft attitude control system with a closed-loop control allocation scheme

Theorem 8.2 *For the spacecraft attitude system with closed-loop control allocation as shown in Fig. 8.2, the closed-loop CA system can be described as:*

$$\begin{aligned}&(\boldsymbol{I} - z^{-1}(\boldsymbol{I} - \boldsymbol{L}))\boldsymbol{U}_{\text{act}}(z)\\&= (\boldsymbol{L} + \boldsymbol{DG} - z^{-1}\boldsymbol{DG})\boldsymbol{U}_{\text{d}}(z) + \boldsymbol{DF}z^{-1}(1 - z^{-1})\boldsymbol{\Gamma}_{\text{act}}(z)\end{aligned} \tag{8.15}$$

where $\boldsymbol{DED}^{\dagger} = \boldsymbol{L}$*. Then the eigenvalues of* $\mathbf{L}$*, defined as* ℓ_i *in the domain of* (0, 1)*, can be guaranteed. That is to say, the system Eq.* (8.15) *is asymptotically stable for a bounded virtual control demand.*

Proof In order to analyze the stability of the closed-loop system including this presented closed-loop CA, the actual actuator torques can be described in discrete-time mode as follows:

$$\begin{aligned}\boldsymbol{\tau}_{\text{act}}(k) &= \boldsymbol{E}\boldsymbol{\tau}_{\text{CA}}(k) + \boldsymbol{F}\boldsymbol{\tau}_{\text{act}}(k-1) + \boldsymbol{G}\boldsymbol{u}_d(k)\\&= \boldsymbol{ED}^{\dagger}\boldsymbol{u}_{\text{CA}}(k) + \boldsymbol{F}\boldsymbol{\tau}_{\text{act}}(k-1) + \boldsymbol{G}\boldsymbol{u}_{\text{d}}(k)\end{aligned} \tag{8.16}$$

where

$$\boldsymbol{u}_{\text{CA}}(k) = \boldsymbol{u}_{\text{d}}(k) + \boldsymbol{e}(k-1) = \boldsymbol{u}_{\text{d}}(k) + \boldsymbol{u}_{\text{CA}}(k-1) - \boldsymbol{u}_{\text{act}}(k-1) \tag{8.17}$$

From Eqs. (8.16) and (8.17), one can obtain

$$\boldsymbol{u}_{\text{act}}(k) = \boldsymbol{D}\boldsymbol{\tau}_{\text{act}}(k) = \boldsymbol{DED}^{\dagger}\boldsymbol{u}_{\text{CA}}(k) + \boldsymbol{DF}\boldsymbol{\tau}_{\text{act}}(k-1) + \boldsymbol{DG}\boldsymbol{u}_{\text{d}}(k) \tag{8.18}$$

$$\begin{aligned}\boldsymbol{DED}^{\dagger}\boldsymbol{u}_{\text{CA}}(k) =&\boldsymbol{DED}^{\dagger}\boldsymbol{u}_{\text{d}}(k) + \boldsymbol{DED}^{\dagger}\boldsymbol{u}_{\text{CA}}(k-1)\\&- \boldsymbol{DED}^{\dagger}\boldsymbol{u}_{\text{act}}(k-1)\end{aligned} \tag{8.19}$$

Subsequently, defining $\boldsymbol{DED}^{\dagger} = \boldsymbol{L}$ and substituting Eq. (8.18)–(8.19), one can obtain

$$\begin{aligned}\boldsymbol{L}\boldsymbol{u}_{\mathrm{CA}}(k) =& \boldsymbol{L}\boldsymbol{u}_{\mathrm{d}}(k) + \boldsymbol{u}_{\mathrm{act}}(k-1) - \boldsymbol{D}\boldsymbol{F}\boldsymbol{\tau}_{\mathrm{act}}(k-2) - \boldsymbol{D}\boldsymbol{G}\boldsymbol{u}_{\mathrm{d}}(k-1)\\ &- \boldsymbol{L}\boldsymbol{u}_{\mathrm{act}}(k-1)\end{aligned} \tag{8.20}$$

and further with

$$\begin{aligned}\boldsymbol{u}_{\mathrm{act}}(k) - (\boldsymbol{I}-\boldsymbol{L})\boldsymbol{u}_{\mathrm{act}}(k-1) =& (\boldsymbol{L}+\boldsymbol{D}\boldsymbol{G})\boldsymbol{u}_{\mathrm{d}}(k) - \boldsymbol{D}\boldsymbol{G}\boldsymbol{u}_{\mathrm{d}}(k-1)\\ &+ \boldsymbol{D}\boldsymbol{F}(\boldsymbol{\tau}_{\mathrm{act}}(k-1) - \boldsymbol{\tau}_{\mathrm{act}}(k-2))\end{aligned} \tag{8.21}$$

Utilizing the standard Z-transform theory, one can obtain the discrete-time model of the closed-loop CA system as follows:

$$\begin{aligned}\boldsymbol{U}_{\mathrm{act}}(z) - z^{-1}(\boldsymbol{I}-\boldsymbol{L})\boldsymbol{U}_{\mathrm{act}}(z) =& (\boldsymbol{L}+\boldsymbol{D}\boldsymbol{G})\boldsymbol{U}_{\mathrm{d}}(z) - z^{-1}\boldsymbol{D}\boldsymbol{G}\boldsymbol{U}_{\mathrm{d}}(z)\\ &+ \boldsymbol{D}\boldsymbol{F}(z^{-1}\boldsymbol{\Gamma}_{\mathrm{act}}(z) - z^{-2}\boldsymbol{\Gamma}(z))\end{aligned}$$

that is,

$$(\boldsymbol{I} - z^{-1}(\boldsymbol{I}-\boldsymbol{L}))\boldsymbol{U}_{\mathrm{act}}(z) = (\boldsymbol{L}+\boldsymbol{D}\boldsymbol{G} - z^{-1}\boldsymbol{D}\boldsymbol{G})\boldsymbol{U}_{\mathrm{d}}(z) + \boldsymbol{D}\boldsymbol{F}z^{-1}(1-z^{-1})\boldsymbol{\Gamma}_{\mathrm{act}}(z)$$

Since the discrete-time model of the closed-loop CA system is obtained as Eq. (8.15), the dynamic properties of the closed-loop system with the SPD and closed-loop CA ought to be studied subsequently. Then the characteristic equation of the closed-loop CA system from Eq. (8.15) can be obtained as following [14, 23]:

$$\left|\boldsymbol{I} - z^{-1}(\boldsymbol{I}-\boldsymbol{L})\right| = 0$$

that is

$$|z\boldsymbol{I} - (\boldsymbol{I}-\boldsymbol{L})| = 0 \tag{8.22}$$

Based on the fundamental theorem of the discrete-time system, the necessary and sufficient condition for a discrete- time system to be stable is that the norms of all the characteristic roots are smaller than one [27]. Obviously, if the eigenvalues of $\boldsymbol{I}-\boldsymbol{L}$ defined as ς_i are in the domain of (0,1), that is to say, the eigenvalues of $\boldsymbol{L}$ defined as $\ell_i = 1 - \varsigma_i$ are in the domain of (0,1), the closed-loop control allocation system described as Eq. (8.15) is stable. Based on the similar rationale as proposed in [14–16, 22], then the overall closed-loop system is also stable since the SPD-based baseline controller has been designed to guarantee the stability of the attitude dynamic system of the spacecraft.

The eigenvalues of the characteristic equation of the closed-loop control allocation system can be obtained as follows:

$$\begin{aligned}\boldsymbol{L} = \boldsymbol{D}\boldsymbol{E}\boldsymbol{D}^{\dagger} &= \boldsymbol{D}(\boldsymbol{I}-\boldsymbol{G}\boldsymbol{D})\boldsymbol{W}^{-2}(\boldsymbol{R}_1\boldsymbol{W}_1)^2\boldsymbol{D}^{\dagger}\\ &= \boldsymbol{D}\boldsymbol{W}^{-1}(\boldsymbol{I} - (\boldsymbol{D}\boldsymbol{W}^{-1})^{\dagger}\boldsymbol{D}\boldsymbol{W}^{-1})\boldsymbol{W}^{-1}(\boldsymbol{R}_1\boldsymbol{W}_1)^2\boldsymbol{D}^{\dagger}\end{aligned} \tag{8.23}$$

Let the singular value decomposition of $\boldsymbol{D}$ and $\boldsymbol{D}\boldsymbol{W}^{-1}$ be represented as:

$$\boldsymbol{D}=\boldsymbol{U}_1\boldsymbol{\Sigma}_1\boldsymbol{V}_1^{\mathrm{T}}=\boldsymbol{U}_1\begin{bmatrix}\boldsymbol{\Sigma}_{1\mathrm{r}} & 0\end{bmatrix}\begin{bmatrix}\boldsymbol{V}_{1\mathrm{r}}^{\mathrm{T}}\\ \boldsymbol{V}_{10}^{\mathrm{T}}\end{bmatrix}=\boldsymbol{U}_1\boldsymbol{\Sigma}_{1\mathrm{r}}\boldsymbol{V}_{1\mathrm{r}}^{\mathrm{T}}$$

$$\boldsymbol{D}\boldsymbol{W}^{-1}=\boldsymbol{U}_2\boldsymbol{\Sigma}_2\boldsymbol{V}_2^{\mathrm{T}}=\boldsymbol{U}_2\begin{bmatrix}\boldsymbol{\Sigma}_{2\mathrm{r}} & 0\end{bmatrix}\begin{bmatrix}\boldsymbol{V}_{2\mathrm{r}}^{\mathrm{T}}\\ \boldsymbol{V}_{20}^{\mathrm{T}}\end{bmatrix}=\boldsymbol{U}_2\boldsymbol{\Sigma}_{2\mathrm{r}}\boldsymbol{V}_{2\mathrm{r}}^{\mathrm{T}}$$

where $\boldsymbol{U}_1$, $\boldsymbol{V}_1$, $\boldsymbol{U}_2$ and $\boldsymbol{V}_2$ are unitary matrixes, $\boldsymbol{\Sigma}_1$ and $\boldsymbol{\Sigma}_2$ are diagonal matrixes with strictly positive diagonal entries and their dimensions are equal to the rank of $\boldsymbol{D}$ and $\boldsymbol{D}\boldsymbol{W}^{-1}$, respectively. Then these will yield

$$\boldsymbol{L}=\boldsymbol{U}_1\boldsymbol{\Sigma}_1\boldsymbol{V}_1^{\mathrm{T}}\boldsymbol{W}^{-1}\boldsymbol{V}_{20}\boldsymbol{V}_{20}^{\mathrm{T}}\boldsymbol{W}^{-1}(\boldsymbol{R}_1\boldsymbol{W}_1)^2\boldsymbol{V}_1\boldsymbol{\Sigma}_1^{-1}\boldsymbol{U}_1^{\mathrm{T}}$$

Utilizing the basic results and techniques of matrix theory [28], the nonzero eigenvalues of matrix product $\boldsymbol{M}\boldsymbol{N}$, the following equation can be obtained

$$\ell_{nz}(\boldsymbol{M}\boldsymbol{N})=\ell_{nz}(\boldsymbol{N}\boldsymbol{M})$$

where ℓ_{nz} denote the nonzero eigenvalues of the matrix.

From above analysis, one can further obtain that

$$\begin{aligned}\ell_{\mathrm{nz}}(\boldsymbol{L})&=\ell_{\mathrm{nz}}(\boldsymbol{V}_{20}^{\mathrm{T}}\boldsymbol{W}^{-1}(\boldsymbol{R}_1\boldsymbol{W}_1)^2\boldsymbol{V}_1\boldsymbol{\Sigma}_1^{-1}\boldsymbol{U}_1^{\mathrm{T}}\boldsymbol{U}_1\boldsymbol{\Sigma}_1\boldsymbol{V}_1^{\mathrm{T}}\boldsymbol{W}^{-1}\boldsymbol{V}_{20})\\&=\ell_{\mathrm{nz}}(\boldsymbol{V}_{20}^{\mathrm{T}}\boldsymbol{W}^{-1}(\boldsymbol{R}_1\boldsymbol{W}_1)^2\boldsymbol{W}^{-1}\boldsymbol{V}_{20})\end{aligned}$$

Next, following result can be achieved from the definition of matrix singular values

$$\ell_{\mathrm{nz}}(\boldsymbol{V}_{20}^{\mathrm{T}}\boldsymbol{W}^{-1}(\boldsymbol{R}_1\boldsymbol{W}_1)^2\boldsymbol{W}^{-1}\boldsymbol{V}_{20})=\sigma^2(\boldsymbol{R}_1\boldsymbol{W}_1\boldsymbol{W}^{-1}\boldsymbol{V}_{20})>0 \tag{8.24}$$

which states that the nonzero eigenvalues of $\boldsymbol{L}$ are real and positive. That is, $\ell_{\mathrm{nz}}(\boldsymbol{L})>0$ can be guaranteed.

On the other hand, what remains to be shown is that the eigenvalues of $\boldsymbol{L}$ are bounded by 1, and let the maximum eigenvalue of $\boldsymbol{L}$ be $\ell_{\max}(\boldsymbol{L})$, namely

$$\begin{aligned}\ell_{\max}(\boldsymbol{L})&=\sigma_{\max}^2(\boldsymbol{R}_1\boldsymbol{W}_1\boldsymbol{W}^{-1}\boldsymbol{V}_{20})=\left\|\boldsymbol{R}_1\boldsymbol{W}_1\boldsymbol{W}^{-1}\boldsymbol{V}_{20}\right\|^2\\&\le\left\|\boldsymbol{R}_1\boldsymbol{W}_1\boldsymbol{W}^{-1}\right\|^2\|\boldsymbol{V}_{20}\|^2\le\left\|\boldsymbol{R}_1\boldsymbol{W}_1\boldsymbol{W}^{-1}\right\|^2\end{aligned}$$

which using the property of singular decomposition, that is $\|\boldsymbol{V}_{20}\|^2=\ell_{\max}(\boldsymbol{V}_{20}^T\boldsymbol{V}_{20})=1$. Then

$$\ell_{\max}(\boldsymbol{L}) \le \left\|\boldsymbol{R}_1\boldsymbol{W}_1\boldsymbol{W}^{-1}\right\|^2 = \sup_{x\neq 0}\frac{\boldsymbol{x}^{\mathrm{T}}\boldsymbol{W}^{-1}(\boldsymbol{R}_1\boldsymbol{W}_1)^2\boldsymbol{W}^{-1}\boldsymbol{x}}{\boldsymbol{x}^{\mathrm{T}}\boldsymbol{x}} \tag{8.25}$$

Defining $\boldsymbol{W}^{-1}\boldsymbol{x} = \boldsymbol{y}$, then $\boldsymbol{W}\boldsymbol{y} = \boldsymbol{x}$, so Eq. (31) can be re-described as:

$$\begin{aligned}\ell_{\max}(\boldsymbol{L}) &\le \sup_{y\neq 0}\frac{\boldsymbol{y}^{\mathrm{T}}(\boldsymbol{R}_1\boldsymbol{W}_1)^2\boldsymbol{y}}{\boldsymbol{y}^{\mathrm{T}}\boldsymbol{W}^2\boldsymbol{y}}\\ &= \sup_{y\neq 0}\frac{\boldsymbol{y}^{\mathrm{T}}(\boldsymbol{R}_1\boldsymbol{W}_1)^2\boldsymbol{y}}{\boldsymbol{y}^{\mathrm{T}}((\boldsymbol{R}_0\boldsymbol{W}_0)^2 + (\boldsymbol{R}_1\boldsymbol{W}_1)^2 + (\boldsymbol{W}_2)^2)\boldsymbol{y}}\\ &\le \sup_{y\neq 0}\frac{\boldsymbol{y}^{\mathrm{T}}(\boldsymbol{R}_1\boldsymbol{W}_1)^2\boldsymbol{y}}{\boldsymbol{y}^{\mathrm{T}}(\boldsymbol{R}_1\boldsymbol{W}_1)^2\boldsymbol{y}} = 1\end{aligned} \tag{8.26}$$

since $\boldsymbol{W}_0$, $\boldsymbol{W}_1$, and $\boldsymbol{W}_2$ denote the diagonal positive-defined weighting matrices, so $\boldsymbol{y}^{\mathrm{T}}((\boldsymbol{R}_0\boldsymbol{W}_0)^2 + (\boldsymbol{W}_2)^2)\boldsymbol{y} > 0$ for any $\boldsymbol{y} \neq 0$. From Eqs. (8.24) to (8.26), the eigenvalues of $\boldsymbol{L}$, defined as ℓ_i in the domain of (0,1), can be achieved. That is to say, the system Eq. (8.25) is asymptotically stable. □

Remark 8.3 From the above stability analysis, one can see that the discrete-time method was utilized for the closed-loop CA system. On the one hand, the operating and working signals from the computer are discrete-time in practical engineering environment such as the orbiting spacecraft. On the other hand, it is also convenient for stability analysis of the considered CA. In view of the whole process of the CA in Sect. 8.3.2.2, the discrete-time analysis was useful and frequently-used because of the special formation of the proposed CA algorithm formulated as a constrained optimal quadratic programming problem described as Eq. (8.11), so utilization of the standard Z-transform theory for the stability analysis is necessary and reasonable.

Remark 8.4 The formal proof of the overall closed-loop stability of the system (such as SPD+CA) does not exist in literature, and that is why the research achievements for this difficult problem have been done. In other words, most of the existing CA schemes are studied and worked as an open-loop system in some sense, and the stability/stabilization issue of closed-loop system under combination of CA with the baseline controller is not considered due to the inherent difficulty for such a design task. From this point of view, the developed control scheme in this paper is novel and has significant contribution towards such a practical engineering application for spacecraft ACS, and also other over-actuated systems.

8.4 Simulation Results

To verify the effectiveness and performance of the proposed attitude stabilization control and the closed-loop constrained optimal control allocation scheme, numerical simulations have been carried out using the rigid spacecraft system given in Eqs.

(2.17) and (2.20). It should be noted that, for orbiting spacecraft, there are more than three actuators aligned with the three body-fixed axes of the spacecraft. Therefore, a typical application case with an over-actuated dynamic system. The common configuration with four RWs can be referred in Chap. 3 and is chosen as the spacecraft actuators in this chapter. In such a configuration, the rotational axes of three reaction wheels (such as RWs 1, 2, and 3) are orthogonal to the spacecraft's orthogonal shaft and the fourth one is installed with the equiangular direction with the orthogonal-to-each other three axes. However, in practice, the orthogonal configuration of actuators will never be perfect. Whether due to limited manufacturing tolerances or warping of the spacecraft structure during launch, some alignment errors can always exist. Referring to Chap. 4 and to model such uncertainties, it is assumed that the RW mounted on X-axis is tilted over nominal direction with constant angles, $\Delta\alpha_1$ and $\Delta\beta_1$; also other RWs mounted left are assumed to be tilted over nominal direction with $\Delta\alpha_2$, $\Delta\beta_2$, $\Delta\alpha_3$, $\Delta\beta_3$, $\Delta\alpha_4$ and $\Delta\beta_4$, respectively. In this section, the actuators' misalignment will be taken consideration to verify the effectiveness and superiority of the proposed control schemes. The normal value and uncertainty of the CA or configuration matrix ($\boldsymbol{D}$ and $\Delta\boldsymbol{D}$) are analyzed and presented in Chap. 3.

In addition, the external disturbances are assumed to be as follows:

$$\boldsymbol{d}(t) = 0.2 \times 10^{-3} \begin{bmatrix} 3\cos(10\omega_{\mathrm{d}}t) + 4\sin(3\omega_{\mathrm{d}}t) - 10 \\ -1.5\sin(2\omega_{\mathrm{d}}t) + 3\cos(5\omega_{\mathrm{d}}t) + 15 \\ 3\sin(10\omega_{\mathrm{d}}t) - 8\sin(4\omega_{\mathrm{d}}t) + 10 \end{bmatrix} \mathrm{N.m}$$

For comparison purposes, performances of the attitude stabilization control system with/without actuator uncertainties and external disturbances are discussed respectively, and the simulation parameters are provided in Tables 8.1 and 8.2 for all numerical simulations carried out in the MATLAB/Simulink software environment.

8.4.1 Attitude Control without Actuator Uncertainties and Disturbances

In this simulation case, the proposed SPD controller incorporating the proposed CLCA, noted as SPD+CLCA, is applied to the rigid spacecraft attitude stabilization control system. Figures 8.3 and 8.4 show the time responses of the attitude MRPs and angular velocity, respectively, without actuator uncertainties and external disturbances. As it can be seen, the SPD+CLCA could achieve a fine performance. The spacecraft attitude system reaches the commanded target smoothly with a settling time less than 30s. Moreover, the transformed quaternion and Euler angle in Figs. 8.5 and 8.6 further illustrate the effectiveness of the proposed scheme. The time response of demanded/virtual control torques and the actual actuator/reaction flywheel torques are shown in Figs. 8.7 and 8.8, respectively. Note that, the restriction

Table 8.1 Main parameters of a rigid spacecraft

Mission	Attitude stabilization
Inertia moments (kg.m^2)	
Principal moments of inertia	$J_1 = 20,\ J_2 = 17,\ J_3 = 15$
Products of inertia	$J_{12} = 0,\ J_{23} = 0,\ J_{13} = 0.9$
Attitude	
Attitude control type	Three-axis control by four reaction fly-wheels
Initial attitude quaternion	$\boldsymbol{Q}_0 = [q_0, \boldsymbol{q}_v^{\mathrm{T}}]^{\mathrm{T}} = [0.9; -0.3; 0.26; 0.18]$
Initial angular velocity (rad/s)	$\boldsymbol{w}_0$=[0.1; -0.1; -0.05]
Reaction fly-wheels	
The maximum torque (N.m)	0.5
The assembling locations/angles (deg)	α_4 =35.26, β_4 =45
The misalignment error angles (rad)	$\Delta\boldsymbol{\alpha}$ =[0.3; 0.3; 0.2; 0.2], $\Delta\boldsymbol{\beta}$ =[0.3; 0.3; 0.2; 0.2]

Table 8.2 Control parameters chosen for numerical analysis

Control schemes	Control gains
Proposed SPD controller	$u_{\max} = 10,\ \beta = 0.908,\ \alpha^2 = 0.1$
Closed-loop control allocator	$\tau_{\min} = -0.5,\ \tau_{\max} = 0.5,$ $\eta_{\min} = -0.002,$ $\eta_{\max} = 0.002,\ \boldsymbol{W}_0 = \mathrm{diag}\begin{pmatrix} 10 & 5 & 1 & 0.1 \end{pmatrix},$ $\boldsymbol{W}_1 = \mathrm{diag}\begin{pmatrix} 10 & 5 & 1 & 2 \end{pmatrix},$ $\boldsymbol{W}_2 = \mathrm{diag}\begin{pmatrix} 10 & 5 & 1 & 0.1 \end{pmatrix}$

on the actuator output torque magnitude 0.5 N.m is considered explicitly, and thanks to both the SPD controller design and the closed-loop constrained optimal control allocator, the SPD+CLCA can achieve the requirements of actuators' constraints.

The time responses of the closed-loop CA errors are shown in Fig. 8.9. It states clearly that the errors of the closed-loop CA converge to zero within 20s, and further expounds and proves the stability of the closed-loop CA system. Additionally, fuel or electrical energy saving for spacecraft in orbit is of crucial importance to prolong the working life and to achieve the targeted missions. Therefore the energy consumption optimization problem should also be considered for spacecraft ACS. To analyze the system energy consumption, the performance function which is derived from the optimal CA problem is defined as: $\boldsymbol{E} = \frac{1}{2}\int_0^T ||\boldsymbol{\tau}||^2 \mathrm{d}t$, where T denotes the simulation time, and $T = 50$ s is chosen in the simulation. Figure 8.10 shows the time response of the energy consumption performance of the proposed strategy without actuator uncertainties and external disturbances. From Fig. 8.10, it can be seen clearly that the energy consumption using SPD+CLCA presented in this chapter is small and acceptable for real application.

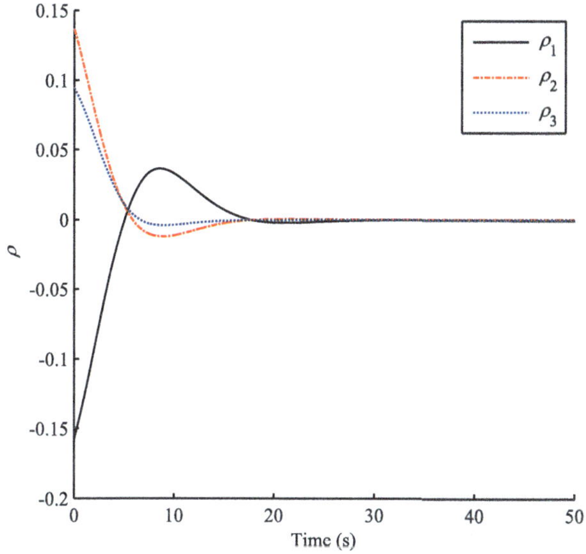

Fig. 8.3 Time responses of MRPs

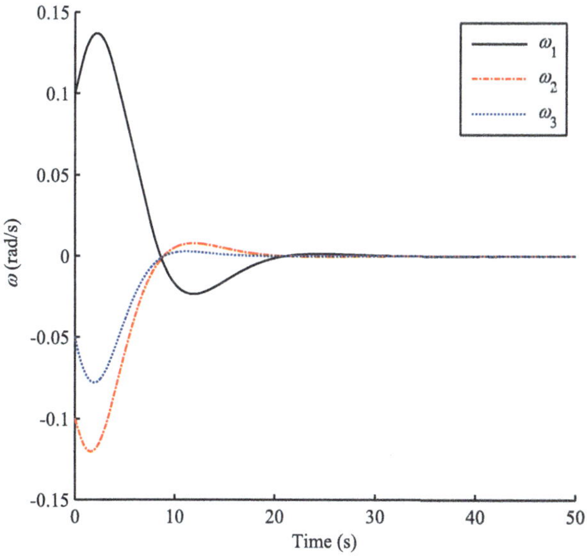

Fig. 8.4 Time responses of angular velocity

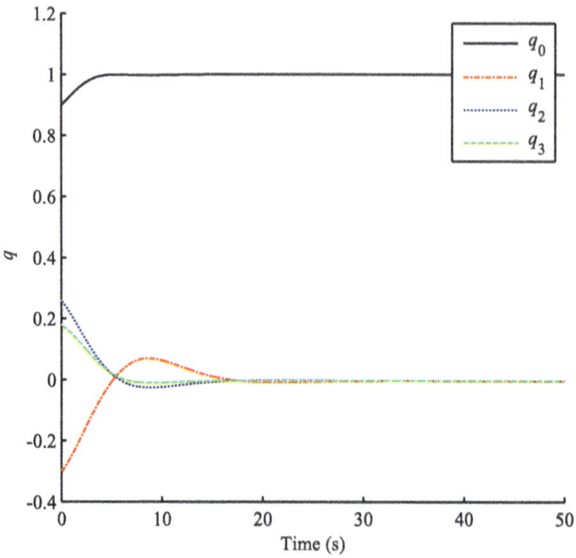

Fig. 8.5 Time responses of quaternion

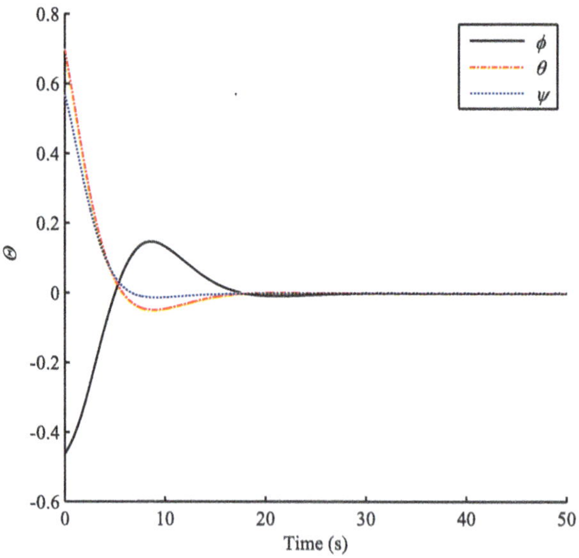

Fig. 8.6 Time responses of Euler angle

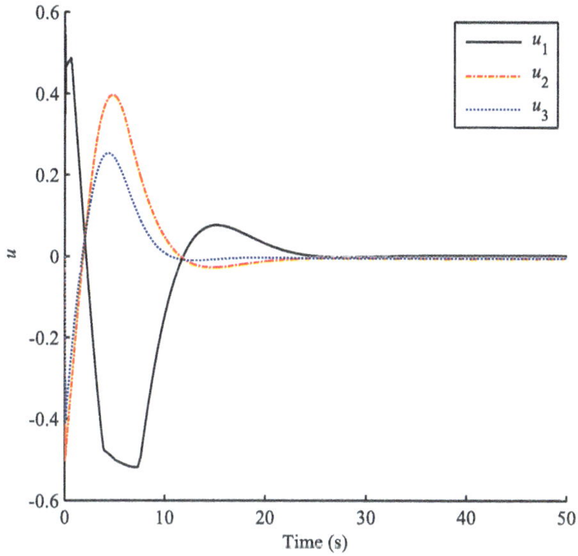

Fig. 8.7 Time responses of control torques

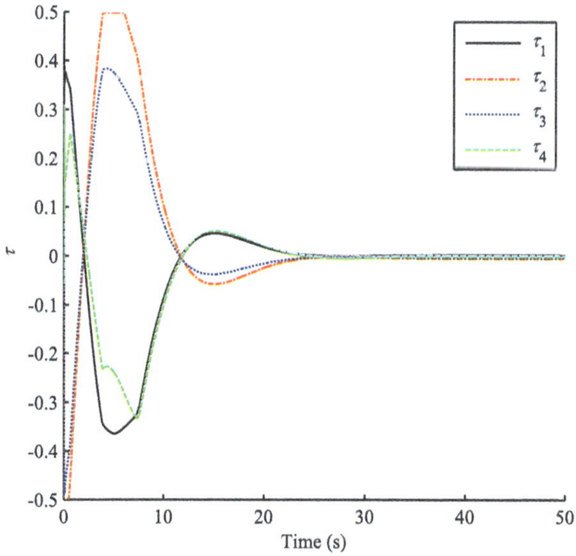

Fig. 8.8 Time responses of the actual actuator control torques

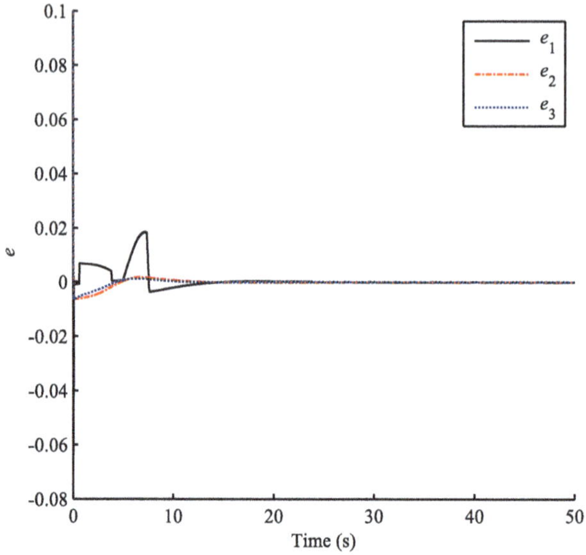

Fig. 8.9 Time responses of closed-loop control allocation errors

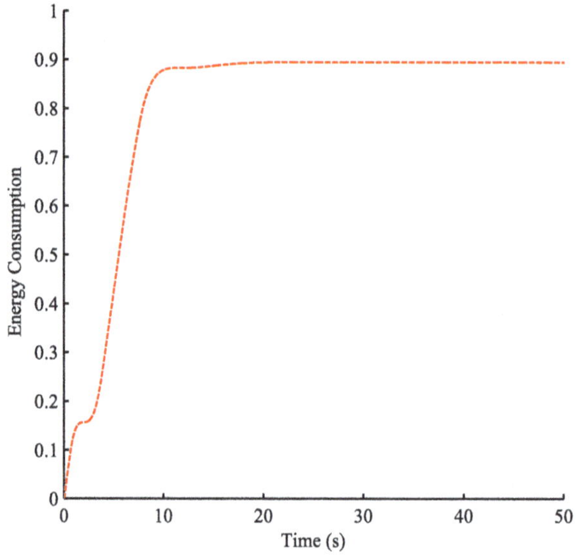

Fig. 8.10 Time responses of energy consumption

8.4.2 Attitude Control with Actuator Uncertainties and Disturbances

Actuator uncertainties and external disturbances are considered for the rigid spacecraft attitude stabilization control system in this subsection and results are presented subsequently. Firstly, Figs. 8.11 and 8.12 represent time responses of the attitude MRPs and angular velocities, respectively. As it can be observed that the SPD+CLCA achieves a good performance. The spacecraft attitude system reaches the commanded stabilization target smoothly with a settling time less than 20 s with a high accuracy in steady-state error. The transformed quaternion and Euler angle in Figs. 8.13 and 8.14 further illustrated the effectiveness of the proposed SPD+CLCA scheme. The time response of demanded/virtual control torques and the actual actuator/reaction wheel torques are shown in Figs. 8.15 and 8.16, respectively. In addition, Fig. 8.17 shows the time responses of the closed-loop CA errors. More specially, the time response of the energy consumption performance of the proposed strategy with actuator uncertainties and external disturbances are shown in Fig. 8.18, and it can be seen clearly that the energy consumption of SPD+CLCA presented in this chapter is also acceptable for real applications with actuator uncertainties and external disturbances. Such an improvement is benefitted and achieved by the closed-loop constrained optimal control allocator proposed in this paper and it is more robust to uncertainties and disturbances and more suitable to meet the practical requirements for orbiting spacecraft.

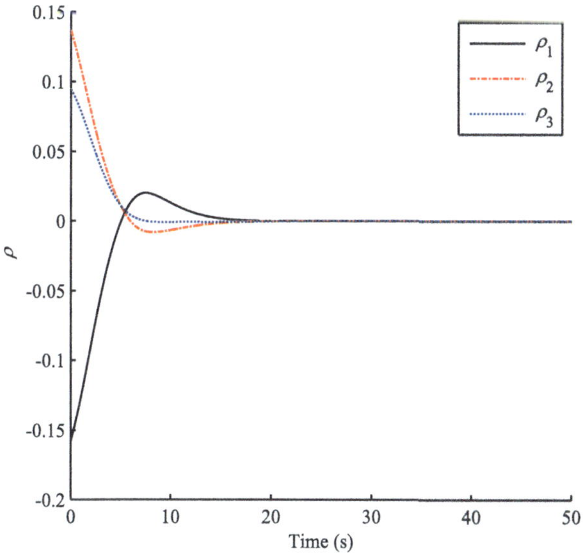

Fig. 8.11 Time responses of MRPs

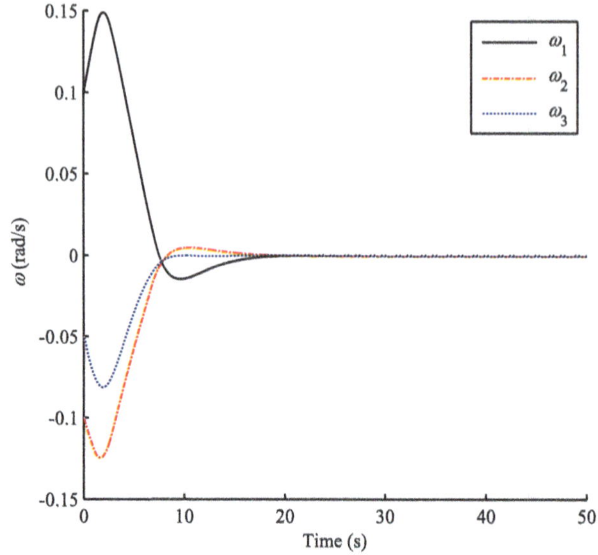

Fig. 8.12 Time responses of angular velocity

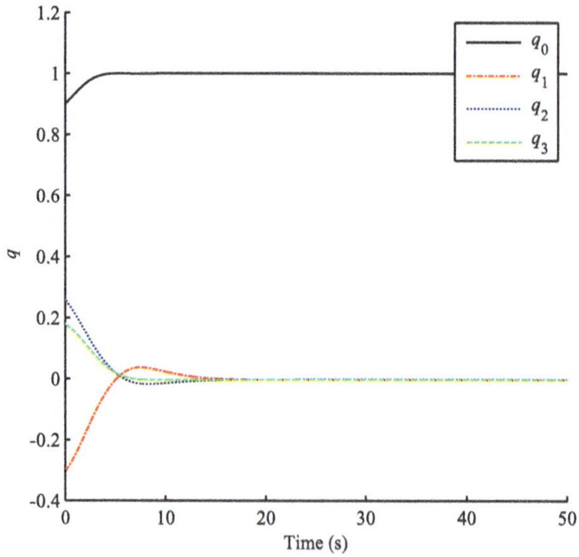

Fig. 8.13 Time responses of quaternion

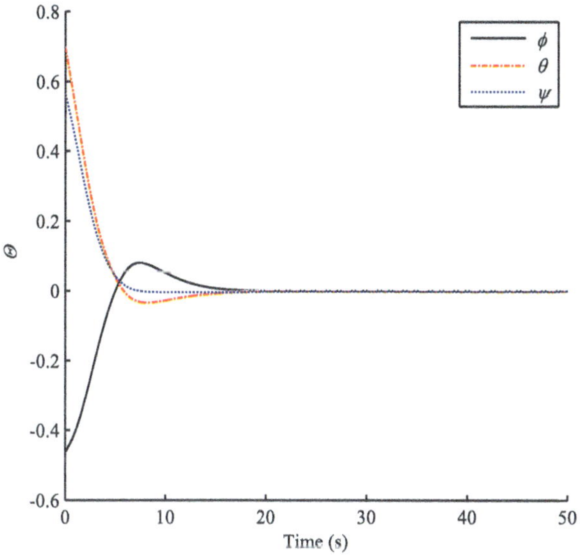

Fig. 8.14 Time responses of Euler angle

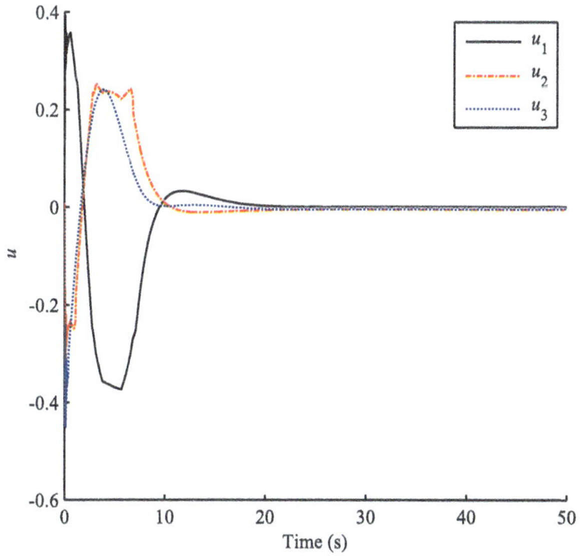

Fig. 8.15 Time responses of control torques

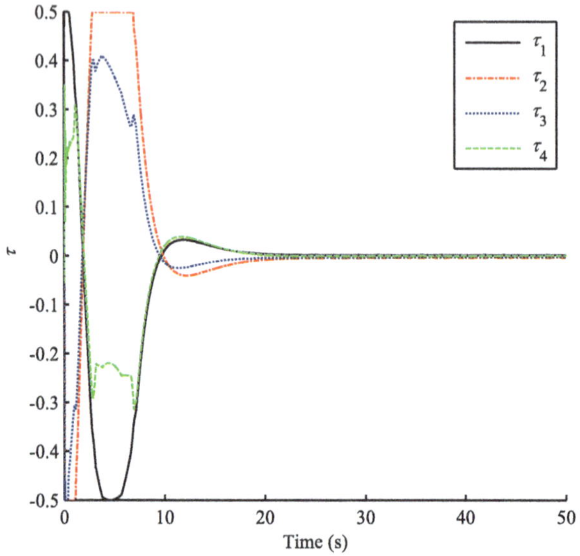

Fig. 8.16 Time responses of the actual actuator control torques

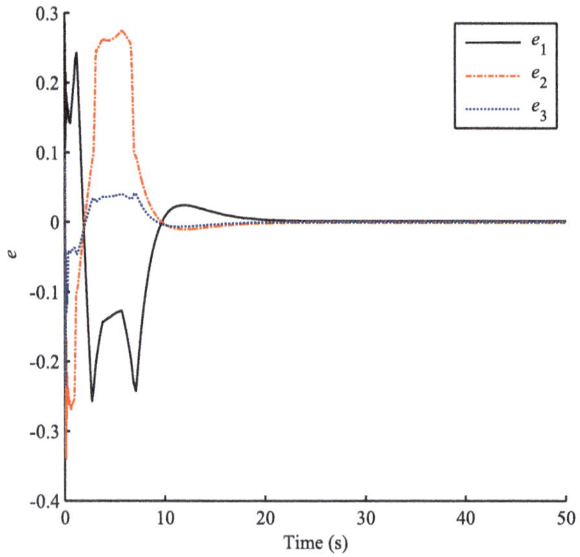

Fig. 8.17 Time responses of the closed-loop control allocation errors

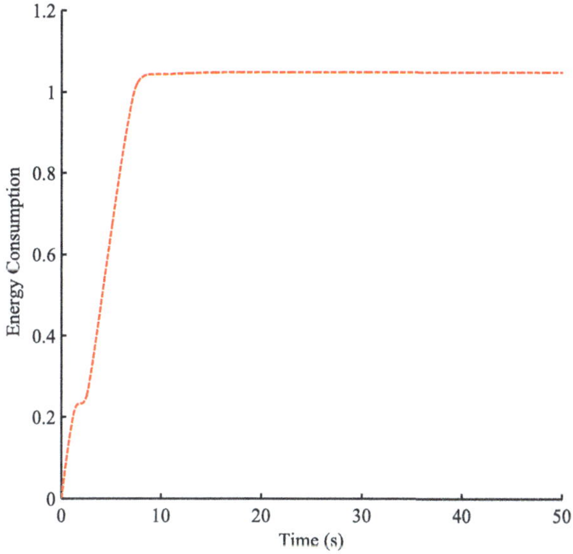

Fig. 8.18 Time responses of energy consumption

Summarizing all the results presented previously, the proposed control scheme SPD+CLCA can achieve a fine performance, whether the actuator uncertainties and external disturbances are considered or not. Most significantly, the proposed SPD+CLCA can be proved theoretically for ensuring system to be stable with a specified trajectory. Extensive simulations were also carried out using different control parameters and disturbances. These results show that in the closed-loop system, attitude stabilization control and energy saving have been accomplished in spite of these undesired uncertainties in the system. Moreover, the flexibility in the choice of controller and control allocator parameters can be utilized to obtain desirable performance while meeting the position and rate constraints of the actuators. These control approaches provide the theoretical foundations for the practical application of the advanced control theory to spacecraft ACSs design.

8.5 Conclusions

In this chapter, a SPD nonlinear control strategy has been incorporated into a novel closed-loop constrained optimal CA framework to solve the attitude stabilization control problem of a rigid spacecraft. Specifically, a SPD nonlinear control law is firstly designed as the baseline/virtual feedback controller to achieve attitude stabilization explicitly, considering input saturation. Then, a closed-loop constrained optimal control allocation scheme is further developed by considering the actuator position and rate constraints simultaneously to suitably distribute the virtual control command into the available actuators. Furthermore, the asymptotical stability of the closed-loop system has been proved and guaranteed. Finally, numerical simulations of the proposed novel control strategy were also presented to demonstrate the advantages and improvements with respect to the time responses and energy consumption saving over existing methods, especially, by taking into account the actuator uncertainties and external disturbances as well. It should be pointed out that actuators faults such as loss of effectiveness or stuck are not considered directly in this chapter, which will be addressed in the next chapter.

References

1. Krstic, M., Tsiotras, P. (1999) Inverse optimal stabilization of a rigid spacecraft. IEEE Transactions on Automatic Control 44(5): 1042–1049
2. Xing, G.Q., Parvez, S.A. (2001) Nonlinear attitude state tracking control for spacecraft. Journal of Guidance, Control, and Dynamics 24(3): 624–626
3. Wie, B., Lu, J. (1995) Feedback control logic for spacecraft eigenaxis rotations under slew rate and control constraints. Journal of Guidance, Control, and Dynamics 18(6): 1372–1379
4. Junkins, J.L., Akella, M.R., Robinert, R.D. (1997) Nonlinear adaptive control of spacecraft maneuvers. Journal of Guidance, Control, and Dynamics 20(6): 1104–1110
5. Hu, Q.L. (2009) Robust adaptive sliding mode attitude maneuvering and vibration damping of three-axis-stabilized flexible spacecraft with actuator saturation limits. Nonlinear Dynamics 55(4): 301–321
6. Boskovic, J.D., Li, S. M., Mehra, R. K. (2004) Robust tracking control design for spacecraft under control input saturation. Journal of Guidance, Control, and Dynamics 27(4): 627–633
7. Wen, J.T.Y., Kreutz-Delgado, K. (1991) The attitude control problem. IEEE Transactions on Automatic Control 36(10): 1148–1162
8. Tsiotras, P. (1998) Further passivity results for the attitude control problem. IEEE Transactions on Automatic Control 43(11): 1597–1600
9. Xiao, B., Hu, Q.L., Zhang, Y.M. (2011) Angular velocity-free control for rigid spacecraft attitude stabilization under input constraint. IFAC Proceedings Volumes 44(1): 8491–8496
10. Ali, I., Radice, G., Kim, J. (2010) Backstepping control design with actuator torque bound for spacecraft attitude maneuver. Journal of Guidance, Control, and Dynamics 33(1): 254–259
11. Boskovic, J.D., Li, S., Mehra, R.K. (2001) Robust adaptive variable structure control of spacecraft under control input saturation. Journal of Guidance, Control, and Dynamics 24(1): 14–22
12. Su, Y.X., Zheng, C.H. (2011) Globally asymptotic stabilization of spacecraft with simple saturated proportional derivative control. Journal of Guidance, Control, and Dynamics 31(6): 1932–1935
13. Page, A., Steinberg, M. (2000) A closed-loop comparison of control allocation methods. In: AIAA Guidance, navigation, and Control Conference and Exhibit, AIAA-2000-4538
14. Zhang, Y., Chen, Z. (2012) A closed-loop control allocation method for satellite precision pointing. In: The 10th International Conference on Industrial Informatics, Beijing, China, pp 1108–1112
15. Buffington, J.M., Enns, D.F. (1996) Lyapunov stability analysis of daisy chain control allocation. Journal of Guidance, Control, and Dynamics 19(6): 1226–1230
16. Buffington, J.M., Enns, D.F., Teel, A.R. (1998) Control allocation and zero dynamics. Journal of Guidance, Control, and Dynamics 21(3): 458–464
17. Wallsgrove, R.J., Akella, M.R. (2005) Globally stabilizing saturated attitude control in the presence of bounded unknown disturbances. Journal of Guidance, Control, and Dynamics 28(5): 957–963

18. Slotine, J.J.E., Li, W. (1991) Applied Nonlinear Control. Prentice–Hall, New Jersey
19. Bodson, M. (2002) Evaluation of optimization methods for control allocation. Journal of Guidance, Control, and Dynamics 25(4): 703–711
20. Durham, W. C. (1993) Constrained control allocation. Journal of Guidance, Control, and Dynamics 16(4): 717–725
21. Zhang, Y.M., Suresh, V.S., Jiang, B., Theilliol, D. (2007) Reconfigurable control allocation against aircraft control effector failures. In: 2007 IEEE International Conference on Control Applications, IEEE, pp 1197–1202
22. Yildiz, Y., Kolmanovsky, I. (2011) Stability properties and cross-coupling performance of the control allocation scheme CAPIO. Journal of Guidance, Control, and Dynamics 34(4): 1190–1196
23. Härkegård, O. (2004) Dynamic control allocation using constrained quadratic programming. Journal of Guidance, Control, and Dynamics 27(6): 1028–1034
24. Härkegård, O., Glad, S.T. (2005) Resolving actuator redundancy–optimal control vs. control allocation. Automatica 41(1): 137–144
25. Golub, G.H., Van Loan, C.F. (2012) Matrix computations, vol 3. JHU press
26. Shi, Z., Li, B., Hu, Q.L., Gao, Q.J. (2012) Dynamic control allocation based-on robust adaptive backstepping control for attitude maneuver of spacecraft with redundant reaction wheels. In: 2012 24th Chinese Control and Decision Conference (CCDC), IEEE, pp 1354–1359
27. Franklin, G.F., Powell, J.D., Workman, M.L. (1998) Digital control of dynamic systems. Addison-wesley Menlo Park, CA
28. Zhang, F.Z. (2011) Matrix theory: Basic results and techniques. Springer Science and Business Media

Chapter 9
Closed-Loop Based Control Allocation for Spacecraft Attitude Stabilization with Actuator Faults

9.1 Introduction

During the aerospace missions for the on-orbiting spacecraft, actuator faults are frequently encountered such that the control system performance may be degraded, even the unstable ACS may result in the complete collapse, immeasurable losses or catastrophic consequences. Thus, the spacecraft ACS should be capable of dealing with or tolerating possible actuator faults, while achieving an acceptable control system performances in the presence of actuator faults at the same time. There exist some main approaches to improve the stability and reliability of the spacecraft ACS, such as high reliable control system design with redundant devices, FDD based FTC system design, CA based FTC system design and so on. Recently, CA techniques have drawn significant interests in the aircraft and spacecraft control [1–13], especially in the FTC system design. Gui et al. [14] introduced an adaptive SMC law to dynamically compensate the perturbations due to actuator faults and other uncertainties, in which the real time CA algorithm delivers the control command to actuators. In [15], a novel dynamic near-optimal CA scheme with combination of a saturated baseline controller for spacecraft attitude control using single-gimbal control moment gyros. This dynamic control allocation is implemented with online update law, which has a modest computational complexity. All these methods can primarily generate commands to control effectors to achieve some demand moment or acceleration. Some of them take in account the position and rate limits, while some also allow for satisfying a secondary objective. Generally speaking, all the aforementioned CA can be implemented practically to address the concerned constraints and objectives. However, the authors in [16] illustrated that different CA algorithms can lead to different closed-loop maneuver performances owing to the CA errors between the designed signal and actual input. Usually, this CA error is treated as small and neglectable in the control design, which, in fact, may lead to a degraded performance or even potential closed-loop instability. In all the aforementioned references, it is assumed

Q. Hu et al., *Control Allocation for Spacecraft Under Actuator Faults*,
https://doi.org/10.1007/978-981-16-0439-3_9

that the system state trajectories are governed by the baseline controller, and CA algorithms have no/little effect on the closed-loop performance.

However, the separation-property assumption is impracticable in most situations, especially considering the actuator dynamics, actuator faults, and the uncertainties of the control distribution matrix. The mismatch between the actual and demand control effects has been considered in [17–23]. In [17], an online sliding-mode CA scheme is proposed for a linear system with actuator faults. Although the closed-loop stability has been proved in the presence of a bounded mismatch between the actual and reconstructed faults, the CA algorithm is a static linear mapping that cannot optimize its solution to the fault-reconstruction errors. In [18], Doman et al. proposed a quantized control allocator, which acts as a quantization element that converts continuous torque-command vectors into quantized torque vectors. Utilizing the stability analysis for quantized control systems, a sufficient condition is given on the state stability region in the presence of large reaction CA errors. In [19, 20], direct commands are directly given to individual actuators.Yang et al. [24] cast the CA as a nonlinear convex programming problem which depicts the desired objective function subject to three-axis moment demands and the limits of effectors' movement demands and the limits of effectors movement. The method can analytically and deterministically give one optimal solution for CA and prove this optimal solution to be guaranteed within a certainly maximal computation time. In [25], a novel fault tolerant integral sliding mode CA schemes for a class of over-actuated affine nonlinear system is developed. The online control allocation reroutes the control effort to the healthy actuators using knowledge of the actuator effectiveness level estimates. Wang et al. [26] introduced a novel adaptive sliding mode based control allocation scheme is proposed for a modified helicopter to accommodate simultaneous actuator faults. The low-level control allocation/re-allocation module is used to distribute the control signals among the required actuators, which can also reconfigure the distribution of the control signals in the presence of actuator faults. Although the system stability analysis was given rigorously, the controller employed the static PICA. In [21, 22], the robust CA problem was considered with uncertainties in the distribution matrix. The proposed algorithms were designed to minimize the gap between the designed control signals and the actual control effects. In [27], a new actuator-dynamics-based dynamic CA scheme is developed for flight control of a small fixed-wing UAV with direct lift control to compensate unknown, external disturbances. Nevertheless, the closed-loop stability with the robust CA remains to be proved. Moreover, these robust CA methods still conjecture the existence of a separation property. In [23], a closed-loop CA algorithm is proposed with the stability of both the nominal system and the CA part guaranteed separately. In summary, the stability analysis of a closed-loop system under the combination of CA and baseline controller remains a problem, and has not been satisfactorily solved at present especially for the control system in the presence of actuators' faults.

This chapter continues the study of the closed-loop CA issue for spacecraft attitude stabilization in the previous chapter. Compared with the Chap. 8, the actuator faults are taken into consideration simultaneously. More specifically, a PD-like control law designed by the dynamic attitude-gain technique is first presented as a virtual or baseline control law to achieve the spacecraft attitude stabilization control and it is robust to the bounded fault-estimation uncertainties. Then, a robust CA method is developed in this chapter, and an analytical solution is obtained. The proposed CA algorithm ensures a precisely quantifiable degree of robustness with respect to the actuator faults and its estimation uncertainties. Unlike the existing robust CA algorithms, the control scheme given in this chapter considers the faults which lead to the unavailability of the control power, and improves the system robustness in regard to the fault and uncertainty. In addition, the approach presented here provides asymptotic stability backed up by a rigorous closed-loop-stability analysis. The closed-loop stability with online control allocator is authenticated via a Lyapunov-like analysis and further evaluated through simulations.

(*Remark: The main control schemes and contents of this chapter are from the published journal paper "Qinglei Hu, Xiao Tan and Maruthi R. Akella, Closed-loop-based control allocation for spacecraft attitude stabilization with actuator fault, Journal of Guidance, Control, and Dynamics, 2018, 41(4): 944-953". The authors appreciate the permission from the American Institute of Aeronautics and Astronautics to reuse partly the results published in the relevant Journal.*)

9.2 Problem Formulation

9.2.1 System Schematic Diagram

Consider the scenario when certain actuator faults occur. In particular, consider the situation in which redundant actuators are available, but a subset of those actuators suffer complete or partial loss of control power. In this scenario, the relationship between the actual control input and the command signal is modeled as

$$\begin{aligned} \boldsymbol{\tau}_{actual} &= \boldsymbol{D}\boldsymbol{u}_{actual} \\ \boldsymbol{u}_{actual} &= \boldsymbol{E}\boldsymbol{u}_{command} \end{aligned} \tag{9.1}$$

where $D \in \mathbf{R}^{3\times m}$ is the actuator distribution matrix; $\boldsymbol{E} = \text{diag}(e_1, e_2, \ldots, e_m)$ denotes the effectiveness matrix with every diagonal element $0 \le e_i \le 1$. To facilitate the CA algorithm implementation, redundancy among actuators is required, that is, rank $(\boldsymbol{DE}) = 3$ is assumed during the maneuver.

In order to compensate for the actuator faults, it is assumed that the actuator fault information has been successfully obtained from a fault detection and diagnosis (FDD) scheme, which generally includes fault detection, isolation, and identification [28–31]. However, it should be noted that no matter what kind of FDD scheme is employed, there often exists nontrivial estimation error concerning the faults. In this chapter, we model the relationship between actual faults E and their estimated values $\hat{E}$ in a fractional form $\boldsymbol{E} = (\boldsymbol{I} + \Delta\boldsymbol{E})\,\hat{\boldsymbol{E}}$, where the estimated values $\hat{\boldsymbol{E}} = \text{diag}\left(\hat{e}_1, \hat{e}_2, \ldots, \hat{e}_m\right)$ and error matrix $\Delta\boldsymbol{E} = \text{diag}\left(\Delta e_1, \Delta e_2, \ldots, \Delta e_m\right)$ are diagonal. The following assumptions are made on the estimated faults and errors

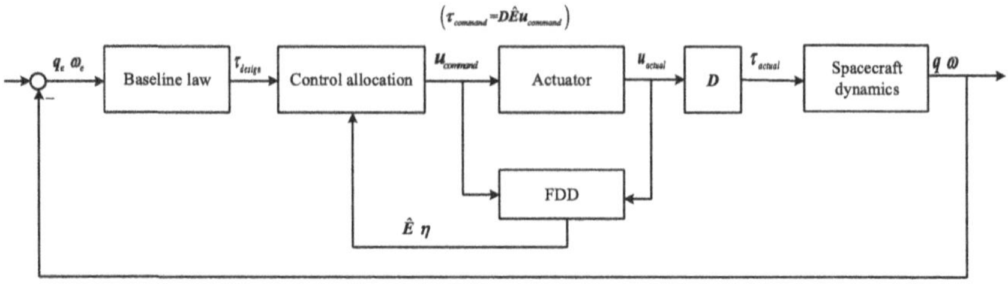

Fig. 9.1 Active fault-tolerant control schematic diagram

$$0 \le \hat{e}_i, \; -\eta \le \Delta e_i \le \eta, \text{ for } i = 1, 2, \ldots, m \tag{9.2}$$

The positive constant scalar η represents the level of FDD estimation imprecision and is within a known bound to designers.

The overall structure of the attitude controller incorporating online CA is synthesized into Fig. 9.1. The notation $\boldsymbol{\tau}_{design}$ represents the torque designed by the baseline controller. $\boldsymbol{u}_{command}$ denotes the command signal to each individual actuator from the CA scheme. $\boldsymbol{\tau}_{command}$ is then defined as $\boldsymbol{\tau}_{command} \triangleq \boldsymbol{D}\hat{\boldsymbol{E}}\boldsymbol{u}_{command}$; $\boldsymbol{u}_{actual}$ is the actual torque/force actuators effect on the vehicle, $\boldsymbol{\tau}_{actual}$ represents the actual total torque that actuates the attitude maneuver. When any particular actuator undergoes a partial effectiveness fault, the FDD scheme estimates the faulty information $\hat{\boldsymbol{E}}$ with uncertainty level η . The red blocks show where the discrepancies between $\boldsymbol{\tau}_{design}$ and $\boldsymbol{\tau}_{actual}$ originate. Due to position/rate limits, $\boldsymbol{\tau}_{design}$ is not necessarily equal to $\boldsymbol{\tau}_{command}$. In the fault-actuator scenario, considering that the FDD cannot provide an exact information about the actuator faults, $\boldsymbol{\tau}_{command}$ is usually not equal to $\boldsymbol{\tau}_{actual}$. For brevity, we use notations $\boldsymbol{\tau}_d$, $\boldsymbol{\tau}_a$, $\boldsymbol{u}_c$ and $\boldsymbol{u}_a$ for $\boldsymbol{\tau}_{design}$, $\boldsymbol{\tau}_{actual}$, $\boldsymbol{u}_{commnad}$ and $\boldsymbol{u}_{actual}$ in the subsequent sections, respectively.

9.2.2 Problem Formulation

Considering the FDD estimation uncertainty mentioned above, the mismatch between the control demand and the actual control input $\boldsymbol{\tau}_{actual} - \boldsymbol{\tau}_{design}$ can only be minimized but not fully eliminated. Theoretically, the discrepancy may lead to a degradation of pointing accuracy or even system instability followed by catastrophic outcomes. Thus, the main motivation for this work is to provide a rigorous assurance of closed-loop stability. The control objective is to design a baseline controller combined with robust CA algorithms that ensure closed-loop stability in the presence of actuator faults and FDD estimation uncertainty.

In this chapter, the notation $\|\cdot\|$ refers to Euclidean norm and $\lambda(\boldsymbol{A})$ denotes the eigenvalues of $\boldsymbol{A} \in \mathbf{R}^{n \times n}$. Abbreviation $\boldsymbol{A} > 0$ is used to describe a real-valued symmetric positive definite matrix and $\boldsymbol{A} \ge 0$ denotes a positive semi-definite matrix.

9.3 Control Scheme Design and Stability Analysis

CA is used for the control of redundantly actuated systems. Generally speaking, it is possible and reasonable to split the controller design into two steps: (1) Design a baseline control law specifying the total control effort to be produced; and (2) Design a control allocator that maps the total control demand onto individual actuator settings. In this section, we propose a modified proportional-derivative (PD) type baseline control law and a robust least-squares allocator. A detailed closed-loop stability analysis then follows.

9.3.1 *Baseline Control Law Design*

Examining the existing literature on spacecraft attitude control, one will find extensive control schemes based on various approaches, such as PID control [32, 33], nonlinear feedback control [34], SMC [35] and adaptive control [36]. While these control schemes are effective in different scenarios, PD control stands out for its simplicity, practicability, and is easy-to-tune. In the subsequent design, we will show how a simple PD-type controller is modified to be implemented to achieve closed-loop stability.

The PD-type baseline law is typically designed as

$$\boldsymbol{\tau}_d = -k\left(\boldsymbol{\omega} + k_1 \boldsymbol{q}_v\right) \tag{9.3}$$

where gain k is a positive constant. In this work, utilizing the dynamic attitude-gain technique [37], the attitude gain k_1 satisfies the following first-order dynamics

$$\dot{k}_1 = -\gamma_k \left(1 + \frac{1}{\mu}\right) k k_1 \boldsymbol{q}_v^{\mathrm{T}} \boldsymbol{q}_v \tag{9.4}$$

where γ_k, μ are user-selected positive constants and $k_1(0) > 0$. A detailed analysis of the behavior of $k_1(t)$ is presented in the following subsection.

The stability of the nominal system in Eqs. (2.14) (2.20) using the baseline controller in Eq. (9.3) can be shown through widely documented techniques from the literature. For example, refer to [32] for more details.

9.3.2 *Online Robust Control Allocation Design*

From previous analysis, although there exist rich results on CA algorithms, the stability of the overall closed-loop control system under the combination of a baseline controller and CA algorithm is still an open problem, especially under a faulty actuator scenario.

To find the optimal control command minimizing the discrepancies between the designed control demand and actual control input, a robust least-squares CA problem (RobCA) is considered as

$$\boldsymbol{u}_c = \arg\min_{u} \max_{-\eta \le \Delta e_i \le \eta} \left\| \boldsymbol{D}\hat{\boldsymbol{E}}\boldsymbol{u} + \boldsymbol{D}\Delta\boldsymbol{E}\hat{\boldsymbol{E}}\boldsymbol{u} - \boldsymbol{\tau}_d \right\| \tag{9.5}$$

It should be noted that if $\eta = 0$, this problem reduces to a standard least-squares problem with a well-known pseudo-inverse solution. A similar min-max optimization problem is solved in [38].

To reduce the notational burden, let $\boldsymbol{A} \triangleq \boldsymbol{D}\hat{\boldsymbol{E}}$, $\Delta\boldsymbol{A} \triangleq \boldsymbol{D}\Delta\boldsymbol{E}\hat{\boldsymbol{E}}$. Note that $\|\Delta\boldsymbol{A}\| \le \eta \|\boldsymbol{A}\|$. Then the worst-case residual is

$$r(\boldsymbol{u}) = \max_{-\eta \le \Delta e_i \le \eta} \|(\boldsymbol{A}\boldsymbol{u} - \boldsymbol{\tau}_d) + \Delta\boldsymbol{A}\boldsymbol{u}\|$$

Using the triangle inequality, one can write

$$r(\boldsymbol{u}) \le \|\boldsymbol{A}\boldsymbol{u} - \boldsymbol{\tau}_d\| + \eta \|\boldsymbol{A}\| \|\boldsymbol{u}\|$$

Now choose $\Delta\boldsymbol{A} = \frac{\eta\|\boldsymbol{A}\|}{\|\boldsymbol{u}\|} \upsilon \boldsymbol{u}^{\mathrm{T}}$ with

$$\upsilon = \begin{cases} \frac{\boldsymbol{A}\boldsymbol{u} - \boldsymbol{\tau}_d}{\|\boldsymbol{A}\boldsymbol{u} - \boldsymbol{\tau}_d\|}, & \text{if } \boldsymbol{A}\boldsymbol{u} \ne \boldsymbol{\tau}_d \\ \text{any unit norm vector}, & \text{otherwise} \end{cases}$$

Then, in the direction of υ, the worst-case residual can be rewritten as

$$\|(\boldsymbol{A}\boldsymbol{u} - \boldsymbol{\tau}_d) + \Delta\boldsymbol{A}\boldsymbol{u}\| = \|\boldsymbol{A}\boldsymbol{u} - \boldsymbol{\tau}_d\| + \eta \|\boldsymbol{A}\| \|\boldsymbol{u}\|$$

This problem can be formulated as the second-order cone programming (SOCP)

$$\begin{aligned} &\min_{\boldsymbol{u},\,\alpha,\,\beta} \alpha \\ &\text{subject to } \|\boldsymbol{A}\boldsymbol{u} - \boldsymbol{\tau}_d\| \le \alpha - \beta, \\ &\qquad \eta \|\boldsymbol{A}\| \|\boldsymbol{u}\| \le \beta \end{aligned} \tag{9.6}$$

where α is the optimization variable to be minimized. Then the following results can be stated.

Theorem 9.1 *Considering the RobCA problem in Eq. (9.5), the optimal solution* $\boldsymbol{u}_c$ *is analytically expressed by*

$$\boldsymbol{u}_c = \begin{cases} \boldsymbol{A}^{\mathrm{T}}\left(\varepsilon \boldsymbol{I} + \boldsymbol{A}\boldsymbol{A}^{\mathrm{T}}\right)^{-1}\boldsymbol{\tau}_d, & \text{for } \varepsilon = \eta^2 \|\boldsymbol{A}\|^2 \frac{\alpha-\beta}{\beta} > 0 \\ \boldsymbol{A}^{\mathrm{T}}\left(\boldsymbol{A}\boldsymbol{A}^{\mathrm{T}}\right)^{-1}\boldsymbol{\tau}_d, & \text{else} \end{cases} \tag{9.7}$$

where $\varepsilon > 0$, $\boldsymbol{A} \triangleq \boldsymbol{D}\hat{\boldsymbol{E}}$, α *and* β *are the optimal solution to the problem in Eq. (9.6).*

Proof: The dual problem of SOCP (9.6) is

$$\begin{aligned} &\text{maximize } \boldsymbol{\tau}_d^{\mathrm{T}} z_1 \\ &\text{subject to } \boldsymbol{A}^{\mathrm{T}} z_1 + z_2 = 0, \\ &\qquad \|z_1\| \le 1, \\ &\qquad \|z_2\| \le \eta \|\boldsymbol{A}\| \end{aligned} \tag{9.8}$$

where $z_1 \in R^3$, $z_2 \in R^n$ are the dual variables.

Since both the primal and dual problems are feasible, there exist optimal points for both of them. If $\alpha = \beta$ at the optimum, then $\boldsymbol{A}\boldsymbol{u} = \boldsymbol{\tau}_d$, and

$$\alpha = \eta \|\boldsymbol{A}\| \|\boldsymbol{u}\|$$

In this case, the optimal $\boldsymbol{u}$ is the minimum norm solution to $\boldsymbol{A}\boldsymbol{u} = \boldsymbol{\tau}_d$ (i.e., $\boldsymbol{u} = \boldsymbol{A}^{\dagger}\boldsymbol{\tau}_d$).

Now assume $\alpha > \beta$. According to the optimization goal, the primal and dual optimal objectives are rewritten as

$$\|\boldsymbol{A}\boldsymbol{u} - \boldsymbol{\tau}_d\| + \eta \|\boldsymbol{A}\| \|\boldsymbol{u}\| = \alpha = \boldsymbol{\tau}_d^{\mathrm{T}} z_1 = -\left(\boldsymbol{A}\boldsymbol{u} - \boldsymbol{\tau}_d\right)^{\mathrm{T}} z_1 - \boldsymbol{u}^{\mathrm{T}}\left(-\boldsymbol{A}^{\mathrm{T}} z_1\right)$$

For $\boldsymbol{A}^{\mathrm{T}} z_1 + z_2 = 0$, one has $\|\boldsymbol{A}\boldsymbol{u} - \boldsymbol{\tau}_d\| + \eta \|\boldsymbol{A}\| \|\boldsymbol{u}\| = -\left(\boldsymbol{A}\boldsymbol{u} - \boldsymbol{\tau}_d\right)^{\mathrm{T}} z_1 - \boldsymbol{u}^{\mathrm{T}} z_2$. With $\|z_1\| \le 1$ and $\|z_2\| \le \eta \|\boldsymbol{A}\|$ in mind, one can further have

$$z_1 = -\frac{\boldsymbol{A}\boldsymbol{u} - \boldsymbol{\tau}_d}{\|\boldsymbol{A}\boldsymbol{u} - \boldsymbol{\tau}_d\|}, \qquad z_2 = -\eta \|\boldsymbol{A}\| \frac{\boldsymbol{u}}{\|\boldsymbol{u}\|}$$

Using these values in $\boldsymbol{A}^{\mathrm{T}} z_1 + z_2 = 0$, the following expression can be further obtained as

$$\boldsymbol{A}^{\mathrm{T}}\boldsymbol{A}\boldsymbol{u} + \eta \|\boldsymbol{A}\| \frac{\|\boldsymbol{A}\boldsymbol{u} - \boldsymbol{\tau}_d\|}{\|\boldsymbol{u}\|}\boldsymbol{u} = \boldsymbol{A}^{\mathrm{T}}\boldsymbol{\tau}_d$$

Then an analytical expression of the optimal solution is obtained

$$\boldsymbol{u}_c = \left(\varepsilon \boldsymbol{I} + \boldsymbol{A}^{\mathrm{T}}\boldsymbol{A}\right)^{-1}\boldsymbol{A}^{\mathrm{T}}\boldsymbol{\tau}_d$$

in which

$$\varepsilon \triangleq \eta \|\boldsymbol{A}\| \frac{\|\boldsymbol{A}\boldsymbol{u} - \boldsymbol{\tau}_d\|}{\|\boldsymbol{u}\|} = \eta^2 \|\boldsymbol{A}\|^2 \frac{\alpha - \beta}{\beta} > 0$$

In addition, since $\varepsilon > 0$, both $\left(\varepsilon \boldsymbol{I} + \boldsymbol{A}^{\mathrm{T}}\boldsymbol{A}\right)$ and $\left(\varepsilon \boldsymbol{I} + \boldsymbol{A}\boldsymbol{A}^{\mathrm{T}}\right)$ are invertible. By simple matrix transformation, one can further obtain

$$\begin{gathered}\left(\varepsilon \boldsymbol{I} + \boldsymbol{A}^{\mathrm{T}}\boldsymbol{A}\right)\boldsymbol{A}^{\mathrm{T}} = \boldsymbol{A}^{\mathrm{T}}\left(\varepsilon \boldsymbol{I} + \boldsymbol{A}\boldsymbol{A}^{\mathrm{T}}\right) \\ \Updownarrow \\ \boldsymbol{A}^{\mathrm{T}}\left(\varepsilon \boldsymbol{I} + \boldsymbol{A}\boldsymbol{A}^{\mathrm{T}}\right)^{-1} = \left(\varepsilon \boldsymbol{I} + \boldsymbol{A}^{\mathrm{T}}\boldsymbol{A}\right)^{-1}\boldsymbol{A}^{\mathrm{T}}\end{gathered}$$

This completes the proof of Theorem 9.1. □

The algorithm for the RobCA problem is summarized as follows:

(1) Solve the SOCP problem (9) by software (YALMIP, etc) to obtain the optimal α and β.

(2) Calculate $\varepsilon = \eta^2 \left\|\boldsymbol{A}\right\|^2 \frac{\alpha-\beta}{\beta}$.

(3) Obtain the solution of the RobCA problem by calculating $\boldsymbol{u}_c$ in Eq. (9.7).

9.3.3 Online Pseudo-Inverse Control Allocation

Typically, under healthy actuator conditions, the CA algorithm with more than three actuators for three-axis stabilization can be derived using a least-squares method, that is

$$\begin{gathered}\min_{\boldsymbol{u}} \ \|\boldsymbol{u}\| \\ \text{subject to } \boldsymbol{D}\boldsymbol{u} = \boldsymbol{\tau}_d\end{gathered}$$

which is equivalent to the well-known pseudo-inverse solution $\boldsymbol{u} = \boldsymbol{D}^{\dagger}\boldsymbol{\tau}_d$, where $\boldsymbol{D}^{\dagger} = \boldsymbol{D}^{\mathrm{T}}\left(\boldsymbol{D}\boldsymbol{D}^{\mathrm{T}}\right)^{-1}$. In the faulty scenario, since the actuator effectiveness matrix $\boldsymbol{E}$ is unknown and estimated to be $\hat{\boldsymbol{E}}$ by the online FDD, a corresponding pseudo-inverse control allocation (PICA) algorithm is proposed

$$\boldsymbol{u}_c = \boldsymbol{A}^{\mathrm{T}}\left(\boldsymbol{A}\boldsymbol{A}^{\mathrm{T}}\right)^{-1}\boldsymbol{\tau}_d \tag{9.9}$$

9.3.4 Closed-Loop Stability with Online Control Allocation Analysis

Though the CA-based control schemes have been studied by many researchers [5, 6, 10], most of the preceding CA schemes are studied as an open-loop system, and the stability issue of a closed-loop system is not considered due to the inherent difficulty for such a control-design task. In this subsection, we provide a complete stability proof of the overall closed-loop system with the PD-type baseline control law and the control allocators.

As a stepping stone, the closed-loop stability incorporating the online PICA algorithm is analyzed first, which is useful for analyzing the RobCA scheme in the next step.

Theorem 9.2 *Consider the rigid spacecraft attitude system given by Eqs. (2.14) and (2.20) with loss of actuator effectiveness faults expressed in Eq. (9.1). With application of the baseline control law in Eqs. (9.3) (9.4), and the PICA algorithm in Eq. (9.9), suppose that is a sufficiently small constant satisfying the condition in Eq. (9.14)in Appendix A, and the fault estimation error satisfies*

$$\eta \leq \frac{\xi}{\left\|\boldsymbol{A}\boldsymbol{A}^{\mathrm{T}}\right\| \left\|\left(\boldsymbol{A}\boldsymbol{A}^{\mathrm{T}}\right)^{-1}\right\|}$$

during the maneuver, where $\xi < 1$ is a user-chosen positive constant. Then, all of the signals in the resulting system are bounded, and for all initial conditions $\boldsymbol{q}_v(0)$ and $\boldsymbol{\omega}(0)$, the attitude and the angular velocity asymptotically converge to zero, i.e.,

$$\lim_{t\to\infty} \boldsymbol{q}_v(t) = \mathbf{0}$$
$$\lim_{t\to\infty} \boldsymbol{\omega}(t) = \mathbf{0}$$

Proof: Please refer to Appendix A for details. □

Now the proposed RobCA algorithm is taken into consideration. This time, the closed-loop stability for a PD controller incorporating the RobCA is analyzed.

Theorem 9.3 *Consider the rigid spacecraft attitude system given by Eqs. (2.14) and (2.20) with loss of actuator effectiveness faults expressed in Eq. (9.1). With application of the baseline control law in Eqs. (9.3)–(9.4) and the RobCA algorithm in Eq. (9.7), suppose that γ_k is a sufficiently small constant satisfying the condition in Eq. (9.18), and the fault estimation error satisfies*

$$\eta \leq \frac{\xi - \varepsilon\left\|\left(\varepsilon\boldsymbol{I} + \boldsymbol{A}\boldsymbol{A}^{\mathrm{T}}\right)^{-1}\right\|}{\left\|\boldsymbol{A}\boldsymbol{A}^{\mathrm{T}}\right\| \left\|\left(\varepsilon\boldsymbol{I} + \boldsymbol{A}\boldsymbol{A}^{\mathrm{T}}\right)^{-1}\right\|}$$

during the maneuver, where $\xi < 1$ is a user-chosen constant. Then, all of the signals in the resulting system are bounded, and asymptotic stability of the closed-loop attitude system is guaranteed. That is, for all initial conditions $\boldsymbol{q}_v(0)$ and $\boldsymbol{\omega}(0)$, the attitude and the angular velocity converge to zero, i.e.,

$$\lim_{t\to\infty} \boldsymbol{q}_v(t) = \mathbf{0}$$
$$\lim_{t\to\infty} \boldsymbol{\omega}(t) = \mathbf{0}$$

Proof please refer to Appendix B for details. □

Remark 9.1 In the stability analysis, the FDD estimation imprecision η is assumed to be within a state-dependent constraint. This assumption explicitly demands the estimation accuracy of the employed FDD scheme and provides a guideline for FDD design. From another perspective, the assumption explicitly states the fault-tolerant capability of the baseline controller.

Remark 9.2 It is worth noting that in [37], the attitude error can be regulated asymptotically to zero in the disturbance-free scenario, regardless of the behavior of the time-varying attitude gain. This feature is mainly attributed to the additional "leakage term" in the control law. However, this technique is not applicable in this scenario. The main obstruction is that to ensure $\dot{\omega} \to \mathbf{0}$ with Barbalat's lemma, we need to show $\ddot{\omega} \in L_{\infty}$. Considering the dynamics for attitude motion with actuator faults, it leads to the assumption that $\boldsymbol{E}(t)$ is a uniform continuous function. However, the actuator faults are usually stochastic and happen abruptly, so the entries in $\boldsymbol{E}(t)$ jump from one value to another.

Remark 9.3 The most fascinating part of the proposed fault-tolerant control strategy is that, with an independently designed baseline controller, FDD scheme and online control allocator, the overall closed-loop stability is still theoretically retained. Based on this control framework, one can design or improve the baseline controller, FDD and CA algorithm separately without worrying about closed-loop stability.

9.4 Simulation Results

To verify the effectiveness of the proposed control scheme, simulations are applied to a rigid spacecraft with four RWs. The four wheels are mounted in a pyramid configuration as shown in Chap. 3 three of which rotate parallel to the satellite's body axes and the fourth is installed with the equiangular direction with the three principal axes. In this case, the RW configuration matrix is

$$\boldsymbol{D} = \begin{bmatrix} 1\,0\,0\,\sqrt{3}/3 \\ 0\,1\,0\,\sqrt{3}/3 \\ 0\,0\,1\,\sqrt{3}/3 \end{bmatrix}$$

The inertia matrix of the rigid spacecraft is

$$\boldsymbol{J} = \begin{bmatrix} 20 & 0 & 0.9 \\ 0 & 17 & 0 \\ 0.9 & 0 & 15 \end{bmatrix}$$

During the maneuver, spacecraft actuators are subjected to random partial effectiveness loss. The time-varying faults can be estimated by FDD schemes in real time. Whatever FDD scheme is employed, random faults are always estimated with imprecision. The degree of estimation imprecision is denoted by the estimation-uncertainty

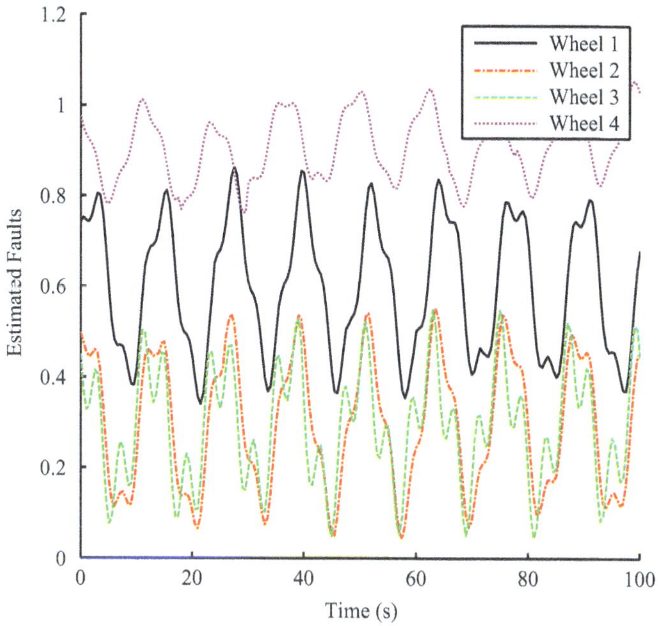

Fig. 9.2 FDD generated fault estimation information $\hat{E}$

upper bound η. To simplify the simulation, the estimated faulty information $\hat{\boldsymbol{E}}$ from FDD is arbitrarily generated by the following function:

$$\hat{e}_i(t) = \begin{cases} 0.6 + 0.20\text{rand}(t_i) + 0.20\sin(0.5t + \pi/6) + 0.05\cos(0.5\pi t + \pi/6), & i = 1 \\ 0.3 + 0.02\text{rand}(t_i) + 0.20\sin(0.5t + \pi/3) + 0.05\cos(0.5\pi t + \pi/3), & i = 2 \\ 0.3 + 0.02\text{rand}(t_i) + 0.15\sin(0.5t + \pi/2) + 0.10\cos(0.5\pi t + \pi/2), & i = 3 \\ 0.9 + 0.05\text{rand}(t_i) + 0.10\sin(0.5t + 2\pi/3) + 0.02\cos(0.5\pi t + 2\pi/3), & i = 4 \end{cases}$$

where $\text{rand}(\cdot) \in (-1, 1)$ is a random number generator, $t_i = \text{mod}(t + \delta t_i, T)$, $\delta t_i = 2i - 2$ and sampling time $T = 12$ s. The terms in $\hat{e}_i(t)$ simulate the average part, the 'jump' part and the continuous part in $\hat{\boldsymbol{E}}$. Figure 9.2 demonstrates the value of the estimated faulty information. Note that the estimated effectiveness value of Wheel 4 occasionally exceeds a value of 1, which is possible because, for example, the iterative-learning-observer (ILO) based FDD in [29] does not put a hard saturation on the estimation value.

For a fair comparison, we carry out two different cases: (1) the uncertainty $\eta = 5\%$; (2) the uncertainty $\eta = 33\%$. In the former case, assumptions in Eqs. (9.10) and (9.17) are satisfied and from the earlier theoretical analysis, the baseline controller with RobCA or PICA can handle case 1; we shall verify that claim later. In the latter case, assumptions in Eqs. (9.10) and (9.17) are no longer satisfied and the effectiveness of the proposed RobCA is compared with the conventional PICA. Con-

Table 9.1 Parameters of a rigid spacecraft

Parameters	Values
Initial quaternion	$\left[q_0, \boldsymbol{q}_v{}^{\mathrm{T}}\right]^{\mathrm{T}} = [0.9, -0.3, 0.26, 0.18]$
Closed-loop control	$\tau_{\min} = -0.5\,\mathrm{N.m}$, $\tau_{\max} = 0.5\,\mathrm{N.m}$,
Initial angular velocity	$\omega = [0.1, -0.1, -0.05]^{\mathrm{T}}\,\mathrm{rad/s}$
Controller parameters	$k = 3$, $k_1\,(0) = 0.2$, $\xi = 0.8$, $\mu = 0.8$, $\gamma_k = 0.004$
Actuator fault type	Efficiency loss

troller parameters are shown in Table 9.1 below. Note that these parameters remain the same throughout all the simulations.

The RobCA algorithm described in this paper has been implemented in YALMIP, a MATLAB-based optimization environment, and SeDuMi, a general toolbox for quadratic and semidefinite problems, for the following numerical results. For practical applications, a customized SOCP solver should be developed that can take advantage of various techniques such as iterating from a warm start, choosing a more suitable numerical method and tuning optimal parameters in that method. For more information on real-time implementation, one can refer to [39].

9.4.1 Performance with FDD Estimation Uncertainty Bound $\eta = 5\%$

During CA based spacecraft attitude stabilization control scheme design, the proposed control schemes can handle the estimation uncertainty and ensure closed-loop stability. In this scenario, the performance of proposed strategies is examined. Figures 9.3, 9.4, 9.5, 9.6, 9.7, 9.8 show the simulation results. There are three sub-figures on each figure, one for each of the PICA algorithm with $\eta = 5\%$, the RobCA with $\eta = 5\%$, and the PICA algorithm with $\eta = 0\%$. The PICA results with $\eta = 0\%$ (PICAU) serve as a baseline.

Figures 9.3, 9.4, 9.5 shows the performance of several different control schemes. System performance of the PICAU scheme shows that, in the uncertainty-free nominal condition, the attitude can be effectively controlled by the baseline controller. It is also observed that, although there exists FDD imprecision in PICA and RobCA schemes, the closed-loop stability is still ensured. The desired maneuver can still be performed with a high steady-state pointing accuracy, which is comparable to the nominal scenario. The time histories of the control torque and actual actuator torques are depicted in Figs. 9.6 and 9.7. Generally speaking, the actual torque of the RobCA and PICA algorithms follow the reference signal with minimal perturbation. To examine more features of the aforementioned control strategies, an energy-consumption index is introduced as $\boldsymbol{E} = \sum_{i=1}^{4} \int_0^T |u_{actual,\,i}(t)|^2 \mathrm{d}t$. The time response of the energy

Fig. 9.3 Time response of quaternion with $\eta = 5\%$

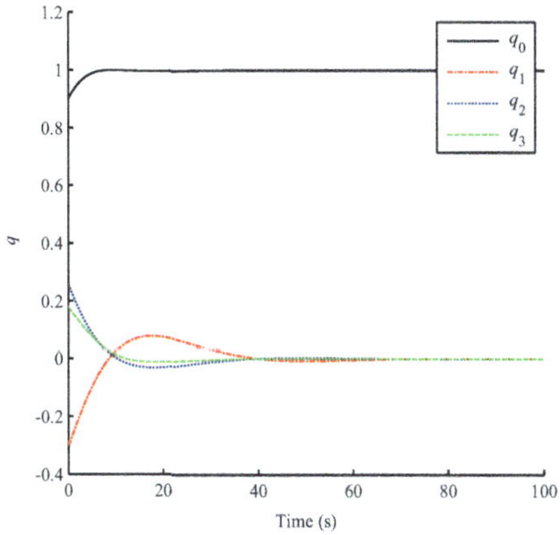

(a) PICA

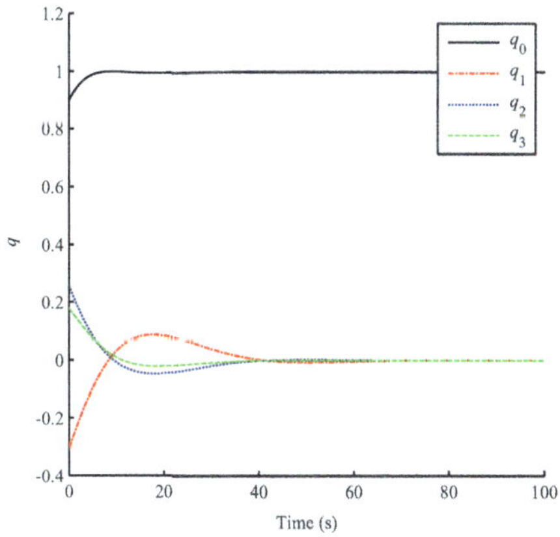

(b) PICAU

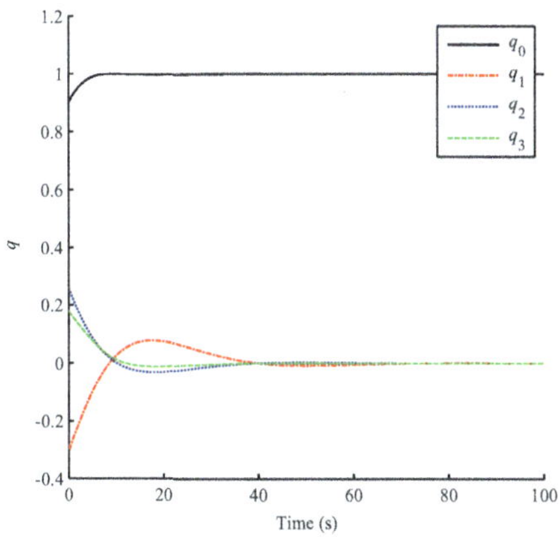

(c) RobCA

Fig. 9.4 Time response of angular velocity with $\eta = 5\%$

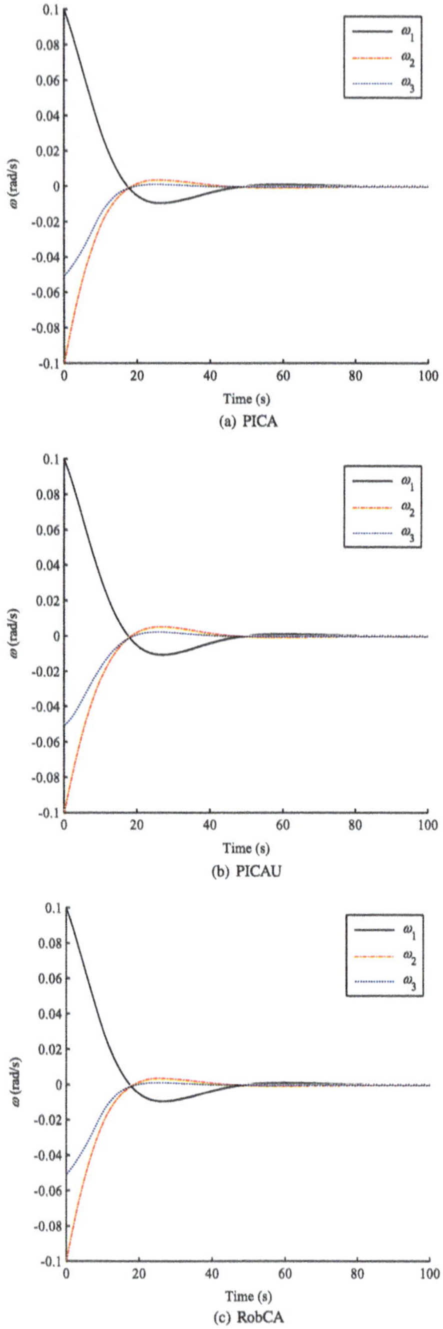

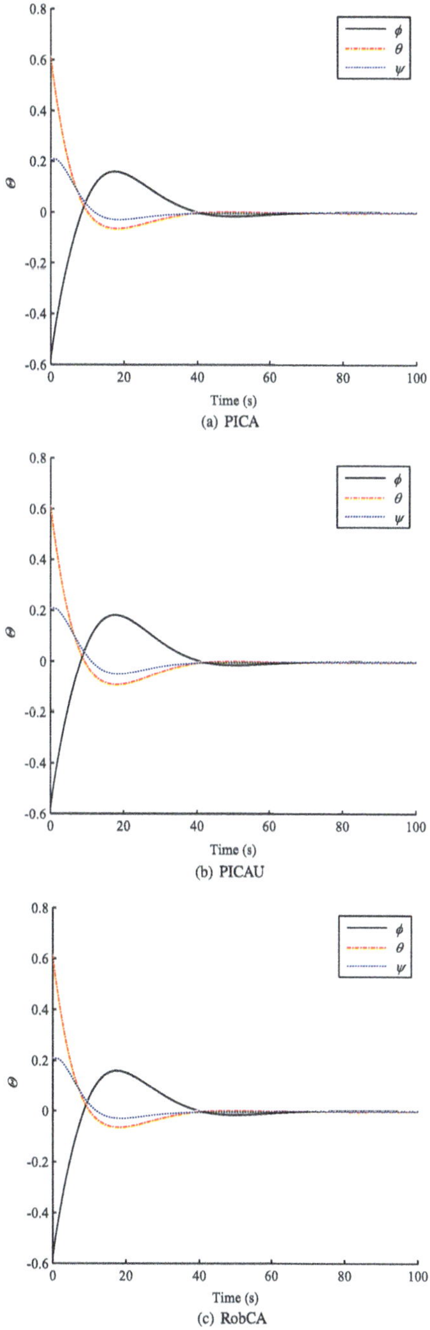

(a) PICA

(b) PICAU

(c) RobCA

Fig. 9.5 Time response of Euler angle with $\eta = 5\%$

Fig. 9.6 Time response of control torque with $\eta = 5\%$

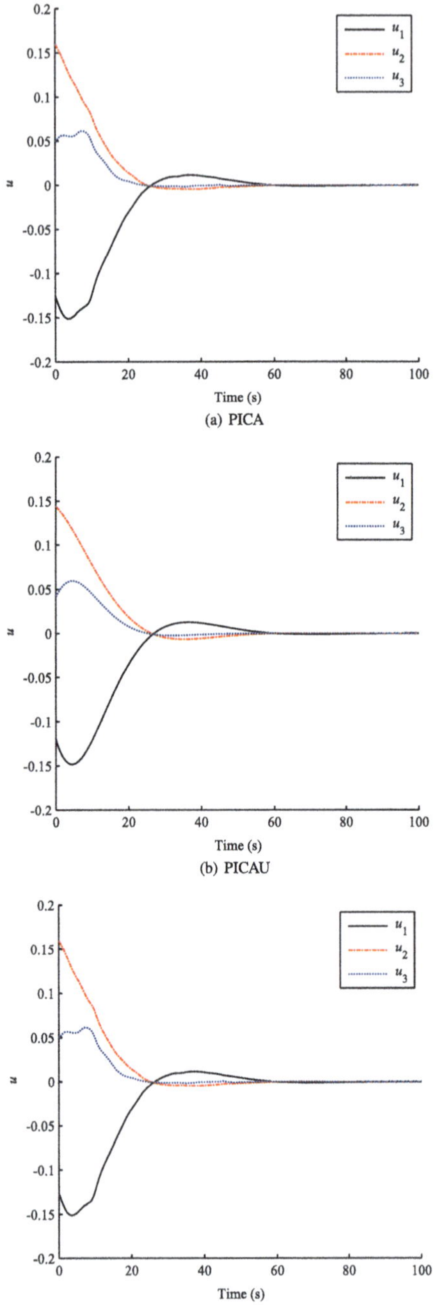

(a) PICA

(b) PICAU

(c) RobCA

Fig. 9.7 Time response of actual actuation torque with $\eta = 5\%$

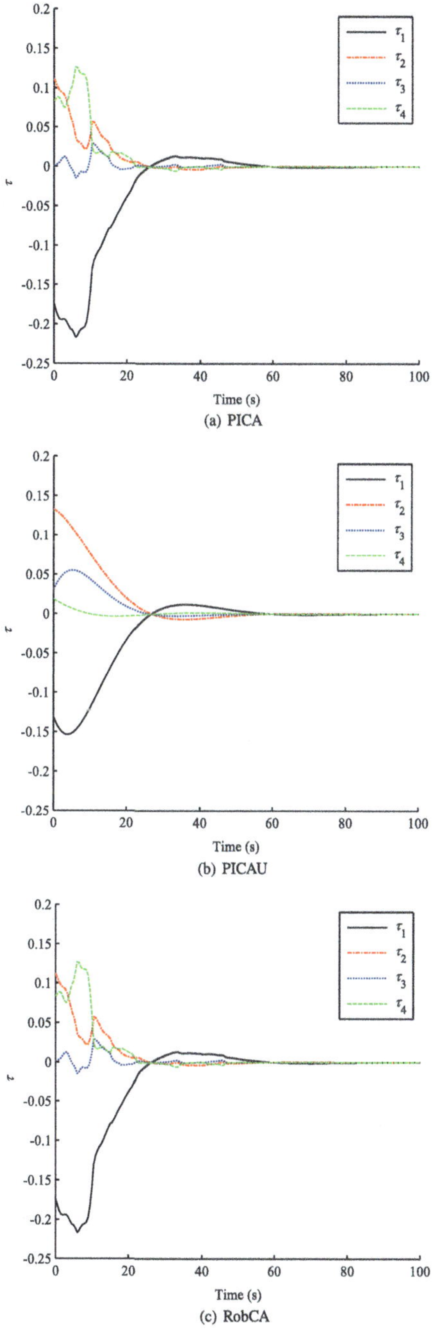

(a) PICA

(b) PICAU

(c) RobCA

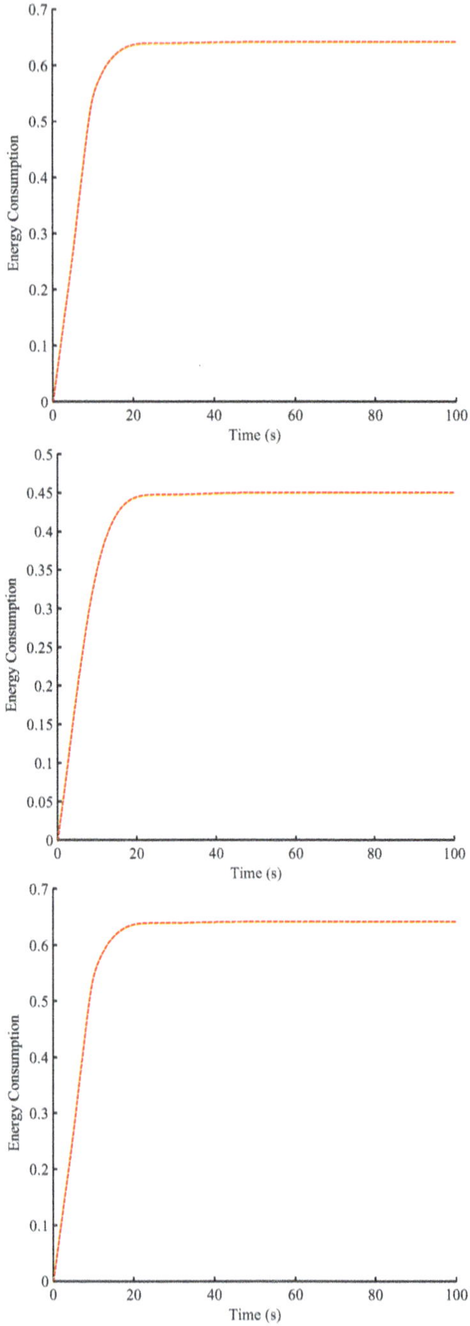

Fig. 9.8 Time response of energy consumption with $\eta = 5\%$

consumption is shown in Fig. 9.8. In the next subsection, we introduce a larger estimation uncertainty to facilitate the illustration of desirable properties of the RobCA scheme.

9.4.2 Performance with FDD Estimation Uncertainty Bound $\eta = 33\%$

In this case, the FDD estimation imprecision is assumed to be 33%. Note that the nominal system (PICAU) is simulated without estimation uncertainty and only used for illustrative purpose. The control performance are presented in Figs. 9.9, 9.10, 9.11, 9.12, 9.13, 9.14. Referring to Figs. 9.9, 9.10, 9.11, both the PICA and RobCA have degraded from the nominal performance under such severe FDD imprecision. Compared to the PICA scheme, the time responses of attitude and velocity of the RobCA have fewer oscillations and less overshoot. Figures 9.12 and 9.13 show the time response of the commanded control torques and actual output torques. The time response of the energy-consumption of the three schemes are shown in Fig. 9.14. One can obtain that the energy-consumption of the RobCA scheme is always smaller than that of the PICA scheme, which accounts for the better closed-loop performance in Figs. 9.9, 9.10, 9.11. It also reflects the superiority of the RobCA scheme from an energy perspective.

In summary, with the proposed control scheme (PD-type controller + online CA) applied, closed-loop attitude stabilization is achieved when bounded uncertainty exists in the FDD estimation. In addition, if the FDD imprecision η is too large to satisfy the constraints in Eqs. (9.10) and (9.17), the proposed RobCA algorithm performs better than PICA algorithm. That is to say, the RobCA is robust to a larger FDD estimation uncertainty.

9.5 Conclusions

In this chapter, a modified PD type baseline controller incorporating various CA schemes has been proposed for attitude stabilization in the presence of actuator faults and fault-estimation uncertainty. In the case of fractional actuation power loss, a RobCA algorithm is designed to reduce the worst-case discrepancies between the demand signal and the actual control input, which gains a better performance than a pseudo-inverse control allocator. As for the controller design, a PD baseline controller is proposed with its attitude-gain time varying, which provide a certain degree of robustness against fault-estimation uncertainty. Although the CA algorithm and the baseline controller are designed separately, the overall control system is rigorously proved to retain closed-loop asymptotic convergence irrespective of bounded

Fig. 9.9 Time response of quaternion with $\eta = 33\%$

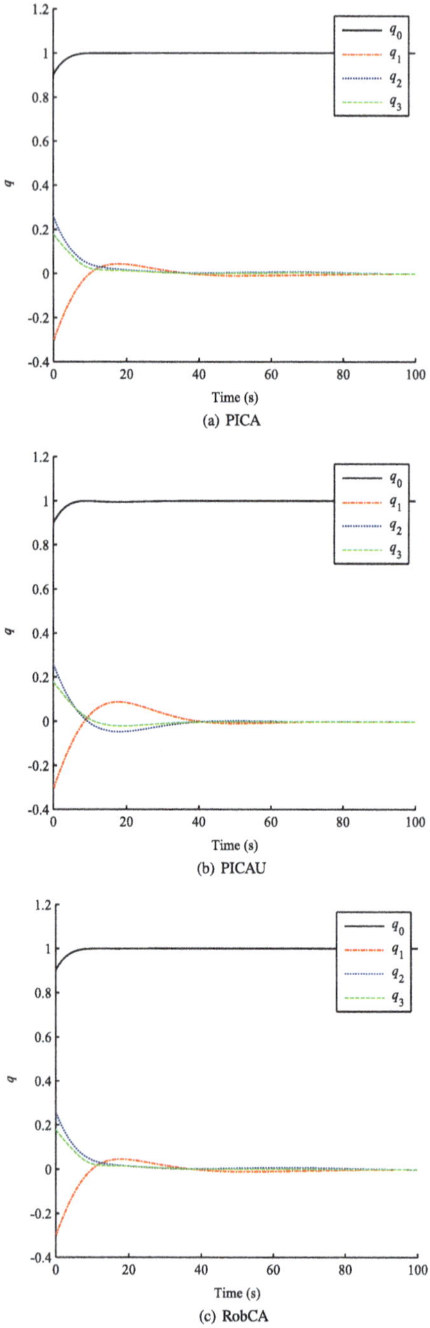

Fig. 9.10 Time response of angular velocity with $\eta = 33\%$

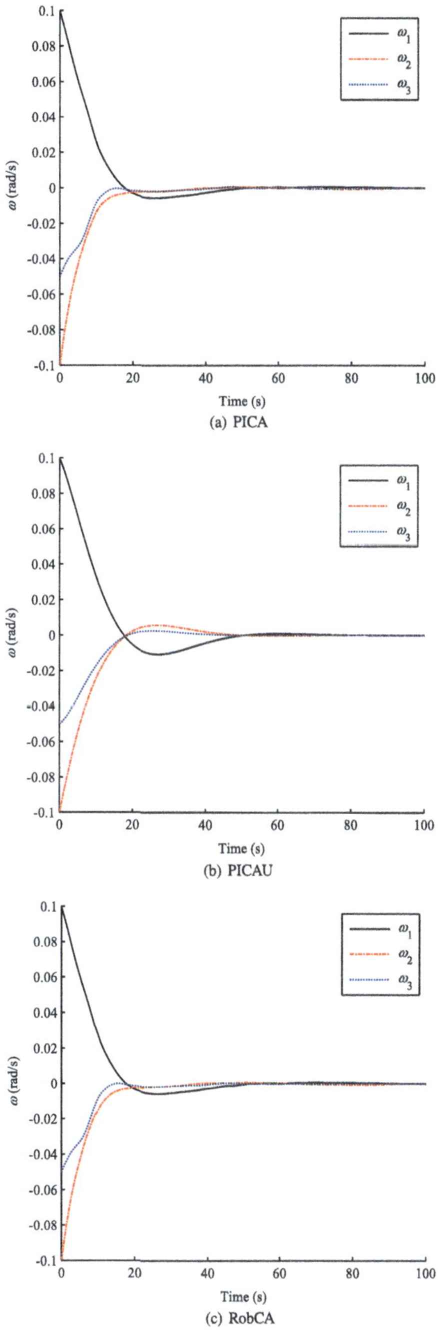

Fig. 9.11 Time response of Euler angle with $\eta = 33\%$

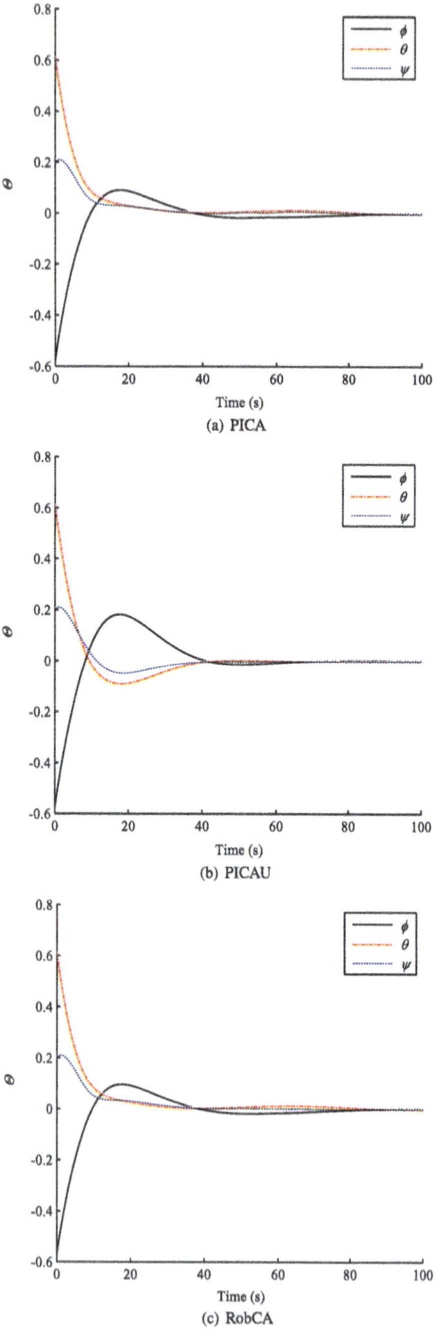

Fig. 9.12 Time response of control torque with $\eta = 33\%$

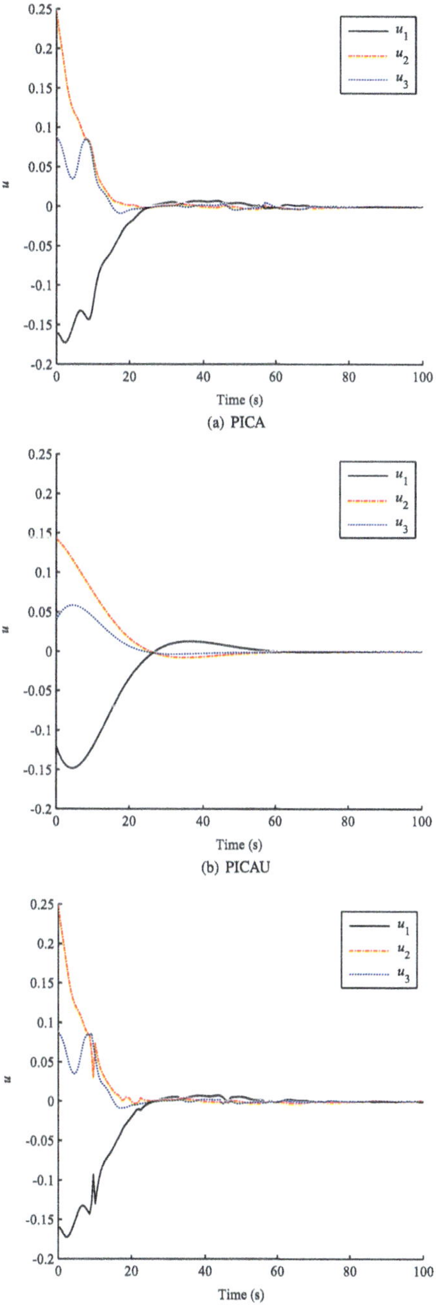

Fig. 9.13 Time response of actual actuation torque with $\eta = 33\%$

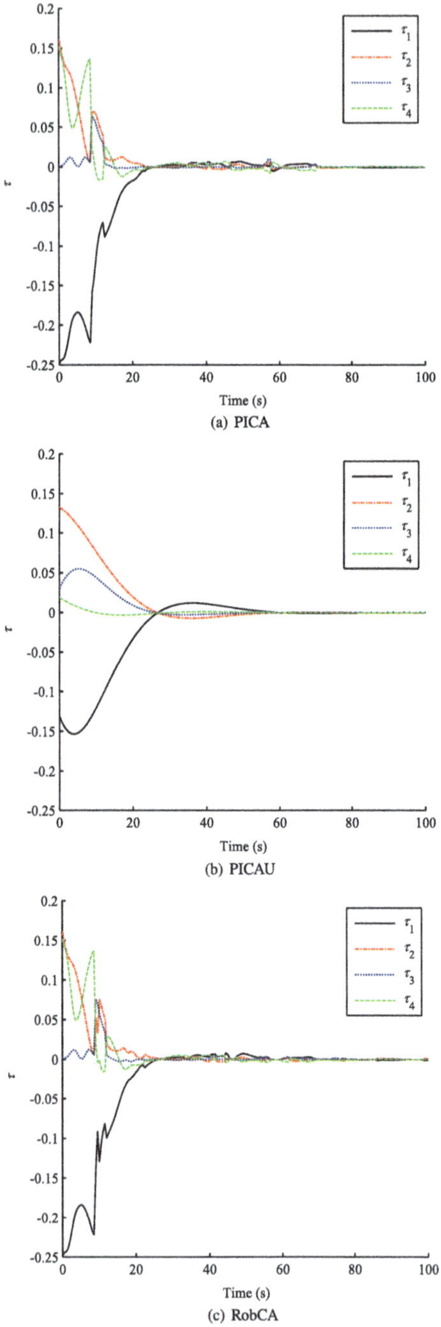

Fig. 9.14 Time response of energy consumption with $\eta = 33\%$

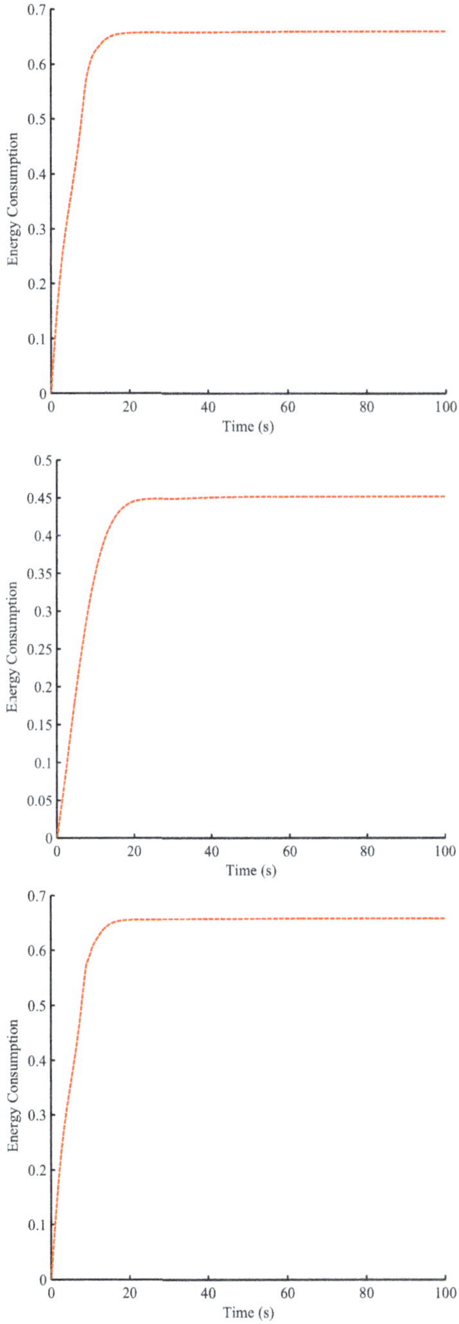

fault-estimation uncertainty. Numerical simulation studies validate the feasibility and effectiveness of the proposed RobCA algorithm and show the stability of the closed-loop system in the presence of imprecise fault estimation.

Appendices

Appendix A

Proof of Theorem 9.3

With PICA in Eq. (9.9) applied, we have

$$\boldsymbol{\tau}_{actual} = \boldsymbol{D}\boldsymbol{E}\boldsymbol{u}_c = \boldsymbol{D}(1 + \Delta\boldsymbol{E})\hat{\boldsymbol{E}}\boldsymbol{u}_c = \left[\boldsymbol{I} + \boldsymbol{A}\Delta\boldsymbol{E}\boldsymbol{A}^{\mathrm{T}}\left(\boldsymbol{A}\boldsymbol{A}^{\mathrm{T}}\right)^{-1}\right]\boldsymbol{\tau}_{design}$$

We abbreviate $\boldsymbol{\tau}_a \triangleq (\boldsymbol{I} + \boldsymbol{B}\boldsymbol{C})\boldsymbol{\tau}_d$, where $\boldsymbol{B} = \boldsymbol{A}\Delta\boldsymbol{E}\boldsymbol{A}^{\mathrm{T}}$ and $\boldsymbol{C} = \left(\boldsymbol{A}\boldsymbol{A}^{\mathrm{T}}\right)^{-1}$.

Choose a Lyapunov function candidate

$$V_1 = \frac{\boldsymbol{\omega}^{\mathrm{T}}\boldsymbol{J}\boldsymbol{\omega}}{2} + kk_1\boldsymbol{q}_v^{\mathrm{T}}\boldsymbol{q}_v + kk_1(1 - q_0)^2 + \mu\frac{k_1^2}{2\gamma_k} = \frac{\boldsymbol{\omega}^{\mathrm{T}}\boldsymbol{J}\boldsymbol{\omega}}{2} + 2kk_1(1 - q_0) + \mu\frac{k_1^2}{2\gamma_k}$$

Differentiation of V_1 with respect to time yields

$$\begin{aligned}\dot{V}_1 &= \boldsymbol{\omega}^{\mathrm{T}}(\boldsymbol{\omega} \times \boldsymbol{J}\boldsymbol{\omega} + \boldsymbol{\tau}_a) + kk_1\boldsymbol{q}_v^{\mathrm{T}}\boldsymbol{\omega} + 2k(1 - q_0)\dot{k}_1 + \mu\tfrac{k_1\dot{k}_1}{\gamma_k} \\ &= \boldsymbol{\omega}^{\mathrm{T}}(1 + \boldsymbol{B}\boldsymbol{C})\boldsymbol{\tau}_d + kk_1\boldsymbol{q}_v^{\mathrm{T}}\boldsymbol{\omega} + 2k(1 - q_0)\dot{k}_1 + \mu\tfrac{k_1\dot{k}_1}{\gamma_k}\end{aligned}$$

Substituting the baseline controller $\boldsymbol{\tau}_d$ and the adaptive law $\dot{k}_1$ in Eqs. (9.3)–(9.4), we have

$$\begin{aligned}\dot{V}_1 &= -k\boldsymbol{\omega}^{\mathrm{T}}(\boldsymbol{I} + \boldsymbol{B}\boldsymbol{C})\boldsymbol{\omega} - kk_1\boldsymbol{\omega}^{\mathrm{T}}\boldsymbol{B}\boldsymbol{C}\boldsymbol{q}_v + 2k(1 - q_0)\dot{k}_1 + \mu\frac{k_1\dot{k}_1}{\gamma_k} \\ &= -k\boldsymbol{\omega}^{\mathrm{T}}(\boldsymbol{I} + \boldsymbol{B}\boldsymbol{C})\boldsymbol{\omega} - kk_1\boldsymbol{\omega}^{\mathrm{T}}\boldsymbol{B}\boldsymbol{C}\boldsymbol{q}_v \\ &\quad - 2\gamma_k k^2 k_1\left(1 + 1/\mu\right)(1 - q_0)\boldsymbol{q}_v^{\mathrm{T}}\boldsymbol{q}_v - (1 + \mu)kk_1^2\boldsymbol{q}_v^{\mathrm{T}}\boldsymbol{q}_v \\ &\le -k\left[(1 - \|\boldsymbol{B}\boldsymbol{C}\|)\|\boldsymbol{\omega}\|^2 - k_1\|\boldsymbol{B}\boldsymbol{C}\|\|\boldsymbol{\omega}\|\left\|\boldsymbol{q}_v\right\| + \mu k_1^2\left\|\boldsymbol{q}_v\right\|^2\right] \\ &\quad - 2\gamma_k k^2 k_1\left(1 + 1/\mu\right)(1 - q_0)\boldsymbol{q}_v^{\mathrm{T}}\boldsymbol{q}_v - kk_1^2\boldsymbol{q}_v^{\mathrm{T}}\boldsymbol{q}_v\end{aligned}$$

To ensure that the first term in the Lyapunov derivative is negative requires

$$(k_1\|\boldsymbol{B}\boldsymbol{C}\|)^2 \le 4(1 - \|\boldsymbol{B}\boldsymbol{C}\|)\mu k_1^2$$

Note that $\|\boldsymbol{BC}\| = \left\|\boldsymbol{A}\Delta\boldsymbol{E}\boldsymbol{A}^{\mathrm{T}}\left(\boldsymbol{A}\boldsymbol{A}^{\mathrm{T}}\right)^{-1}\right\| \le \eta\left\|\boldsymbol{A}\boldsymbol{A}^{\mathrm{T}}\right\|\left\|\left(\boldsymbol{A}\boldsymbol{A}^{\mathrm{T}}\right)^{-1}\right\|$. Assume $\eta\left\|\boldsymbol{A}\boldsymbol{A}^{\mathrm{T}}\right\| \left\|\left(\boldsymbol{A}\boldsymbol{A}^{\mathrm{T}}\right)^{-1}\right\| \le \xi$, where $\xi < 1$ is a user-chosen positive constant. Then, the constraint on the fault-estimation error from the FDD scheme is derived

$$\eta \le \frac{\xi}{\left\|\boldsymbol{A}\boldsymbol{A}^{\mathrm{T}}\right\|\left\|\left(\boldsymbol{A}\boldsymbol{A}^{\mathrm{T}}\right)^{-1}\right\|} \tag{9.10}$$

This assumption puts a constraint on the estimation uncertainty, which explicitly states the fault-tolerant capability of the proposed scheme and the accuracy requirement of the corresponding FDD scheme.

Let $\mu \stackrel{\Delta}{=} \frac{\xi^2}{4(1-\zeta)}$. Note that

$$(1 - \|\boldsymbol{BC}\|)\|\boldsymbol{\omega}\|^2 \ge (1-\xi)\|\boldsymbol{\omega}\|^2$$

$$2\sqrt{\mu(1-\xi)}\|\boldsymbol{\omega}\|\left\|\boldsymbol{q}_v\right\| = \xi\|\boldsymbol{\omega}\|\left\|\boldsymbol{q}_v\right\| \ge \|\boldsymbol{BC}\|\|\boldsymbol{\omega}\|\left\|\boldsymbol{q}_v\right\|$$

then

$$\begin{aligned} \dot{V}_1 \le & -k\left[\sqrt{(1-\xi)}\|\boldsymbol{\omega}\| - k_1\sqrt{\mu}\left\|\boldsymbol{q}_v\right\|\right]^2 \\ & - 2\gamma_k\left(1+1/\mu\right)k^2k_1(1-q_0)\boldsymbol{q}_v^{\mathrm{T}}\boldsymbol{q}_v - kk_1^2\boldsymbol{q}_v^{\mathrm{T}}\boldsymbol{q}_v \\ \le & \ 0 \end{aligned} \tag{9.11}$$

which implies that V_1, ω, and k_1 are bounded for all time.

Now we need to prove the convergence of the spacecraft attitude and its angular rate. Note that $2\gamma_k\left(1+1/\mu\right)k^2(1-q_0)\boldsymbol{q}_v^{\mathrm{T}}\boldsymbol{q}_v k_1 \ge 0$, $kk_1^2\boldsymbol{q}_v^{\mathrm{T}}\boldsymbol{q}_v \ge 0$ for all states. Integrating both sides of Eq. (9.11), one has

$$2\gamma_k\left(1+1/\mu\right)k^2\int_0^{+\infty}(1-q_0)\boldsymbol{q}_v^{\mathrm{T}}\boldsymbol{q}_v k_1 dt + k\int_0^{+\infty}k_1^2\boldsymbol{q}_v^{\mathrm{T}}\boldsymbol{q}_v dt \le V_1(0) - V_1(\infty) \tag{9.12}$$

Since

$$\frac{d}{dt}\left((1-q_0)\boldsymbol{q}_v^{\mathrm{T}}\boldsymbol{q}_v k_1\right) = -\gamma_k k(1-q_0)\left(1-q_0^2\right)^2 k_1 + \left(1/2 + q_0 - 3q_0^2/2\right)\boldsymbol{q}_v^{\mathrm{T}}\boldsymbol{\omega}k_1$$

is bounded, $(1-q_0)\boldsymbol{q}_v^{\mathrm{T}}\boldsymbol{q}_v k_1$ is uniformly continuous with a finite integral. Then, Barbalat's lemma dictates that $(1-q_0)\boldsymbol{q}_v^{\mathrm{T}}\boldsymbol{q}_v k_1$ converges to 0 as $t \to \infty$.

Similarly, since $\left[\sqrt{(1-\zeta)}\|\boldsymbol{\omega}\| - k_1\sqrt{\mu}\left\|\boldsymbol{q}_v\right\|\right]^2$ is also uniformly continuous with a finite integral, Barbalat's lemma further shows that the first term of Eq. (9.11) converges to zero such that

$$\left.\begin{array}{l}(1-q_0)\,\boldsymbol{q}_v^{\mathrm{T}}\boldsymbol{q}_v k_1 \to 0 \\ \left[\sqrt{(1-\xi)}\,\|\omega\| - k_1\sqrt{\mu}\,\|\boldsymbol{q}_v\|\right]^2 \to 0\end{array}\right\} \Rightarrow \left\{\begin{array}{l}k_1\boldsymbol{q}_v \to 0 \\ \omega \to 0\end{array}\right. \tag{9.13}$$

To meet the overall control objective, the stability analysis therefore reduces to show that $k_1\boldsymbol{q}_v \to 0 \Rightarrow \boldsymbol{q}_v \to 0$, meaning that $k_1(t)$ does not converge to zero unless q_v converges to the origin. From the internal dynamics of k_1 in Eq. (9.4), $\dot{k}_1 = -\gamma_k\left(1+\frac{1}{\mu}\right)kk_1\boldsymbol{q}_v{}^{\mathrm{T}}\boldsymbol{q}_v \geq -\gamma_k\left(1+\frac{1}{\mu}\right)kk_1$ is derived. Because the lower bound on $k_1(t)$ is described simply by a first-order dynamic, $k_1(t)$cannot cross zero or tend to zero in finite time, but only at infinity. Since $\dot{k}_1(t) \leq 0$, then $k_1(t)$ is monotonously decreasing with upper bound $k_1(0)$. Just like in [40], here we carefully examine the behavior of $k_1(t)$ to show there exists a condition on γ_k such that $k_1(t)$ has a positive lower bound.

For a given $\gamma_k > 0$, from Eq. (9.12), there exists a positive function $c(\gamma_k)$ such that

$$k\int_0^{+\infty} k_1{}^2\boldsymbol{q}_v{}^{\mathrm{T}}\boldsymbol{q}_v dt \leq c(\gamma_k)$$

Assertion 9.1 For the adjustment law in Eq. (9.4), for any given $k_1(0) > 0$ and $\overline{k}_1 > 0$ where $\overline{k}_1 < k_1(0)$, if there exists a $\gamma_k > 0$ such that

$$\gamma_k\left(1+\frac{1}{\mu}\right)c(\gamma_k) \leq \left(k_1^2(0) - \overline{k}_1^2\right)/2 \tag{9.14}$$

then $k_1(t) \geq \overline{k}_1$ for all $t \geq 0$.

Proof: Taking a closer inspection of the first-order dynamic, we can further show that

$$k_1\dot{k}_1 = -\gamma_k\left(1+\frac{1}{\mu}\right)kk_1{}^2\boldsymbol{q}_v{}^{\mathrm{T}}\boldsymbol{q}_v$$

Upon integrating this equation on time interval $[0, +\infty)$, one obtains $k_1{}^2(\infty) = k_1{}^2(0) - 2\gamma_k\left(1+\frac{1}{\mu}\right)k\int_0^{+\infty}k_1{}^2\boldsymbol{q}_v{}^{\mathrm{T}}\boldsymbol{q}_v dt$. Hence if a $\gamma_k > 0$ that satisfies Eq. (9.14) is found, then $k_1(t)$ is bounded below by $\overline{k}_1$. This completes the proof of Assertion 9.1.

Assuming that a sufficiently small $\gamma_k > 0$ is found that satisfies Eq. (9.14), then $k_1(t) \geq \overline{k}_1 > 0$ for all time. In such a case, because $\lim\limits_{t\to\infty} k_1(t)\boldsymbol{q}_v(t) = 0$ and $k_1(t) \geq \overline{k}_1$ for all time, it follows that $\lim\limits_{t\to\infty}\boldsymbol{q}_v(t) = 0$.

This completes the proof of Theorem 9.2.

Appendix B

Proof of Theorem 9.3 For the RobCA algorithm, the optimal solution can be synthesized in a uniform expression, that is

$$\boldsymbol{u}_{command} = \boldsymbol{A}^{\mathrm{T}}\left(\varepsilon \boldsymbol{I} + \boldsymbol{A}\boldsymbol{A}^{\mathrm{T}}\right)^{-1}\boldsymbol{\tau}_{design}, \varepsilon \geq 0$$

To streamline the analysis, the actual torque can be calculated as

$$\begin{aligned}
\boldsymbol{\tau}_{actual} &= \boldsymbol{D}\left(\boldsymbol{I} + \Delta \boldsymbol{E}\right)\hat{\boldsymbol{E}}\boldsymbol{u}_c \\
&= \boldsymbol{A}\left(\boldsymbol{I} + \Delta \boldsymbol{E}\right)\boldsymbol{A}^{\mathrm{T}}\left(\varepsilon \boldsymbol{I} + \boldsymbol{A}\boldsymbol{A}^{\mathrm{T}}\right)^{-1}\boldsymbol{\tau}_{design} \\
&= \left(\boldsymbol{I} + \left(\boldsymbol{A}\Delta \boldsymbol{E}\boldsymbol{A}^{\mathrm{T}} - \varepsilon \boldsymbol{I}\right)\left(\varepsilon \boldsymbol{I} + \boldsymbol{A}\boldsymbol{A}^{\mathrm{T}}\right)^{-1}\right)\boldsymbol{\tau}_{design}
\end{aligned}$$

We abbreviate $\boldsymbol{\tau}_a \triangleq \left(1 + \boldsymbol{B}'\boldsymbol{C}'\right)\boldsymbol{\tau}_d$, where $\boldsymbol{B}' = \boldsymbol{A}\Delta\boldsymbol{E}\boldsymbol{A}^{\mathrm{T}} - \varepsilon\boldsymbol{I}$, $\boldsymbol{C}' = \left(\varepsilon\boldsymbol{I} + \boldsymbol{A}\boldsymbol{A}^{\mathrm{T}}\right)^{-1}$. In this case, the following Lyapunov function candidate is considered

$$\begin{aligned}
V_2 &= \frac{\boldsymbol{\omega}^{\mathrm{T}}\boldsymbol{J}\boldsymbol{\omega}}{2} + kk_1\boldsymbol{q}_v^{\mathrm{T}}\boldsymbol{q}_v + kk_1(1 - q_0)^2 + \mu\frac{k_1^2}{2\gamma_k} \\
&= \frac{\boldsymbol{\omega}^{\mathrm{T}}\boldsymbol{J}\boldsymbol{\omega}}{2} + 2\left(1 - q_0\right)kk_1 + \mu\frac{k_1^2}{2\gamma_k}
\end{aligned} \tag{9.15}$$

A procedure similar to the preceding analysis applies

$$\begin{aligned}
\dot{V}_2 &= \boldsymbol{\omega}^{\mathrm{T}}\left(-\boldsymbol{\omega} \times \boldsymbol{J}\boldsymbol{\omega} + \boldsymbol{\tau}_a\right) + kk_1\boldsymbol{q}_v^{\mathrm{T}}\boldsymbol{\omega} + 2k\left(1 - q_0\right)\dot{k}_1 + \mu_1\frac{k_1\dot{k}_1}{\gamma_k} \\
&= \boldsymbol{\omega}^{\mathrm{T}}\left(1 + \boldsymbol{B}'\boldsymbol{C}'\right)\boldsymbol{\tau}_d + kk_1\boldsymbol{q}_v^{\mathrm{T}}\boldsymbol{\omega} + 2k\left(1 - q_0\right)\dot{k}_1 + \mu_1\frac{k_1\dot{k}_1}{\gamma_k}
\end{aligned}$$

Substituting the baseline controller and the adaptive law in Eqs. (9.3) (9.4) yields

$$\begin{aligned}
\dot{V}_1 &= -k\boldsymbol{\omega}^{\mathrm{T}}\left(\boldsymbol{I} + \boldsymbol{B}'\boldsymbol{C}'\right)\boldsymbol{\omega} - kk_1\boldsymbol{\omega}^{\mathrm{T}}\boldsymbol{B}'\boldsymbol{C}'\boldsymbol{q}_v \\
&\quad - 2\gamma_k k^2 k_1\left(1 + 1/\mu_1\right)\left(1 - q_0\right)\boldsymbol{q}_v^{\mathrm{T}}\boldsymbol{q}_v - \left(1 + \mu_1\right)kk_1^2\boldsymbol{q}_v^{\mathrm{T}}\boldsymbol{q}_v \\
&\leq -k\left[\left(1 - \left\|\boldsymbol{B}'\boldsymbol{C}'\right\|\right)\left\|\boldsymbol{\omega}\right\|^2 - k_1\left\|\boldsymbol{B}'\boldsymbol{C}'\right\|\left\|\boldsymbol{\omega}\right\|\left\|\boldsymbol{q}_v\right\| + \mu_1 k_1^2\left\|\boldsymbol{q}_v\right\|^2\right] \\
&\quad - 2\gamma_k k^2 k_1\left(1 + 1/\mu_1\right)\left(1 - q_0\right)\boldsymbol{q}_v^{\mathrm{T}}\boldsymbol{q}_v - kk_1^2\boldsymbol{q}_v^{\mathrm{T}}\boldsymbol{q}_v
\end{aligned} \tag{9.16}$$

To ensure $\dot{V}_2 \leq 0$, one requires $\left(k_1\left\|\boldsymbol{B}'\boldsymbol{C}'\right\|\right)^2 \leq 4\left(1 - \left\|\boldsymbol{B}'\boldsymbol{C}'\right\|\right)\mu_1 k_1^2$. Assume that the fault-estimation error from the FDD scheme satisfies

$$\eta \leq \frac{\xi - \varepsilon \left\| \left(\varepsilon \boldsymbol{I} + \boldsymbol{A}\boldsymbol{A}^{\mathrm{T}}\right)^{-1} \right\|}{\left\| \boldsymbol{A}\boldsymbol{A}^{\mathrm{T}} \right\| \left\| \left(\varepsilon \boldsymbol{I} + \boldsymbol{A}\boldsymbol{A}^{\mathrm{T}}\right)^{-1} \right\|} \tag{9.17}$$

for a user-defined positive constant $\xi < 1$. From

$$\begin{aligned} \left\| \boldsymbol{B}'\boldsymbol{C}' \right\| &\leq \eta \left\| \boldsymbol{A}\boldsymbol{A}^{\mathrm{T}} \right\| \left\| \left(\varepsilon \boldsymbol{I} + \boldsymbol{A}\boldsymbol{A}^{\mathrm{T}}\right)^{-1} \right\| + \varepsilon \left\| \left(\varepsilon \boldsymbol{I} + \boldsymbol{A}\boldsymbol{A}^{\mathrm{T}}\right)^{-1} \right\| \\ &\leq \xi \end{aligned}$$

and $\mu_1 \triangleq \frac{\xi^2}{4(1-\xi)}$, we have

$$\begin{aligned} \left(k_1 \left\| \boldsymbol{B}'\boldsymbol{C}' \right\|\right)^2 &\leq k_1^2 \xi^2 \\ &\leq k_1^2 \frac{\xi^2}{1-\xi} \left(1 - \left\| \boldsymbol{B}'\boldsymbol{C}' \right\|\right) \\ &= 4\mu_1 k_1^2 \left(1 - \left\| \boldsymbol{B}'\boldsymbol{C}' \right\|\right) \end{aligned}$$

Thus, the first term of the right side of Eq. (9.16) is guaranteed to be non-positive. Note that

$$\left(1 - \left\| \boldsymbol{B}'\boldsymbol{C}' \right\|\right) \|\boldsymbol{\omega}\|^2 \geq (1-\xi) \|\boldsymbol{\omega}\|^2$$

$$2\sqrt{\mu_1 (1-\xi)} \|\boldsymbol{\omega}\| \left\| \boldsymbol{q}_v \right\| = \xi \|\boldsymbol{\omega}\| \left\| \boldsymbol{q}_v \right\| \geq \left\| \boldsymbol{B}'\boldsymbol{C}' \right\| \|\boldsymbol{\omega}\| \left\| \boldsymbol{q}_v \right\|$$

Then the following inequality on the derivative of V_2 can be obtained:

$$\begin{aligned} \dot{V}_2 \leq &- k\left[\sqrt{(1-\xi)} \|\boldsymbol{\omega}\| - k_1\sqrt{\mu_1} \left\| \boldsymbol{q}_v \right\|\right]^2 \\ &- 2\gamma_k \left(1 + 1/\mu_1\right) k^2 k_1 (1 - q_0) \boldsymbol{q}_v^{\mathrm{T}} \boldsymbol{q}_v - k k_1^2 \boldsymbol{q}_v^{\mathrm{T}} \boldsymbol{q}_v \\ \leq &\, 0 \end{aligned}$$

which implies that V_2, ω, k_1 are bounded for all time.

Now, a justification of the upper bound on the FDD estimation error η is provided. Denote the eigenvalues of $\boldsymbol{A}\boldsymbol{A}^{\mathrm{T}}$ as $\lambda\left(\boldsymbol{A}\boldsymbol{A}^{\mathrm{T}}\right) \triangleq \{\lambda_1, \lambda_2, \lambda_3\}$, where $0 < \lambda_1 \leq \lambda_2 \leq \lambda_3$. Since $\left(\varepsilon \boldsymbol{I} + \boldsymbol{A}\boldsymbol{A}^{\mathrm{T}}\right) > 0$, there exists an orthogonal matrix $\boldsymbol{P}$ such that $\left(\varepsilon \boldsymbol{I} + \boldsymbol{A}\boldsymbol{A}^{\mathrm{T}}\right) = \boldsymbol{P}\mathrm{diag}\left(\varepsilon + \lambda_1, \varepsilon + \lambda_2, \varepsilon + \lambda_3\right) \boldsymbol{P}^{\mathrm{T}}$. Thus $\left(\varepsilon \boldsymbol{I} + \boldsymbol{A}\boldsymbol{A}^{\mathrm{T}}\right)^{-1} = \boldsymbol{P}\mathrm{diag}\left(\frac{1}{\varepsilon+\lambda_1}, \frac{1}{\varepsilon+\lambda_2}, \frac{1}{\varepsilon+\lambda_3}\right) \boldsymbol{P}^{\mathrm{T}}$ and $\left\| \left(\varepsilon \boldsymbol{I} + \boldsymbol{A}\boldsymbol{A}^{\mathrm{T}}\right)^{-1} \right\|$ is calculated as $\frac{1}{\varepsilon+\lambda_1}$. From Eqs. (9.6) and (9.7), we know ε is a small positive variable. Since ξ is a positive constant satisfying $\xi < 1$, it is reasonable to assume there exists ξ such that $\xi - \frac{\varepsilon}{\varepsilon+\lambda_1} > 0$, i.e., the upper bound on η is a positive quantity.

A procedure similar to that of Eqs. (9.11)–(9.13) applies to $\dot{V}_2$ and shows that

$$\lim_{t\to\infty} k_1(t)\,\boldsymbol{q}_v(t) = 0,\ \lim_{t\to\infty} \boldsymbol{\omega}(t) = 0 \tag{9.18}$$

The fact that $k_1(t)\,\boldsymbol{q}_v(t)$ tends to zero does not guarantee that $\boldsymbol{q}_v$ will tend to zero. The objective now is to show that there exists a condition on $\gamma_k > 0$ such that $k_1(t) \geq \overline{k}_1 > 0$ for all time.

Note that, in this case, for a given $\gamma_k > 0$, there exists another positive function $c_1(\gamma_k)$ such that

$$k\int_0^{+\infty} {k_1}^2 {\boldsymbol{q}_v}^{\mathrm{T}} \boldsymbol{q}_v dt \leq c_1(\gamma_k)$$

Assertion 9.2 For the adjustment law in Eq. (9.4), for any given $k_1(0) > 0$ and $\overline{k}_1 > 0$ where $\overline{k}_1 < k_1(0)$, if there exists a $\gamma_k > 0$ such that

$$\gamma_k\left(1 + 1/\mu_1\right)c_1(\gamma_k) \leq \left({k^2}_1(0) - {\overline{k}^2}_1\right)/2 \tag{9.19}$$

then $k_1(t) \geq \overline{k}_1$ for all $t \geq 0$.

The proof process of Assertion 9.2 is similar to that of Assertion 9.1, and is omitted here.

Assuming that a sufficiently small $\gamma_k > 0$ is found, it follows that $k_1(t)$ is bounded below by $\overline{k}_1 > 0$. With the fact that $\lim_{t\to\infty} k_1(t)\,\boldsymbol{q}_v(t) = 0$, we have $\lim_{t\to\infty} \boldsymbol{q}_v(t) = 0$. Thus, the asymptotic convergence of $\boldsymbol{q}_v$ and $\boldsymbol{\omega}$ is assured. This completes the proof of Theorem 9.3.

References

1. Zhang, Y.M., Jiang, J. (2008) Bibliographical review on reconfigurable fault-tolerant control systems. Annual Reviews in Control 32(2): 229–252
2. Durham, W.C. (1994) Constrained control allocation - three-moment problem. Journal of Guidance, Control, and Dynamics 17(2): 330–336
3. Chang, H.T., Huang, P.F., Zhang, Y.Z., Meng, Z.J., Liu, Z.X. (2018) Distributed control allocation for spacecraft attitude takeover control via cellular space robot. Journal of Guidance, Control, and Dynamics 41(11): 2499–2506
4. Durham, W. (1994) Attainable moments for the constrained control allocation problem. Journal of Guidance, Control, and Dynamics 17(6): 1371–1373
5. Buffington, J.M., Enns, D.F. (1996) Lyapunov stability analysis of daisy chain control allocation. Journal of Guidance, Control, and Dynamics 19(6): 1226–1230
6. Eberhardt, R., Ward, D. (1999) Indirect adaptive flight control of a tailless fighter aircraft. In: Guidance, Navigation, and Control Conference and Exhibit, pp 466–476
7. Jin, J. (2005) Modified pseudoinverse redistribution methods for redundant controls allocation. Journal of Guidance, Control, and Dynamics 28(5): 1076–1079
8. Ikeda, Y., Hood, M. (2000) An application of L1 optimization to control allocation. In: AIAA Guidance, Navigation, and Control Conference and Exhibit, AIAA-2000–4566

9. Doman, D., Ngo, A. (2002) Dynamic inversion-based adaptive/reconfigurable control of the X-33 on ascent. Journal of Guidance, Control, and Dynamics 25(2): 275–284
10. Burken, J., Lu P., Wu, Z., Bahm, C. (2001) Two reconfigurable flight-control design methods: Robust servomechanism and control allocation. Journal of Guidance, Control, and Dynamics 24(3): 482–493
11. Enns, D. (1998) Control allocation approaches. In: Guidance, Navigation, and Control Conference and Exhibit, AIAA-1998–4109
12. Härkegård, O. (2004) Dynamic control allocation using constrained quadratic programming. Journal of Guidance, Control, and Dynamics 27(6): 1028–1034
13. Bodson, M. (2002) Evaluation of optimization methods for control allocation. Journal of Guidance, Control, and Dynamics 25(4): 703–711
14. Gui, H.C., De Ruiter, A.H.J. (2019) Adaptive fault-tolerant spacecraft pose tracking with control allocation. IEEE Transactions on Control Systems Technology 27(2): 479–494
15. Hu, Q.L., Tan, X (2020) Dynamic near-optimal control allocation for spacecraft attitude control using a hybrid configuration of actuators. IEEE Transactions on Aerospace and Electronic Systems 56(2): 1430–1443
16. Page, A., Steinberg, M. (2000) A closed-loop comparison of control allocation methods. In: AIAA Guidance, navigation, and Control Conference and Exhibit, AIAA-2000–4538
17. Alwi, H., Edwards, C. (2008) Fault tolerant control using sliding modes with online control allocation. Automatica 44(7): 1859–1866
18. Doman, D.B., Gamble, B.J., Ngo, A.D. (2009) Quantized control allocation of reaction control jets and aerodynamic control surfaces. Journal of Guidance, Control, and Dynamics 32(1): 13–24
19. Cai, W.C., Liao, X.H., Song, D.Y. (2008) Indirect robust adaptive fault-tolerant control for attitude tracking of spacecraft. Journal of Guidance, Control, and Dynamics 31(5): 1456–1463
20. Hu, Q.L., Tan, X., Akella, M. (2017) Finite-time fault-tolerant spacecraft attitude control with torque saturation. Journal of Guidance, Control, and Dynamics 40(10): 2524–2537
21. Cui, L., Yang, Y. (2011) Disturbance rejection and robust least-squares control allocation in flight control system. Journal of Guidance, Control, and Dynamics 34(6): 1632–1643
22. Shen, Q., Wang, D., Zhu, S., Poh, E.K. (2017) Robust control allocation for spacecraft attitude tracking under actuator faults. IEEE Transactions on Control Systems Technology 25(3): 1068–1075
23. Hu, Q.L., Li, B., Zhang, Y.M. (2014) Nonlinear proportional-derivative control incorporating closed-loop control allocation for spacecraft. Journal of Guidance, Control, and Dynamics 37(3): 799–812
24. Yang, Y., Gao Z. (2020) A new method for control allocation of aircraft flight control system. IEEE Transactions on Automatic Control 65(4): 1413–1428
25. Chen, L., Edwards, C., Alwi, H. (2018) Integral sliding mode fault tolerant control allocation for a class of affine nonlinear system. International Journal of Robust and Nonlinear Control 29(3): 565–582
26. Wang, B., Zhang, Y.M. (2018) An adaptive fault-tolerant sliding mode control allocation scheme for multirotor helicopter subject to simultaneous actuator faults. IEEE Transactions on Industrial Electronics 65(5): 4227–4236
27. Yan, Y.D., Yang, J., Liu, C.J., Coombes, M., S.H. Li, Chen, W.H. (2020) On the actuator dynamics of dynamic control allocation for a small fixed-wing UAV with direct lift control. IEEE Transactions on Control Systems Technology 28(3): 984–991
28. Chen, W., Saif, M. (2007) Observer-based fault diagnosis of satellite systems subject to time-varying thruster faults. Journal of dynamic systems, measurement, and control 129(3): 352–356
29. Hu, Q.L., Li. B., Friswell, M. (2013) Observer-based fault diagnosis incorporating online control allocation for spacecraft attitude stabilization under actuator failures. The Journal of the Astronautical Sciences 60(2): 211–236
30. Edwards, C., Spurgeon, S., Patton, R. (2000) Sliding mode observers for fault detection and isolation. Automatica 36(4): 541–553

31. Henry, D. (2008) Fault diagnosis of microscope satellite thrusters using H-infinity/H-2 filters. Journal of Guidance, Control, and Dynamics 31(3): 699–711
32. Tsiotras, P. (1998) Further passivity results for the attitude control problem. IEEE Transactions on Automatic Control 43(11): 1597–1600
33. Su, Y.X., Zheng, C.H. (2011) Globally asymptotic stabilization of spacecraft with simple saturated proportional derivative control. Journal of Guidance, Control, and Dynamics 31(6): 1932–1935
34. Wen, J.T.Y., Kreutz-Delgado, K. (1991) The attitude control problem. IEEE Transactions on Automatic Control 36(10): 1148–1162
35. Lo, S., Chen, Y. (1995) Smooth sliding mode control for spacecraft attitude tracking maneuvers. Journal of Guidance, Control, and Dynamics 18(6): 1345–1349
36. Guo,J.X., Tao,G., Liu, Y. (2014) Adaptive actuator failure and structural damage compensation of nasa generic transport model. Journal of Dynamic Systems, Measurement, and Control 136(3): 031009
37. Wallsgrove, R., Akella, M. (2005) Globally stabilizing saturated attitude control in the presence of bounded unknown disturbances. Journal of Guidance, Control, and Dynamics 28(5): 957–963
38. Ghaoui, L.E., Lebret, H. (1997) Robust solutions to least squares problems with uncertain data. SIAM Journal on Matrix Analysis and Applications 18(4): 1035–1064
39. Wang, Y., Boyd, S. (2010) Fast model predictive control using online optimization. IEEE Transactions on control systems technology 18(2): 267–278
40. Boskovic, J.D., Li, S. M., Mehra, R. K. (2004) Robust tracking control design for spacecraft under control input saturation. Journal of Guidance, Control, and Dynamics 27(4): 627–633

Chapter 10
Conclusions

10.1 Conclusions

For on-orbiting spacecraft, FTC and CA techniques are generally utilized to ensure the stability, reliability, autonomy and intelligence, and the percentage of successfully accomplished missions of the spacecraft ACS, especially in the presence of external disturbances, actuator constraints and faults. This book focus on spacecraft ACS design including optimization algorithms based CA schemes, optimal control and SMC based controller, FDD and FTC, the stability analysis of the closed-loop CA system, etc. In view of the proposed CA and FTC schemes in the previous chapters, the contributions and achievements of this book are presented in the following aspects.

(1) A SPD based virtual control law incorporated with a null-space-based optimal CA method is proposed for the spacecraft ACS under external disturbances and input saturation.

(2) A NSMC technique based finite-time control law is first proposed to achieve the fast convergence, and then a robust least squares based CA scheme is developed to suitably distribute the desired virtual control signals into the active actuators.

(3) A FSMC based attitude stabilization control law is firstly proposed, and it can be extended to accommodate actuator faults while making use of the remaining active actuating capabilities by utilizing the PICA method.

(4) A finite-time ESO is firstly investigated to estimate and compensate for specified total uncertainty including actuators-misalignments and parametric uncertainties. And then, CLF based optimal control laws are developed to achieve the spacecraft attitude stabilization control.

(5) An ILO based FDD and ISMC based FTC scheme is proposed to achieve the attitude stabilization of rigid spacecraft in the present of actuator uncertainties, external disturbances and input saturation, and then a robust CA algorithm is developed in an optimal manner with fault-tolerant capability.

Q. Hu et al., *Control Allocation for Spacecraft Under Actuator Faults*,
https://doi.org/10.1007/978-981-16-0439-3_10

(6) A closed-loop CA scheme is proposed for dealing with the problem of distributing the virtual control command synthesized by the SPD controller into the individual actuator properly by using constrained optimal quadratic programming algorithms, and the asymptotic stability of the closed-loop CA system is guaranteed theoretically.

(7) Taking further actuators' faults in the design of the closed-loop CA system, a PD-like control law incorporating with a robust online CA method is developed. The asymptotic stability of the closed-loop system with online CA is achieved by a rigorous closed-loop stability analysis.

10.2 Open Problems and Challenges

This book presents systematical and comprehensive descriptions of the CA and FTC issues for over-acutated spacecraft ACS under actuator faults, system uncertainties and disturbances. It provides some interesting solutions for the spacecraft attitude control and CA problems. However, there still exist some open and challenging problems to be addressed for the advanced spacecraft. These problems are listed as follows.

(1) Multiple objective optimization based CA. Most of the existing CA schemes cannot handle the spacecraft ACS with various and different constraints, such as model uncertainties, actuators' deviations, constraints and faults, simultaneously. And that the CA performance index is simple, which is designed without multiple objectives. Thus, the multiple objective optimization based CA should be investigated for the spacecraft ACS to achieve better control performances.

(2) CA-based FTC. One of the advantages of the CA module is that the redundant actuators provide the ability of fault tolerance, by utilizing the available actuators without reconfiguring the control structure or scheme. When actuator faults occur in the spacecraft ACS, CA-based FTC is responsible to distribute the virtual control signals to available actuators such that the ACS achieves the same control system performances or at least the stable response during the aerospace missions. Thus, CA-based FTC is another interesting issue for the spacecraft ACS under uncertainties and actuators' constraints.

(3) Autonomous and intelligent FDD and FTC. As spacecraft system and its components become more and more complex, the corresponding accurate mathematical models are usually difficult to obtain. The model-depended FDD and FTC methodologies would be not able to perform well. Thus, knowledge-based or data-driven or intelligent FDD and FTC methods are also worthy to investigate further by utilizing intelligent and automation techniques comprehensively, such that the spacecraft ACS has a high degree of intelligence and autonomy, which can adjust the FDD-based FTC system online and in real time.

(4) Experimental verification for aerospace engineering. The proposed FTC and CA schemes of spacecraft ACS in this book are only verified with the effectiveness and superiority by utilizing numerical simulations. These are actually a series of theoretical research results, which are difficult to be applied into the practical engi-

neering directly. Therefore, the development and experiment of the semi-physical simulation platform or other verification techniques and evaluation methodologies are also very important issues that need to be investigated for the spacecraft ACS, to promoting the engineering applications of the proposed control system schemes.

The manufacturer's authorised representative in the EU is Springer Nature Customer Service Centre GmbH, Europaplatz 3, 69115 Heidelberg, Germany. If you have any concerns regarding our products, please contact ProductSafety@springernature.com

Printed and bound by CPI Group (UK) Ltd, Croydon, CR0 4YY
30/09/2026
02225294-0005